Statistical Thinking
from Scratch

Statistical Thinking from Scratch

A Primer for Scientists

M. D. EDGE

OXFORD

UNIVERSITY PRESS

OXFORD

UNIVERSITY PRESS

Great Clarendon Street, Oxford, OX2 6DP,
United Kingdom

Oxford University Press is a department of the University of Oxford.
It furthers the University's objective of excellence in research, scholarship,
and education by publishing worldwide. Oxford is a registered trade mark of
Oxford University Press in the UK and in certain other countries

Published in the United States of America by Oxford University Press
198 Madison Avenue, New York, NY 10016, United States of America

British Library Cataloguing in Publication Data
Data available

Library of Congress Control Number: 2019934651

ISBN 978-0-19-882762-7

for Isabel

Contents

Acknowledgments

When one thinks about probability and random processes, one's mind sometimes wanders toward the contingencies inherent in life. Almost any event in one's biography might have happened in a slightly different way, and the unrealized outcomes might have branched into a much-altered story. To give one example out of many from my own life, I took a class in college that shaped my career trajectory, without which this book would not have been written. The person who told me the class would be offered was a friend of my roommate's who happened to visit the day before the application was due—a person I had not met before and haven't seen since. Had my roommate's friend not visited, I would have gone on to live another life that would have followed an autumn term in which I took a different class. Immediately my roommate, his friend, the class instructors, and all the people who set them on a course in which they would interact with me are implicated in the writing of this book. There are thousands of other such events that influenced this book in shallow or deep ways, and each of those events was itself conditional on a chain of contingencies. Taking this view, it is not mystical to say that billions of people—including the living and their ancestors—are connected to this book through a web of interactions. The purpose of an acknowledgments section, then, is not to separate contributors from non-contributors, but to produce a list of the few people whose roles appear most salient according to the faulty memory of the author.

My editors at Oxford University Press, Ian Sherman and Bethany Kershaw, have guided me through the process of improving and finalizing this book. I have always felt that the book has been safe in their hands. Mac Clarke's perceptive copy edits rescued me in more than a few places. The SPi typesetting team ably handled production, managed by Saranya Jayakumar. Alex Walker designed the cover, which has won me many unearned compliments.

Tim Campellone, Paul Connor, Audun Dahl, Carolyn Fredericks, Arbel Harpak, Aaron Hirsh, Susan Johnston, Emily Josephs, Jaehee Kim, Joe Larkin, Jeff Long, Sandy Lwi, Koshlan Mayer-Blackwell, Tom Pamukcu, Ben Roberts, Maimon Rose, William Ryan, Rav Suri, Mike Whitlock, Rasmus Winther, Ye Xia, Alisa Zhao, several anonymous reviewers, and my Psychology 102 students at UC Berkeley all read sections of the book and provided a combination of encouragement and constructive comments. The book is vastly improved because of their input. Arbel Harpak, Aaron Hirsh, and Sandy Lwi warrant special mention for their detailed comments on large portions of the text. My wonderful students also deserve celebration for their bravery in being the first people to rely on this book as a text, as does Rav Suri for being the first instructor (after me) to adopt it for a course.

This book would have remained merely an idea for a book if it were not for Aaron Hirsh and Ben Roberts. Aaron and Ben—along with my wife, Isabel Edge—convinced me that I might be able to write this book and that it was not entirely ridiculous to try. They were also both instrumental in shaping the content and framing, and they guided me as I navigated the possibilities for publication. Along similar lines, Melanie Mitchell helped me arrive at the scheme that kept me writing consistently for years—requiring 1,000 words of new text from myself each week, with the penalty for dereliction a $5 donation to a pro-astrology outfit.

I would not have had the idea to write this book if I had not been able to train as both an empirical researcher and as a developer of statistical methods. My mentors in these fields—including Tricia Clement, Graham Coop, Emily Drabant, Russ Fernald, James Gross, Sheri Johnson, Viveka Ramel, and Noah Rosenberg—made this possible. My interests were seeded by wonderful teachers, including Cari Kaufman, whose introductory class in mathematical statistics gave me a feeling of empowerment that I have wanted to share ever since, Ron Peet, who did the most to shape my interest in math and gave me my first exposure to statistics, and Bin Yu, who taught me more than I thought possible in one semester about working with data. Gary "GH" Hanel taught me calculus, and I have cribbed his sticky phrases conveying the rules for differentiating and integrating polynomials—"out in front and down by one," and "up and under"—after finding that they (and several of his other memory aids) always stay with me, no matter how much calculus I forget and relearn.

Finally, my family has supported me during the process of writing (and of preparing to write). My parents, Chloe and Don, have always supported me in my goals for learning and growth. Isabel, my wife, has helped me believe that my efforts are worthwhile and has been a listener and counselor over years of writing. Maceo, our three-year-old, has not provided any input that I have incorporated into the text, but we like him very much nonetheless.

Doc Edge
Davis, California

Prelude

Practitioners of every empirical discipline must learn to analyze data. Most students' first—and sometimes only—training in data analysis comes from a course offered by their home department. In such courses, the first few weeks are typically spent teaching skills for reading data displays and summarizing data. The remainder of the course is spent discussing a sequence of statistical tests that are relevant for the field in which practitioners are being educated: a course in a psychology department might focus on t-tests and analysis of variance (ANOVA); an economics course might develop linear regression and some extensions aimed at causal inference; future physicians might learn about survival analysis and Cox models. There are at least three advantages to this approach. First, given that students may take only one course in data analysis, it is reasonable to teach the skills they need to be functional data analysts as quickly as possible. Second, courses that focus on procedures useful in the students' major area of study allow instructors to pick relevant examples, encouraging interest. Third, courses like these can teach data analysis while requiring no mathematics beyond arithmetic.

At the same time, the introduction of test after test in the second part of the course comes with major drawbacks. First, as instructors frequently hear from their students, test-after-test litanies can be difficult to understand. The material that conceptually unites the procedures has been squeezed into a short time. As a result, from the students' view, each procedure is a subject unto itself, and it is difficult to develop an integrated view of statistical thinking. Second, for motivated students, standard introductory sequences can give the impression that, though data may be exciting, statistics is uninteresting. For the student, it can seem as though to master statistics is to memorize a vast tree of assumptions and hypotheses, allowing one to draw the appropriate test from a deck when certain conditions are met. Students who learn this style of data analysis cannot be blamed for failing to see that the discipline of statistics is stimulating or even that it is intellectually rooted. Third, the ability to apply a few well-chosen procedures may allow the student to become a functional researcher, but it is an insufficient foundation for future growth as a data analyst. We have taught a set of recipes—and versatile, germane ones—but we have not trained a chef. When new procedures arise, it will be no easier for our student to learn them than it was for him to learn his first set of procedures. That is to say it will require real labor, and success will depend on an exposition written by someone who can translate statistical writing into the language of his field.

Most university statistics departments train future statisticians differently. First, they require their students to take substantial college-level math before beginning their courses in statistics. At minimum, calculus is required, usually with multivariable calculus and linear algebra, and perhaps a course in real analysis. After meeting the mathematical requirements, future statisticians take an entire course—or two—dedicated strictly to probability theory, followed by a course in mathematical statistics. After at least a year of

Statistical Thinking from Scratch: A Primer for Scientists. M. D. Edge, Oxford University Press (2019).
© M. D. Edge 2019. DOI: 10.1093/oso/9780198827627.001.0001

university-level mathematical preparation and a year of statistics courses, the future statistician has never been asked to apply—and perhaps never even heard of—procedures the future psychologist, for example, has applied midway through an introductory course.

At this stage in her development, the well-trained statistics student may not have applied three-way ANOVA, for example, but she deeply understands the techniques she does know, and she sees the interest and coherence of statistics as a discipline. Moreover, should she need to use a three-way ANOVA, she will be able to learn it quickly with little or no outside assistance.

How can the budding researcher—pressed for time, perhaps minimally trained in mathematics, and needing to apply and interpret a variety of statistical techniques—gain something of the comfortable understanding and versatility of the statistician? This book proposes that the research worker ought to learn at least one procedure in depth, "from scratch." This exercise will impart an idea of how statistical procedures are designed, a flavor for the philosophical positions one implicitly assumes when applying statistics in research, and a clearer sense of the strengths and weaknesses of statistical techniques.

Though it cannot turn a non-statistician into a statistician, this book will provide a glimpse of the conceptual framework in which statisticians are trained, adding depth and interest to whatever techniques the reader already knows how to apply. It is perhaps most naturally used as a main or supplementary text in an advanced introductory course—for beginning graduate students or advanced undergraduates, for example—or for a second course in data analysis. I assume the reader already has an interest in understanding the reasoning underlying statistical procedures, a sense of the importance of learning from data, and some familiarity with basic data displays and descriptive statistics. Prior exposure to calculus and programming are helpful but not required—the most relevant concepts are introduced briefly in Chapter 2 and in Appendices A and B. Probability theory is taught as needed and not assumed. In some departments, the book would be suitable for a first course, but the mathematical demands are high enough that instructors may find they prefer to use the book with determined students who are already invested in empirical research. Another possible adjustment is to split the course into two terms, with the Interlude serving as a prelude to the second course, supplementing with data examples from the students' field. The book is also useful as a self-study guide for working researchers seeking to improve their understanding of the techniques they use daily, or for professionals who must interpret results from research studies as part of their work.

There are many excellent statistics textbooks available for non-statisticians, so any new book must make plain how it differs from others. This book has a set of features that are not universal and that in combination might be unique.

First, this book focuses on instruction in exactly one statistical procedure, simple linear regression. The idea is that by learning one procedure from scratch, considering the entire conceptual framework underlying estimation and inference in this one context, one can gain tools, understanding, and intuition that will apply to other contexts. In an era of big data, we are keeping the data small—two variables, and in the data set we use most often, only 11 observations—and thinking hard about it. In saying that we work, "from scratch," I mean that we attempt to take little for granted, exploring as many fundamental questions as possible with a combination of math, simulations, thought experiments, and examples. I chose simple linear regression as the procedure to analyze both because it is mathematically simple and because many of the most widely applied statistical techniques—including t-tests, multiple regression, and ANOVA, as well as machine-learning methods like lasso and ridge regression—can be viewed as special cases or generalizations of simple linear regression. A few of these generalizations are sketched or exemplified in the final chapter.

A second feature is the mathematical level of the book, which is gentler than most texts intended for statisticians but more demanding than most introductory statistics texts for non-statisticians. One goal is to serve as a bridge for students who have realized that they need to learn to read mathematical content in statistics. Learning to read such content unlocks access to advanced textbooks and courses, and it also makes it easier to talk with statisticians when advice is needed. A second reason for including as much math as I have is to increase the reader's interest in statistics—many of the richest ideas in statistics are expressed mathematically, and if one does not engage with some math, it is too easy for a statistics course to become a list of prescriptions. The main mathematical requirement is comfort with algebra (or, at least, neglected algebra skills that can be dusted off and put back into service). The book requires some familiarity with the main ideas of calculus, which are introduced briefly in Appendix A, but it does not require much facility with calculus problems. The mathematical demands increase as the book progresses (up to about Chapter 9), on the theory that the reader gains skills and confidence with each chapter. Nearly all equations are accompanied by explanatory prose.

Third, some of the problems in this book are an integral part of the text. The majority of the included problems are intended to be earnestly attempted by the reader—exceptions are marked as optional. Every solution is included, either in the back of the book or at the book's GitHub repository, github.com/mdedge/stfs/, or companion site, www.oup.co.uk/companion/edge.[1] The problems are interspersed through the text itself and are part of the exposition, providing practice, proofs of key principles, important corollaries, and previews of topics that come later. Many of the problems are difficult, and students should not feel they are failing if they find them tough—the process of making an attempt and then studying the solutions is more important than getting the correct answers. A good approach for students is to spend roughly 75% of their reading time attempting the exercises, referring to the solution for next steps when stuck for more than a few minutes.

Fourth, several of the problems are computational exercises in the free statistical software package R. Many of these problems involve the analysis of simulated data. There are two reasons for this. First, R is the statistical language of choice among statisticians. As of this writing, it is the most versatile statistical software available, and aspiring data analysts ought to gain some comfort in it. Secondly, it is possible to answer difficult statistical questions in R using simulations. When one is practicing statistics, one often encounters questions that are not readily answered by the mathematics or texts with which one is familiar. When this happens, simulation often provides a serviceable resolution. Readers of this book will use simulation in ways that suggest useful answers. All R code to conduct the demonstrations in the text, complete the exercises, and make the figures is available at the book's GitHub repository, github.com/mdedge/stfs/, and there is also an R package including functions specific to the book (installation instructions come at the end of Chapter 2, in Exercise Set 2-2, Problem 3).

The chapters are intended to be read in sequence. Jointly, they introduce two important uses of statistics: estimation, which is to use data to guess the values of parameters that describe a data-generating process, and inference, which is to test hypotheses about the processes that might have generated an observed dataset. (A third key application, prediction, is addressed briefly in the Postlude.) Both estimation and inference rely on the idea that the observed data are representative of many other possible data sets that might have

---

[1] Instructors can obtain separate questions I have used for homework and exams when teaching from this book. Email the author to request additional questions.

been generated by the same process. Statisticians formalize this idea using probability theory, and we will study probability before turning to estimation and inference.

Chapter 1 presents some motivating questions. Chapter 2 is a tutorial on the statistical software package R. (Additional background material is in Appendices A and B, with Appendix A devoted to calculus and Appendix B to computer programming and R.) Chapter 3 introduces the idea of summarizing data by drawing a line. Chapters 4 and 5 cover probability. An Interlude chapter marks the traditional divide between probability theory and statistics. Chapter 6 covers estimation, immediately putting the fourth and fifth chapters to work. Chapter 7 covers inference: what kinds of statements can we make about the world given a model (Chapters 4 and 5) and a sample? Chapters 8, 9, and 10 describe three broad approaches to estimation and inference. These three perspectives share goals and share the framework of probability theory, but they make different assumptions. In the Postlude, I discuss some extensions of simple linear regression and point out a few possible directions for future learning.

CHAPTER 1

Encountering data

Key terms: Computation, Data, Statistics, Simple linear regression, R.

> If we take in our hand any volume, . . . let us ask, Does it contain any abstract reasoning concerning quantity or number? No. Does it contain any experimental reasoning concerning matter of fact and existence? No. Commit it then to the flames: for it can contain nothing but sophistry and illusion.
>
> David Hume, *An Enquiry Concerning Human Understanding* (1748)

In the passage quoted here, Hume claims that there are two types of arguments we should consider accepting: mathematical reasoning and empirical reasoning—reasoning based on observations about the world. The specifics of Hume's claims are beyond our scope, but users of statistics are safe from followers of Hume's dictum. This book will contain some reasoning concerning number, and it will contain examples of "experimental" reasoning about observed facts. Thus, a good Humean need not commit it to the flames.

Hume's statement gives us a way of thinking about the subject of statistics. We want to make claims about the world on the basis of observations. For example, we might want to know whether a corollary of the wave theory of light matches the results of an experiment. We might ask whether a new therapy for diabetes is effective. We might want to know whether people with college degrees earn more than their peers who do not graduate. Whether we are pursuing physical science, biological science, social science, engineering, medicine, or business, we constantly need answers to questions with empirical content.

But what is Hume's "reasoning concerning matter of fact"? Collecting data about the world is one thing, but using those data to make conclusions is another. Consider Figure 1-1, which purports to show data on fertilizer consumption and cereal yields in 11 sub-Saharan African countries.[1] In many sub-Saharan African countries, soil nitrogen limits agricultural yield. One way to increase soil nitrogen is to apply fertilizer. Each of the 11 countries in the dataset is represented by a point on the plot. For each point, the *x* coordinate—that is, the position on the horizontal axis—indicates the country's fertilizer consumption in a given year. The *y* coordinate—the position on the vertical axis—represents the country's yield of cereal grains in that same year. Suppose that on the basis of Figure 1-1, I claim that there is a robust relationship between fertilizer consumption and cereal yield among countries similar to the 11 countries we have sampled. (Note that I have not made claims about any causal

---

[1] These data are fake—more on their source at the end of the chapter—but the question is real. The data in Figure 1-1 do loosely resemble actual data from some sub-Saharan African countries with low grain yields from 2008 to 2010. For example, in 2010, farmers in Mozambique consumed about 9 kg/hectare of fertilizer, and the cereal yield was about 945 kg/hectare. Actual data are available from the World Bank.

Statistical Thinking from Scratch: A Primer for Scientists. M. D. Edge, Oxford University Press (2019).
© M. D. Edge 2019. DOI: 10.1093/oso/9780198827627.001.0001

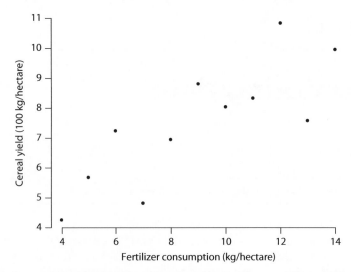

Figure 1-1 Fertilizer consumption and cereal yield in 11 sub-Saharan African countries.

relationships that might explain the relationship; I have merely posited that the relationship "exists" in some sense.)

Suppose you disagree. You might counter that the plot isn't impressive: there aren't many data, the relationship between the variables strikes you as weak, and you don't know the source of the data. These counters are at least potentially legitimate. You and I are at an impasse, Dear Reader: I have made a claim based on data, and you have looked at the same data and made a different claim. Without methods for reasoning about data, it is unclear how to make further progress regarding our disagreement.

How can we develop concepts for reasoning from data? The discipline of statistics provides one answer to this question. Statistics takes Hume's other candidate for non-illusory knowledge—mathematical reasoning—and builds a mathematical framework in which we can set the data.[2] Once we have framed the problem of reasoning about data mathematically, we can make claims by adopting assumptions and then using mathematical reasoning to proceed. The status of the claims we make will usually hinge on the appropriateness of the assumptions we use to get started. As you read about statistical approaches for reasoning about data, consider whether and under what circumstances they are adequate. We will revisit this question in various forms.

In the Prelude, I promised that this book would be only lightly mathematical, yet here, I have proposed that statistics is a way to harness mathematical thinking to reason about data. How will this book help readers strengthen their statistical understanding without engaging in heavy mathematics?

Statistics took shape as a discipline before modern computers were available, with many of the ideas most important to this book appearing in the late nineteenth and early twentieth centuries. Many of the most important statisticians of this era were well-trained mathematicians. With limited computing power but ample mathematical training, they approached the development of their subject mathematically. Today, advances in computing allow those of us with limited mathematical training to answer questions that

---

[2] This is not to suggest that strong mathematical reasoning skills are the sole or even most important qualification of a data analyst. Facility with data and computers, subject-area knowledge, scientific acumen, and common sense are all important.

would be difficult even for a seasoned mathematician to approach directly. We will use computation to answer statistical questions that will not yield to elementary math.

1.1 Things to come

Before we begin in earnest, let's take a moment to anticipate the major topics we'll consider in the rest of the book, motivated by the data in Figure 1-1. We will be focused on understanding simple linear regression, which entails identifying a line that "fits" the data, passing though the cloud of data points in the figure. Linear regression—including both simple linear regression and its generalization, multiple regression—is perhaps the most widely used method in applied statistics, especially when its special cases are considered, including *t*-tests, correlation analysis, and analysis of variance (ANOVA). It only takes a few commands to run a simple linear regression analysis in the statistical software R. (A tutorial in R is coming in the next chapter.) The data are stored in an R object called anscombe, and to fit the linear regression model, we run

```
mod.fit <- lm(y1 ~ x1, data = anscombe)
```

As will be discussed later, the lm() function fits a linear regression model to a dataset. By "fitting" a regression model, we find a line that "best" fits the data shown in Figure 1-1. You can see the data from Figure 1-1 with the "best fit" line drawn in by typing

```
plot(anscombe$x1, anscombe$y1)
abline(mod.fit)
```

The plot with the line drawn in (plus a few improvements to labeling and aesthetics[3]) is shown in Figure 1-2. The plot() function produces a scatterplot—that is, a plot with points located to indicate values of the attributes represented by the *x* and *y* axes.

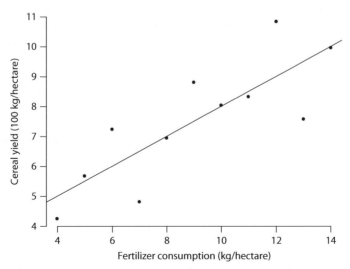

Figure 1-2 The agriculture data from Figure 1-1, with the line of "best" fit from the simple linear regression model.

---

[3] Code for generating all the book's R figures is available at github.com/mdedge/stfs.

The abline() function draws the line implied by the linear regression model. The sense in which this line can be described as the "best" fit is the subject of Chapter 3. As we will see, there are actually many different lines that could be described as "best." The line in Figure 1-2 is best according to a criterion that has a long history in statistics.

The line that's drawn in Figure 1-2 has an equation, meaning that it can be described as $y = a + bx$, where a and b are constant numbers. In words, to find the y coordinate of the line at any value x, one starts with a and then adds the product of b and x. The values of a and b for the pictured line are given in the output of the summary() function:

```
summary(mod.fit)
```

The output is

```
Call:
lm(formula = y1 ~ x1, data = anscombe)
Residuals:
      Min       1Q    Median       3Q       Max
-1.92127 -0.45577 -0.04136 0.70941 1.83882
Coefficients:
             Estimate  Std.  Error    t   value Pr(>|t|)
(Intercept)    3.0001         1.1247      2.667     0.02573 *
x1             0.5001         0.1179      4.241     0.00217 **
- - -
Signif. codes: 0 '***' 0.001 '**' 0.01 '*' 0.05 '.' 0.1 ' ' 1
Residual standard error: 1.237 on 9 degrees of freedom
Multiple R-squared: 0.6665,  Adjusted R-squared: 0.6295
F-statistic: 17.99 on 1 and 9 DF, p-value: 0.00217
```

The key part of the output is the regression table, which is printed in bold here. In the first column of the table, labeled "Estimate," we see the numbers 3 and 0.5. These are the values of the intercept, a, and slope, b, of the line in Figure 1-2. The word "estimate" suggests a way of viewing the line in Figure 1-2 that is different from the one suggested earlier. I initially suggested that the line in Figure 1-2 is the one that "best" fits the data in some sense—in other words, that it is a description of the sample. That's true. The word "estimate" suggests that it is *also* a guess about some unknown quantity, perhaps about a property of a larger population or process that the sample is supposed to reflect. If we assume that data are generated by a particular process, then we can make claims about the types of data that might result. That is the topic of Chapters 4 and 5, on probability theory. And once we have decided that we really do want to make estimates—that is, to use data to learn the parameters of an assumed underlying process—how should we design procedures for estimation? What properties should these procedures have? That is the subject of Chapter 6.

Moving rightward in the table, we see a column labeled "Std. Error," an abbreviation for "standard error." The standard error is an attempt to quantify the precision of an estimate. In a specific sense developed later, the standard error responds to the question, "If we were to sample another dataset from the same population as this, by about how much might we expect the estimate to vary?" So the 0.1179 in the table suggests that if we drew another sample of 11 points generated by the same underlying process—in the example, perhaps agricultural data from 11 other sub-Saharan African countries—we should not be surprised if the slope of the best-fit line differs by ~0.12 from the current estimate. The attempt to identify the precision of estimates is called "interval estimation," and it is one of the topics of Chapter 7.

The final column of the regression table is labeled "Pr(>|t|);" these numbers are called "p values." Their interpretation is subtle and often botched. Loosely, p values measure the plausibility of the data under a specific hypothesis. The hypotheses being tested here are that the data were actually generated by a process described by a line with intercept 0 (first row) or slope 0 (second row). Low p values, like the ones in the table, suggest that (a) the hypotheses are false, or (b) some other assumption entailed by the hypothesis test is wrong, or (c) an unlikely event occurred. Hypothesis testing is the other subject—besides interval estimation—of Chapter 7.

I have alluded to "underlying" assumptions, and the goal of much of the rest of the book is to illustrate the ways in which such modeling assumptions are involved in statistical analysis. Depending on the assumptions that the data analyst can justify, different sets of statistical procedures become available. In Chapters 8, 9, and 10, we apply different sets of assumptions to the dataset to arrive at different procedures for point estimation, interval estimation, and hypothesis testing. The assumptions underlying the standard regression table produced by lm() are the same as those used in Chapter 9. In the Postlude, we consider ways in which the principles developed in the book apply to statistical analyses of other sorts of datasets, ones that are not a natural fit for simple linear regression analysis.

I'd like to close this chapter with an argument for undergoing such an extended meditation on simple linear regression. After all, empirical researchers are busy, and it's possible to teach heuristic interpretations of statistics like those in the regression table. Such teaching is quick, and it allows researchers to get to serviceable answers, at least in the easiest cases. Why spend so much time on statistical thinking? Why not outsource the theory to professional statisticians?

The answer is part honey, part vinegar. On the positive side, data analysis is much more fun and interesting when the analyst has a genuine sense of what she is doing. With some understanding of statistical theory, it's possible to relate scientific claims to their empirical basis, connecting the data to the mathematical framework that justifies the claim. That breeds confidence in researchers, and it also allows for creativity. If you know how the machine works, you can take it apart and repurpose it.

In contrast, rote approaches that rely on heuristics alone can be unsatisfying, anxiety-inducing, creatively stifling, and/or genuinely dangerous. Here is a cautionary anecdote. Suppose you fit a linear model, as we did before, to a new dataset. Whereas before we fit a simple linear regression to the variables y1 and x1 in the anscombe dataset, we now work with variables y3 and x3:

```
mod.fit2 <- lm(y3 ~ x3, data = anscombe)
summary(mod.fit2)
```

The resulting regression table includes

```
Coefficients:
            Estimate  Std. Error  t value   Pr(>|t|)
(Intercept)  3.0025      1.1245      2.670   0.02562 *
x3           0.4997      0.1179      4.239   0.00218 **
```

Each entry in the table is nearly identical to its counterpart table for the earlier model. For a rote data analyst relying on just the model summary, the interpretation of these models would therefore be the same. But look at the data underlying this analysis, shown in Figure 1-3. Whereas the data in Figure 1-2 seem to be randomly scattered around the line, points in Figure 1-3 appear to follow a much more systematic pattern, with the exception of

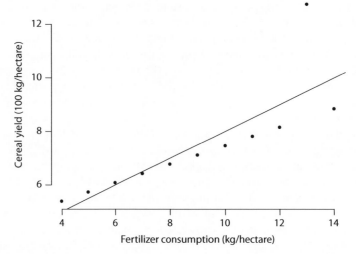

Figure 1-3 The data underlying the analysis of the variables y3 and x3 in the anscombe data set.

one point that departs from it. Whereas the line in Figure 1-2 seems like an appropriate summary of the data, Figure 1-3 suggests a need for serious inspection. What is with that outlying point? Why are the others arranged in a perfectly straight line? These questions are urgent, and the regression table can't be interpreted without knowing their answers.[4] In other words, relying strictly on an automatic response to the regression table will lead to foolish conclusions.

In the remainder of the book, we will provide a basis for a complete interpretation of the numbers in the regression table, of the questions they are trying to answer, and of the ways in which their interpretation depends on the assumptions we are willing to make.

---

[4] These data are drawn from a famous paper by Francis Anscombe, including four fake datasets that give exactly the same regression results but whose plots suggest wildly different interpretations. The paper is "Graphs in statistical analysis" from 1973 in *American Statistician*.

CHAPTER 2

R and exploratory data analysis

Key terms: Dataframe, Exploratory data analysis, `for()` loop, R, R function, Sample mean, Scatterplot, Vector

> I program my home computer
> Beam myself into the future.
>
> Kraftwerk, "Home Computer"

Virtually all statistical computations performed for research purposes are carried out using statistical software. In this book, we will use R, which is a programming language designed for statistics and data analysis.

For many students, R is more difficult to learn than most proprietary statistical software. R uses a command-line interface, which means that the user has to write and enter commands rather than use a mouse to select analyses and options from menus.[1] Once you become comfortable with R, you will find that it is more powerful than the options that are easier to learn.

Why use R if it is more difficult than some of the alternatives? There are several reasons:

(1) Community: R is the computing *lingua franca* of professional and academic statisticians and of many data analysts from other fields. There is an active community generating new content and answers to questions.

(2) Adaptability: R users can write packages that add to R's capabilities. There are thousands of packages available to handle specialized data analysis tasks, to customize graphical displays, to interface with other programs and software, or simply to speed up or ease typical programming tasks. Some of them are wonderful. The adaptability of R means that new statistical techniques are available in R years or even decades before they become available in proprietary packages.

(3) Flexibility: Suppose you want to perform a statistical procedure but modify it slightly. In a proprietary language, this is often difficult—the code used to run the procedure is kept hidden, and the only parameters you are allowed to change are the ones included as options and shown to the user. In R, it is much easier to make changes.

---

[1] There are some graphical user interfaces (GUIs) for R available, such as R Commander. Another way to make R more user-friendly is to use an interactive developer environment (IDE), such as RStudio. In this book, I will assume that you are using R without a GUI or IDE, but you are welcome to use one, and RStudio is recommended.

Statistical Thinking from Scratch: A Primer for Scientists. M. D. Edge, Oxford University Press (2019).
© M. D. Edge 2019. DOI: 10.1093/oso/9780198827627.001.0001

(4) Performance: Compared with many proprietary packages, R is faster and can work with larger datasets. R is not the fastest language available, but it is usually more than fast enough, and when you really need speed, you can program R to interface with faster languages like C++.

(5) Ease of integrating simulation and analysis: In this book, we answer several questions about probability and statistics by simulation. That is, we ask questions like, "What would happen if we used such-and-such estimator on data drawn from such-and-such distribution?" Sometimes, it is possible to answer such questions mathematically, but it is often easier to simulate data of the type we are interested in, apply the technique we are interested in, and see what happens. R provides a framework for carrying out that procedure.

(6) Price: R is free software under the GNU General Public License. The most obvious advantage is that it won't cost you anything, whereas some of the proprietary packages cost hundreds or thousands of dollars per year for a license. More importantly, R's community (point 1) is broader than it would otherwise be because R is free.

I hope that you are convinced that learning R is a good idea for anyone with an interest in statistics or data analysis.[2] In addition, if you have never programmed before, I would suggest that learning one's first programming language is one of the more rewarding intellectual experiences one can have. To program a computer successfully is to be logical, explicit, and correct—few other pursuits force us to take on these qualities in the same way and give us such clear feedback when we have fallen short.

We will focus on a subset of R's features, including simulation, use of R's built-in datasets, basic analyses, and basic plotting. I hope that learning these aspects of R will motivate you to explore its many other features. There are dozens of books and online tutorials that can teach you more about R. A few good ones are listed at the end of the chapter. In the remainder of this chapter, you will install R and complete a tutorial that will prepare you for the exercises in subsequent chapters. More information on basic R commands and object types is available in Appendix B.

Exercise Set 2-1

1) Download and install the current version of R on a computer to which you have regular access. R is available from the Comprehensive R Archive Network (CRAN) at http://www.r-project.org. (For Linux users, the best procedure will depend on your distribution.)
2) Open R. When you have succeeded, you should see a window with the licensing information for R and a cursor ready to take input.
3) (Optional, but recommended) Download and install RStudio. Close your open R session and open RStudio. If you prefer the RStudio interface, then use RStudio to run R for the rest of the exercises in the book.
4) Visit the book's GitHub repository at github.com/mdedge/stfs/ to view additional resources, including R scripts, among which is a script to run all the code in this chapter.

[2] I do not mean to suggest that R has no disadvantages. R cannot return your love:

```
> I love you, R
Error: unexpected symbol in "I love"
```

2.1 Interacting with R

The most important thing you can learn about R is how to get help. The two built-in commands for doing so are `help()` and `help.search()`. The `help()` command directs you to information about other R commands. For example, typing `help(mean)` at the prompt and hitting return will open a web browser and take you to a page with information about how to use the `mean()` function, which we will see a little later. One downside of `help()` is that you have to know the name of the R function you want to use. If you don't know the name of the function you need, you can use `help.search()`. For example, if we didn't know that `mean()` was the function we need to take the mean of a set of numbers, we could try `help.search("mean")`. This brings up a list of functions that match the query `"mean"`. In this specific case, `help.search()` is not too helpful—a lot of functions come up, and the one we want—`base::mean()`—is buried. (We see "`base::mean()`" because `mean()` is in the `base` package, which loads automatically when R is started.) When you don't know the name of the function you need, a web search is usually more helpful.

As you progress in your use of R, you will find `help()` to be increasingly useful. But as you start, you may find the `help()` pages to be hard to understand—one has to learn a little about R before they make sense. This tutorial and the information in Appendix B will help you get comfortable. After that, you can switch to using some of the resources at the end of the chapter. My single favorite resource for beginners is Rob Kabacoff's free website Quick-R (http://www.statmethods.net/), which has help and examples for most of the tasks you'll need to perform frequently in R. If you have a specific question, search for it; it has likely been answered on the R forum or Stack Overflow (www.stackoverflow.com/).

Once you have opened R, you will see a prompt that looks like this:

```
>
```

The simplest way to view R is as a program that responds directly to your commands. You type a command followed by the return key, and R returns an answer. For example, you can ask R to do arithmetic for you:

```
> 3+4
[1] 7

> (9*8*7*sqrt(6))/3
[1] 411.5143
```

Here, the "`[1]`" before the answer means that your answer is the first entry of a *vector* that R returned, which in this case is a vector of length 1. **Vectors** are ordered collections of items of the same type. Both of the commands above are *expressions*—one of the two major types of R commands. R evaluates the expression and prints the answer. The answer is not saved. The other major type of R command is *assignment*, which we will see shortly.

Anything that appears after a "#" sign on a line is treated as a comment and ignored. So, as far as R is concerned,

```
> #The next line gives the sum I need
> 2 + 3 #2 and 3 are important numbers
```

is the same as

```
> 2 + 3
```

Writing good comments is essential to writing readable code. Comments will help you understand and debug code after you have set it aside.

You can use the "up" and "down" arrow keys in the R terminal to return to commands that you have already entered. The "up" arrow brings up your most recently issued command; pressing "up" again brings up the command issued before that, etc. This tactic is especially helpful when you need to re-enter a command with a modification.

You can store values of variables in R. For example, if you wanted to assign a variable called "x" to equal 7, you would type

```
> x <- 7
```

The combination of keystrokes "<-" is used to assign values to variables. The spaces after the x and before the 7 can be omitted without affecting the way the command works. In general, R is flexible about spaces that do not interrupt the names of variables or functions. You can also type

```
> x = 7
```

to do the same thing. Both of the previous two commands are *assignment* commands. In contrast to the values that result from an expression command, the values of assignments are not printed. Instead, they are stored for later use.

To see the value currently assigned to x, you can type

```
> x
[1] 7
```

You can use x in computations in the same way you would use the value assigned to x:

```
> x*7
[1] 49
```

Three important notes arise here. First, R is case-sensitive. If you try to refer to "x" as a capital "X", you will be rewarded with an error message:

```
> X*7
Error: object 'X' not found
```

More generally, R cannot find an object or call a function whose name has been misspelled. This is a frequent source of vexation for beginners, and an occasional one throughout one's R-using life. If R returns errors you do not expect, check carefully for typos.

Second, it is possible to have commands that are spread over multiple lines. If you enter a partial command and then hit return, you will see the typical ">" prompt replaced by a "+" symbol. The "+" indicates that R needs more input before it can evaluate a command. It is easy to miss a closing bracket or parenthesis, for example,

```
> (1+3+5)/(2+4+6
+ )
[1] 0.75
```

When the "+" prompt appears, finish the command. If you don't know how to finish the command, you can get a new prompt with the escape key on a Windows or Mac or with ctrl-c on a Linux machine.

You can see that the particulars of what you type are important. Though you can type all your commands directly into R, it is much better to save your R commands in a separate file—this allows for easier correction of typos and quick reproduction of analyses. When you use a text editor to save your commands, do not include the prompt ">" on each line— the prompt symbol is built into R and is not part of the input.

A plain text editor will work fine for saving your commands. Do not use a program that adds formatting to your text, such as Microsoft Word, because the formatting can interfere with the commands themselves. Many text editors, such as gedit, will helpfully highlight your R code to enhance readability if you save your file with a .R extension. There are also interactive development environments that simultaneously highlight code, track variables, and can feed your code straight into the R console. If you want to use an interactive development environment, RStudio is excellent.

Of course, you will want to use R for more than arithmetic. In the next section, we will work through some data summaries and graphical procedures.

2.2 Tutorial: the *Iris* data

It is time to write some code. Remember to write R code in a separate text editor—you can then paste it into the R console.[3] You ought to execute all the code in this section on your own computer. A script including all the code in this section (as well as other scripts, including all the code in this book) is available at github.com/mdedge/stfs/.

In this tutorial, we will conduct some **exploratory data analysis** of the R dataset iris, which also discussed in Appendix A. The iris dataframe is built into R, along with many other datasets.[4] The iris dataframe includes a set of measurements on 50 iris flowers from each of three different species—*Iris setosa*, *Iris virginica*, and *Iris versicolor*.

The iris dataset is built into R; you do not need to do anything to install it. You can see the whole dataset by entering iris at the command line. It is usually more useful to examine just the first few rows of a dataframe, which you can see using the head() function:

```
> head(iris)
  Sepal.Length  Sepal.Width  Petal.Length  Petal.Width  Species
1          5.1          3.5           1.4          0.2  setosa
2          4.9          3.0           1.4          0.2  setosa
3          4.7          3.2           1.3          0.2  setosa
4          4.6          3.1           1.5          0.2  setosa
5          5.0          3.6           1.4          0.2  setosa
6          5.4          3.9           1.7          0.4  setosa
```

head() is an **R function**. Like mathematical functions (see Appendix A), R functions take inputs, or arguments, and return outputs. In addition to the functions built into R, thousands more are available in add-on packages, and you can also write your own functions.

---

[3] Two other ways to transfer code from the text editor to R: (1) If you are using RStudio, you can highlight code written in the source file and run it by holding ctrl (or, on a Mac, cmd) and hitting the return key. (2) The source() function is another way to run all the commands written in a text file. See help(source).

[4] You can see the names of all the built-in datasets using library(help = "datasets"). Typically, one can refer to a built-in dataset simply by typing its name in the R console, unless that name has been used for something else in the session. The built-in datasets will not be listed by the ls() function (or shown among the objects in the environment in RStudio) unless they are loaded with the data() function, as in data ("iris").

We have explicitly specified the argument iris, indicating in this case that we want to see the first few lines of the object named iris. There is another argument to head() that we can leave unspecified if we are fine with the default values, but which we could also change if we wanted. For example, the command head(iris, n = 10) would produce the first 10 lines of iris, rather than the first 6 lines. Function arguments have names—the equals sign in n = 10 indicates that we are assigning the argument named n the value 10. There is some flexibility in the explicit naming of arguments—for example, we did not write x = iris even though the first argument of head() is named x. The rule is that if the arguments are given in the order the function expects them—in the case of head(), x first and n second—then one does not have to name them. For example, these five commands give the same results

```
> head(iris, 10)
> head(iris, n = 10)
> head(x = iris, 10)
> head(x = iris, n = 10)
> head(n = 10, x = iris)
```

But if the arguments are given in an order different than R expects and not named, you will have problems. The call head(10, iris), for example, gives an error. You can see the arguments that each function expects in order using help().

iris is a **dataframe**, which means that it can hold rows and columns of data, that the rows and columns can be named, and that different columns can store data of different types (numeric, character, etc.). Here, we can see that each row contains an individual flower, and the columns contain measurements of different features of each flower. Each column of the dataframe is a vector, an ordered collection of data of one type.

We can quickly gain some useful information using the summary() function:

```
> summary(iris)
 Sepal.Length     Sepal.Width      Petal.Length     Petal.Width
 Min.   :4.300    Min.   :2.000    Min.   :1.000    Min.   :0.100
 1st Qu.:5.100    1st Qu.:2.800    1st Qu.:1.600    1st Qu.:0.300
 Median :5.800    Median :3.000    Median :4.350    Median :1.300
 Mean   :5.843    Mean   :3.057    Mean   :3.758    Mean   :1.199
 3rd Qu.:6.400    3rd Qu.:3.300    3rd Qu.:5.100    3rd Qu.:1.800
 Max.   :7.900    Max.   :4.400    Max.   :6.900    Max.   :2.500
       Species
 setosa    :50
 versicolor:50
 virginica :50
```

If we would rather not see results for the whole dataset at once, we can probe the individual variables. We refer to individual variables in a dataframe by typing the name of the dataframe first, then $, and then the name of the variable we want to reference. Try entering iris$Sepal.Length into the R terminal. Then try giving that statement as an argument to summary(), as in

```
> summary(iris$Sepal.Length)
   Min. 1st Qu.  Median    Mean 3rd Qu.    Max.
  4.300   5.100   5.800   5.843   6.400   7.900
```

If we would rather not see all the information provided by the summary, we can also ask for specific values, like the **sample mean** or median:[5]

```
> mean(iris$Sepal.Length)
[1] 5.843333
> median(iris$Sepal.Length)
[1] 5.8
```

Histograms provide a visual summary of the data for one variable.

```
> hist(iris$Sepal.Length)
```

produces a basic histogram of the sepal length data. You can improve it by changing the axis label using the xlab argument. For example, the command

```
> hist(iris$Sepal.Length, xlab = "Sepal Length", main = "")
```

produces a plot similar to the one shown in Figure 2-1. (I have made some alterations to the figure for aesthetic reasons; you can view the code used to make Figure 2-1—and all the other R figures in the book—at github.com/mdedge/stfs/.) Notice that the arguments are separated by commas and that when we want to refer to strings of characters rather than named variables, we put text in quotations. You can see other options for hist() using help(hist).

The *Iris* data are well-known because R. A. Fisher—probably the twentieth century's most important statistician—used them in a famous study.[6] When Fisher examined the *Iris* data,

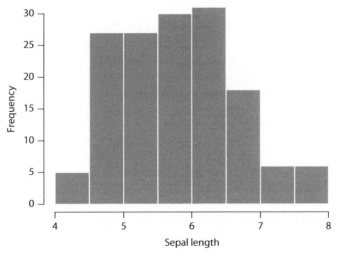

Figure 2-1 Histograms show the empirical distribution of one-dimensional data. The vertical axis shows the count (or proportion) of the observations that fall in the range shown on the horizontal axis.

[5] The mean is the arithmetic average, the sum of the observations divided by the number of observations in the set being summed. A median is a number that falls at the 50th percentile of the data—a number than which half of the observations are greater and half of the observations are smaller.

[6] The study (Fisher, 1936) appeared in the *Annals of Eugenics*. It is distressing to learn but important to acknowledge that the development of statistics in the late nineteenth and early twentieth centuries was

he wanted to find a way to distinguish the three species of iris from each other using measurements of their sepals and petals. We can ask questions similar to those Fisher asked by using a few R commands.

Let's see how we could use sepal measurements to make distinctions between species. First, we need a way of extracting just the data that are associated with a particular species. We will have to make use of the Species variable. Species is a factor. For our purposes, it is sufficient to know that factors are a special type of numeric data for encoding category information, and factor values can be associated with a text label. It is these text labels we see when we examine Species using head(iris). We can use the text labels to extract data associated with a specific species. For example, to see the mean sepal length for *Iris setosa* flowers, we could use:

```
> mean(iris$Sepal.Length[iris$Species=="setosa"])
[1] 5.006
```

The statement in square brackets can be interpreted as a condition that must be met for an observation to be included in the evaluation. In English, a literal rendering of the statement might be the mouthful, "Return the mean of the set of sepal length observations in the iris dataframe that corresponds to entries in the iris dataframe whose value for the species variable label is 'setosa'." Notice that we use the double-equals sign "==" to evaluate whether two objects—in this case, the entries in two vectors—are equal. This is important: a single equals sign *assigns* the object on the left to have the value on the right, but two equals signs *checks* whether the object on the left is equal to the object on the right. To see how this works, see what you get when you enter just iris$Species=="setosa" at the command line, and then try iris$Sepal.Length[iris$Species=="setosa"].

If the mean sepal lengths differ for different species, then sepal length might help distinguish the species. One way to see the mean sepal length values for each species is to use a **for() loop**. Using a for() loop is not the simplest or best way to solve this problem,[7] but for() loops will be a handy way to solve some exercises we will see later, so I'll use this problem as an excuse to introduce them.

A for() statement has two parts: loosely, a statement of what R ought to do, and a statement of how many times R ought to do it. The statement of how many times R ought to repeat the intended action comes first, in parentheses. Following the parentheses, we use curly brackets to enclose the instructions about what R ought to do. (Generally, you can think of curly brackets as a way of grouping together sets of commands.) Here is a simple for() loop that prints a few numbers:

---

connected to eugenics. Statistical innovators including Francis Galton, Karl Pearson, and R. A. Fisher held—and to varying degrees, were motivated by—eugenic views. (Others did not participate in the eugenics movement, such as Yule and Edgeworth.) For one historical perspective on the early connections and later separation of statistical and eugenic thinking, see Louçã (2009).

[7] One superior option is tapply(), part of the apply() family of functions. The function call

```
> tapply(iris$Sepal.Length, iris$Species, mean)
```

gives the means by species, faster (imperceptibly so in this case), with better labels. tapply() is also flexible once you learn it. Another option is

```
> aggregate(iris$Sepal.Length, list(iris$Species), mean)
```

Even though it is often possible to do things more effectively without a for() loop than with one, for() loops are worth learning.

```
> for(i in 1:3){
    print(i)
}
[1] 1
[1] 2
[1] 3
```

Let's break this down. The first part, `for(i in 1:3)`, told R to define a variable i, called an index, as a vector including the numbers 1, 2, and 3. (The `a:b` notation is a shorthand way to tell R to make a vector of all the integers from a to b, inclusive.) The second part, inside the curly brackets, `print(i)`, told R what we want to do with each entry in i—we want to print it to the console. Loosely, in English, the for statements reads, "cycle through each entry in a vector i, containing the integers 1, 2, and 3 [that's the first line], and print the entry [that's the part in brackets]." We could have gotten exactly the same output with the sequence of commands

```
> i <- 1
> print(i)
[1] 1
> i <- 2
> print(i)
[1] 2
> i <- 3
> print(i)
[1] 3
```

In this sequence of commands, we are manually reassigning the value of i and repeating the command to print it each time. That is exactly what R does—it reassigns the value of the index variable and then repeats the commands enclosed in curly brackets—but the `for()` loop is a much less cumbersome way of specifying the sequence, especially when the loop needs to be repeated many times.

We can use `for()` loops to do more interesting things. For example, the following code prints the mean sepal lengths for irises of each of the three species in the dataset:

```
> for(i in unique(iris$Species)){
    print(mean(iris$Sepal.Length[iris$Species == i]))
}
[1] 5.006
[1] 5.936
[1] 6.588
```

Some comments are in order. We started the `for()` statement by writing `for(i in unique (iris$Species))`. The command `unique(iris$Species)` produces

```
[1] setosa versicolor virginica.
```

This is a vector with three entries. Remember that vectors are ordered collections of items of the same type. This vector contains every unique entry in the `Species` variable exactly once. We are instructing R, "Please complete the following action once for every element in the vector `unique(iris$Species)`." Further, we are telling R that in the commands that follow, we will refer to the element of `unique(iris$Species)` that the loop is currently working with as i. Then, in curly brackets, we asked R to print the mean sepal length for flowers of species i.[8] That is, we tell R, "for each species in unique(iris$Species), print the mean sepal length for the observations that are members of that species."[9]

Having examined the mean sepal lengths for each species, we might want to examine the sepal widths. The most straightforward approach would be to re-run the above code, replacing Sepal.Length with Sepal.Width. However, if you find yourself copying and pasting code and replacing variable names, you can often save yourself trouble by writing a custom function. R functions, like mathematical functions, take a set of arguments and return an output. In this case, we want a function that gives mean values of subsets of a vector. In particular, we want mean values of subsets that correspond to species, but we can write our function to be more flexible than that. Let's write a function that prints the mean value of subsets of a vector x, with subsets specified by a user-supplied vector y:

```
conditional.mean <- function(x, y) {
    for(i in unique(y)) {
        print(mean(x[y == i]))
    }
}
```

Notice that inside the first set of curly brackets, we have exactly the same code we used to find the mean value of sepal length for each species, but with iris$Sepal.Length replaced by x and iris$Species replaced by y. We can now run the function conditional.mean(), setting x and y to be vectors we want to examine. To get the same results we got above, we can write

```
> conditional.mean(x = iris$Sepal.Length, y = iris$Species)
[1] 5.006
[1] 5.936
[1] 6.588
```

And now, we can get the mean sepal width value for each species using

```
> conditional.mean(x = iris$Sepal.Width, y = iris$Species)
[1] 3.428
[1] 2.77
[1] 2.974
```

This case is simple enough that using function() doesn't save time,[10] but it is crucial to know how to use function(). It will save time and trouble later.

---

[8] It may be worthwhile to emphasize the distinctions between three types of bracketing characters that have different roles in R: parentheses, square brackets, and curly brackets. Parentheses are used to enclose the arguments to functions—like head(iris, n = 10)—and to indicate order of operations in mathematical expressions. Square brackets are used for data extraction and subsetting. For example, if x is a numeric vector, then x[3] extracts the third element of x, and x[x > 5] extracts all elements of x greater than 5. Curly brackets are used to enclose blocks of commands intended to be executed together, such as the commands that are to be executed during each iteration of a for() loop.

[9] It is customary but not necessary to indent sections of code that occur inside for() or if() statements. The indentations help people read the code.

[10] For example, it would be quicker to use either tapply() or aggregate(), as noted in an earlier footnote. Using R effectively requires both the ability to write your own functions when convenient and the ability to find existing functions that can complete your tasks. Especially while you are learning, it is worth the time to search for functions before writing your own.

Using what we have learned about the mean sepal lengths for each species, we might guess that if we find an iris flower with a long sepal, it is more likely to be from *Iris virginica* than from *Iris setosa*. But to assess how useful sepal length is for species classification, we need to know more. To what extent do the distributions of sepal length overlap? The histogram gives us precise information about the whole distribution of sepal length data for any one species in the study, but it is sometimes an awkward tool for comparing many distributions to each other—it can become difficult to keep track of histograms stacked or overlaid on top of each other. The box plot—though not the most attractive picture[11]—is more to the purpose. We can visualize the entire distribution of sepal widths for several species at once with the box plot. The command to make the appropriate box plot in R is

```
boxplot(iris$Sepal.Length ~ iris$Species)
```

The `boxplot()` function has a feature we have not yet seen. Its argument is written in formula notation. Formula notation has a *dependent* variable on the left and one or more *independent* variables on the right after a tilde. In this case, the independent variable is species, and the dependent variable is sepal length. We will see formula notation again.

We can add appropriate axis labels to produce a display like the one in Figure 2-2.

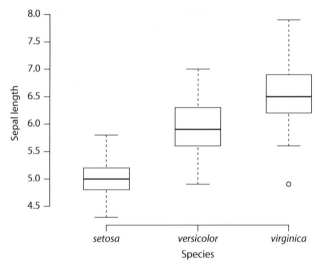

Figure 2-2 Box plots compare distributions of several sets of data. For each flower type on the horizontal axis, the bold horizontal line is the median sepal length. The lower edge of the "box" shows the 25th percentile of the sepal length data, and the upper edge is shows the 75th percentile. The end of the lower "whisker"—whiskers are the dashed lines extending from the box—shows the minimum sepal length, and the end of the upper whisker shows the maximum sepal length. The exception to this occurs when there are points defined to be outliers. Outliers are points that are far removed from the rest of the data, where what is "far removed" is determined as a function of the position and length of the box. When there are outliers, the ends of the whiskers represent the most extreme points that are not outliers, and outliers are drawn in as individual points. There is an outlier for sepal length in the *Iris virginica* data.

---

[11] Prettier alternatives include violin plots, kernel density estimates, and beanplots (Kampstra, 2008). But box plots are both classic and extremely easy to implement.

```
title(xlab = "Species", ylab = "Sepal Length")
```

For each species, the middle 50% of the sepal length observations—that is, the range between the 25th and 75th percentiles,[12] also called the interquartile range—is represented by a box, with the median sepal length represented by a line through the box. The full range of the data is represented by the whiskers extending from the box, with outliers represented by individual points. Figure 2-2 shows that sepal length is informative about species membership. A flower with a sepal length of 7.5 cm is likely to be from *Iris virginica*, and one with a sepal length of 4.5 cm is likely to be from *Iris setosa*. Still, we cannot make perfectly accurate classifications on the basis of sepal length alone. For example, a flower with a sepal length of 5.7 cm could conceivably come from any of the three species. To improve our ability to classify, we can look at multiple variables rather than relying strictly on sepal length.

Scatterplots allow us to examine the joint distribution of two or three variables at once. In R, the plot() function makes scatterplots. Let's use a scatterplot to assess whether we could identify *Iris* species more accurately by considering both sepal width and sepal length. Start with a simple scatterplot:

```
plot(iris$Sepal.Length, iris$Sepal.Width)
```

This doesn't help us with our classification question, though, because we can't tell the species apart. One way to fix the problem is to plot each species using a separate symbol. The pch option of plot() allows us to change the symbol used to plot a point. As of this writing, 25 numbered symbols can be specified with pch.[13] We set *setosa* to be plotted with the symbol corresponding to 1, *versicolor* to be plotted with the second symbol, and *virginica* to be plotted with the third symbol. Adding appropriate axis labels results in the following command:

```
plot(iris$Sepal.Length, iris$Sepal.Width, pch = as.numeric(iris$Species),
xlab = "Sepal Length", ylab = "Sepal Width")
```

A legend makes the plot easier to read:

```
legend("topright", pch = c(1,2,3), legend = c("setosa", "versicolor",
"virginica"))
```

The c() command is for "concatenate," and it joins the comma-separated items in parentheses into a vector. So

```
> c(1,2,3)
[1] 1 2 3
```

gives a vector with entries 1, 2, and 3. Adding the legend results in an image like Figure 2-3.

Figure 2-3 suggests that using sepal width and sepal length alone, it is possible to distinguish *setosa* from the other two species with considerable accuracy. The scatterplot does not

---

[12] The *m*th percentile is a value than which *m* % of the data are smaller. For example, 25% of the observations have values smaller than the 25th percentile, and 75% of the data have values smaller than the 75th percentile. The 1st and 3rd "quartiles" are the 25th and 75th percentiles, respectively.

[13] You can see all the numbered pch options with the command plot(1:25, pch = 1:25) or with a web search for "R pch" or "CRAN pch".

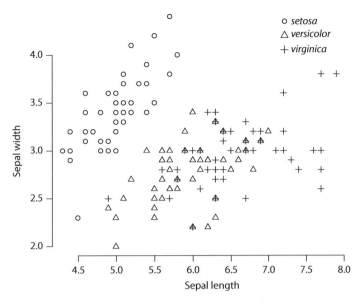

Figure 2-3 A scatterplot of the iris data. Each point represents an observation—in this case, a flower. The *x* coordinate of a point gives the sepal length of the flower, and the *y* coordinate gives the sepal width. The three species are plotted with different symbols.

give us a formal rule for making the classification, but it does suggest that such a rule might be possible—*setosa* flowers are off by themselves in the upper-left corner of the plot, with wide but short sepals. The scatterplot also reveals that sepal width and length are not sufficient to distinguish *Iris versicolor* from *Iris virginica*. The *versicolor* and *virginica* points are interspersed in a way that would make it impossible to separate them completely without further information. R. A. Fisher was in fact able to distinguish all three *Iris* species by jointly using sepal width, sepal length, petal width, and petal length. The method he developed for doing so—linear discriminant analysis—is beyond our scope but closely related to ideas we will encounter later.

You now have a working knowledge of R—you have encountered dataframes, extracted data, wrestled with varying data types, written a for() statement and a function, and made plots. You will have chances to practice all these skills in the exercises and the following chapters.

Exercise Set 2-2

1) Do the same analyses from the tutorial (i.e., histogram, take means, box plots, and scatter-plot) for the petal length and petal width variables from the iris dataset. (This will be easy if you saved your code as you worked through the tutorial.) What do you see?

2) A major reason for R's utility is the wide variety of packages available. In this exercise, we will install and use the gpairs package (Emerson et al., 2013). The gpairs package will allow us to access a lot of the information we gathered in the tutorial using just one command.

   (a) Install gpairs. If you are using R for Mac or Windows, there will be a "Packages" menu, and you can select the "install packages" option. On any operating system, you also have the option of using the command

   ```
   > install.packages("gpairs")
   ```

You will be asked to choose a "mirror"—this is just a server from which you will download the package. Picking the one that is geographically closest to you will usually give the fastest installation. You only need to install a package once.

(b) Load the package. Do this using the library() command:

```
> library(gpairs)
```

Though you only have to *install* the package once, you have to *load* it during every R session in which you want to use it. Also notice that though the quotes around gpairs are required when the package is installed, they are optional when the package is loaded.

(c) Now you have access to the functions and data that are part of the gpairs package. The centerpiece of the gpairs package is the gpairs() function, which produces "generalized pairs plots"—plots that allow simultaneous visualization of pairwise relationships between several variables. Try the following command:

```
> gpairs(iris, scatter.pars = list(col = as.numeric(iris$Species)))
```

Some of the plots should look familiar. What do you see?

(3) Later in the book, we will use a number of original R functions to complete the exercises. Many of the functions are printed here, and one way to define them would be to re-type them (or copy and paste them, if you are reading electronically). But this is tedious and prone to error. A better way is to install the book's companion R package, stfspack. The package is hosted on github.com rather than CRAN, so you will need to follow a slightly different (but just as easy) procedure to acquire it. First, you will need to install and load the devtools package (Wickham & Chang, 2016) from CRAN. The following two lines of code will do the trick:

```
install.packages("devtools")
library(devtools)
```

Installing the devtools package may take a minute or two. With devtools loaded, we can install stfspack from github with

```
install_github("mdedge/stfspack")
```

Now to load the package, use library() as usual

```
library(stfspack)
```

If you restart R and want to use the package, you need only execute the final library() command. That is, you only need to install the package once, but you need to reload it every time you restart R and want to use the functions defined in it. You can also find an R script containing all the functions included in stfspack at the book's GitHub repository, http://github.com/mdedge/stfs/. If you would rather not install stfspack, you can define the functions by running the code in that file.

2.3 Chapter summary

R is a powerful, free software package for performing statistical tasks. We will use it to simulate data, analyze data, and make data displays. More details about R are given in Appendix B.

2.4 Further reading

Resources for newcomers to R include:

Beckerman, A. P., Childs, D. Z., & Petchey, O. L. (2017). *Getting Started with R: An Introduction for Biologists*. Oxford University Press.

A practical guide for getting started with data analysis, with coverage of popular packages for data management (`dplyr`) and visualization (`ggplot2`).

Dalgaard, P. (2008). *Introductory Statistics with R*. Springer, New York.

The first chapter is a great introduction to the R language. The rest of the book covers implementation of the most popular analyses in R.

Kabacoff, R. I. (2010). *R in Action*. Manning, Shelter Island, NY.

Written for people who want to switch to R from packages like SPSS. *R in Action* has short, to-the-point introductions to every basic R function with lots of examples. Quick-R, a website also by Kabacoff, is a favorite resource for R beginners.

There is also an R package called `swirl` that walks the user through several interactive R tutorials.

For readers with some comfort in R who want to deepen their skills, the following are useful:

Matloff, N. (2011). *The Art of R Programming: A Tour of Statistical Software Design*. No Starch Press, San Francisco, CA.

If you want to learn enough R to be able to write software packages for other people to use, this book should be on your shelf. Matloff covers the basics of interacting with R and then provides an in-depth, behind-the-curtain look at its workings. You will want to know the basics of R before starting this book.

Nolan, D., & Lang, D. T. (2015). *Data Science in R: A Case Studies Approach to Computational Reasoning and Problem Solving*. CRC Press, Boca Raton, FL.

Each chapter is a genuine example of a problem solved with data, and all the chapters are written by leading experts in R.

Wickham, H. (2014). *Advanced R*. CRC Press, Boca Raton, FL.

An excellent guide to programming in R. The introduction and foundations sections cover what most users need to know and more. You can read for free online (http://adv-r.had.co.nz/) or pay to own a physical copy.

CHAPTER 3

The line of best fit

Key terms: Best-fit line, Line errors, Line values, Loss function

> And there's all of these rules
> Based on all these predictions
> Makes it so hard to define
> Where do you draw the line?
>
> Joe Walsh, "Half of the Time"

Let's return to the data shown at the beginning of the first chapter. Figure 1-1 is reproduced here as Figure 3-1. Figure 3-1 purports to show, on the x axis, fertilizer consumption from 11 sub-Saharan African countries, and on the y axis, each country's cereal yield.

Just to draw the points in the way they are shown in Figure 3-1 is to do something substantial—we have already raised the possibility that it is useful to consider these two variables jointly rather than separately, and we have found a concise way to represent their relationship visually. However, we will want to know more. Two natural questions are:

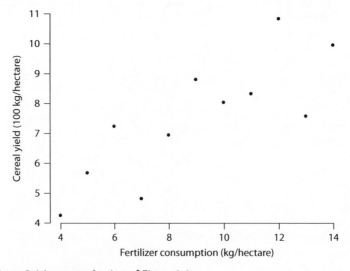

Figure 3-1 Figure 3-1 is a reproduction of Figure 1-1.

Statistical Thinking from Scratch: A Primer for Scientists. M. D. Edge, Oxford University Press (2019).
© M. D. Edge 2019. DOI: 10.1093/oso/9780198827627.001.0001

(1) Is there a way we can summarize these data?

(2) Suppose we knew how much fertilizer is consumed by a country similar to those in our sample. How can we predict the cereal yield in that country?

There are many ways to answer these questions, but one action that responds to both in this case is to draw a line through the data. The line's equation can be used as a summary of the data, and it can be used to generate predictions.

How do we draw such a line? In the case of Figure 3-1, we could simply draw a line with a free hand, "eyeballing" the relationship. We could then examine the line we had drawn, determine an approximate slope and intercept, and write an equation for the line. This might work well enough for some informal purposes, but it is easy to see that it won't do for serious work. One person would draw the line one way, but another would draw it differently. We need a method for drawing the line that leads different data analysts to the same result.

Box 3-1 MATHEMATICS REQUIRED FOR THIS CHAPTER

This chapter assumes that you have had some exposure to math beyond algebraic manipulation of equations, including sum notation, the concept of a function, and calculus (specifically derivatives). Your calculus knowledge might be weathered and dusty—that would be fine—but I'll assume that it can be coaxed back into service with some reminders. If you have never seen some concept necessary for this chapter, that's fine—you may just need to take advantage of some extra resources before proceeding. There is a crash course of the calculus concepts you'll need in Appendix A, and some further calculus resources are listed at the end of that appendix.

In the meantime, Box 3-1 provides a brief reminder of the needed concepts. If the ideas in the box don't sound familiar or you find this chapter difficult, then you may want to do some background work (not too much) before proceeding. The mathematical demands are higher in some of the later chapters, so it is worthwhile to get comfortable now.

Mathematical functions are procedures that take some kind of input—in our case, a number or a set of numbers—and produce output. For example, the function $f(x) = x^2$ takes an input number and squares it, giving for example $f(3) = 3^2 = 9$. The key property of a function is that for any input, the same output always results.

Throughout this book, capital sigma ($\sum$) represents summation. For example, $\sum_{i=1}^{n} f(e_i) = f(e_1) + f(e_2) + \ldots + f(e_n)$. Read this out loud as "the sum of $f(e_i)$ for i from 1 to n." Here, i plays the role of the *index* of summation, and we say in words that we sum "over i" from 1 to n. That is, we add up the terms that result from considering all integer values of i from 1 to n. There are two properties of these sums that we'll use often. First, anything that does not depend on the index over which we sum—in this case, i—can be pulled out of the sum without changing the result. So, if a is a constant that does not depend on i, and a list of numbers that do depend on i are denoted by x_i, then $\sum_{i=1}^{n} ax_i = a\sum_{i=1}^{n} x_i$. (This is just the distributive property, represented slightly more abstractly than you may remember.) Similarly, if there is nothing in the sum that depends on i, then the sum becomes like multiplication by the number of indices being summed over. So if a does not depend on i, then $\sum_{i=1}^{n} a = a + a + a + \ldots + a = na$.

The calculus concept we'll use in this chapter is the *derivative*. Say that there's a function we're interested in called $f(x)$, for example $f(x) = x^2$. The derivative of $f(x)$, which we write

(Continued)

Box 3-1 (Continued)

either as $f'(x)$ or $\frac{d}{dx}f(x)$, is a function that outputs the instantaneous rate of change of $f(x)$. That is, $f'(x)$ tells us how quickly $f(x)$ changes—and in what direction—if we increase the argument of $f(x)$ from x to $x + \Delta x$, where Δx represents an extremely small number.

This notion of instantaneous change can be made much more precise, but for our purposes the point is that derivatives can be used to *optimize* functions—to identify values of a function's arguments that give the largest or smallest possible outputs. For example, we might want to find the value of x for which $f(x) = x^2$ is smallest. One way to find candidate values of x that might optimize $f(x)$ is to find the values of x that satisfy $f'(x) = 0$—the values of x for which the derivative is equal to 0. In the case of $f(x) = x^2$, the derivative is $f'(x) = 2x$ (proven in Appendix A). The equation $f'(x) = 0$ holds only when $x = 0$, and that is also the value of x that minimizes $f(x) = x^2$.

To find any derivative you'll need in this chapter—and any chapter other than Chapter 9—three rules will suffice. (The equation numbers here are those that appear in Appendix A.) First, whenever we have a polynomial function—that is, a function of the form $f(x) = ax^n$, where a and n are constant numbers—the derivative is

$$f'(x) = nax^{n-1}. \tag{A.2}$$

You can memorize the phrase "out in front and down by one," which is what happens to the exponent in a polynomial when the derivative is taken. Second, if a function is multiplied by a constant, then the derivative is multiplied by the same constant, so if $g(x) = af(x)$ with a a constant, then

$$g'(x) = af'(x). \tag{A.3}$$

Third, the derivative of a sum is just the sum of the derivatives of its components. So if $h(x) = f(x) + g(x)$, then the derivative of $h(x)$ is

$$h'(x) = f'(x) + g'(x). \tag{A.4}$$

3.1 Defining the "best" fit

We are looking for a line that best fits the data, also called the **best-fit line**. What do we mean by "best"? There are many possible meanings, and though there are reasonable arguments for privileging some over others, there is usually no sole correct choice. Suppose that we write a line with the equation

$$\tilde{y} = a + bx. \tag{3.1}$$

This is just the generic equation for a line with intercept a and slope b. Notice the "~" over the y, which is used here to distinguish the values along the line going through the data from the actual values of the data points (which are denoted as y, without the ~).

Look at Figure 3-2, which displays the data from Figure 3-1 with a few additions. First, there is a solid line representing a candidate fit line. It was not chosen to be a line of best fit, but it is a line that a reasonable person might draw through the data with a free hand. There are several marks along the line. These marks correspond to the points on the line that are directly above or below the points in the dataset. That is, the marked points share x

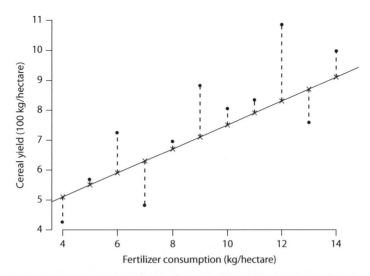

Figure 3-2 We have added a candidate fit line $\widetilde{y} = 3.5 + 0.4\,x$. This is not a line of best fit, but it would be a reasonable line to draw if we were working freehand and simply "eyeballing" the fit. The × marks along the line represent the "line values" of cereal yield. The vertical dashed lines represent the line errors.

coordinates with points from the dataset, but their y coordinates are chosen to fall on the line we have drawn. The y coordinates of the points on the line are what we might call the **line values** of y, and the ith line value is $\widetilde{y}_i$. There are also dashed lines connecting the line values to the observed values. We can write the length of the ith dashed line as e_i, which we will call the ith **line error**.[1,2] Line errors have a positive sign when the dashed line extends up from the solid line and a negative sign when the dashed line extends down from the solid line.[3]

In notation, we can write the relationships defined in the last paragraph as

$$e_i = y_i - \widetilde{y}_i = y_i - (a + bx_i). \tag{3.2}$$

That is, the line errors e_i are the differences between the observed values and the line values, where the line values are the y coordinates of the line at the x coordinates of the observed values. These definitions are helpful because they allow us to think of the line of best fit as the one that, in some sense, makes the line errors as small as possible. Specifically, the line of best fit minimizes the sum of the values that result when we apply some function—a function that we have yet to choose—to the line errors. That is, we minimize

$$\sum_{i=1}^{n} f(e_i), \tag{3.3}$$

where the $f(e_i)$ is the function we choose to apply to the line errors—called a "loss function"—and n is the number of paired observations. The **loss function**—also called a

[1] The word "error" is used here because a perfect line would pass exactly through all the points.
[2] One might imagine defining line errors in terms of distances between the points and the line but not requiring that the distances be vertical. There is a framework for working with distances calculated at other angles called "total least squares"—the two-variable version is called "Deming regression."
[3] Some other sources call line values "predicted values," but we are not yet making any predictions about data we might collect in the future—we are working with data we already have in hand. Other sources call line errors "residuals," but I will also reserve that term for later.

"cost function" or "objective function"—is supposed to describe how bad it is to make a given line error. For example, if we decide that large errors are much more costly than small errors, then we might decide to take the square of the line error as a loss function—in that case, doubling the size of the line error quadruples the output of the loss function (i.e. $(2e_i)^2 = 4e_i^2$). We will call the quantity that we want to minimize, shown in Equation 3.3, the "total loss."

How do we choose the loss function? The option that might occur to you first is to minimize the sum of the line errors themselves, setting the loss function as $f(e_i) = e_i$ and the total loss as $\sum_{i=1}^n e_i$. This will fail. As written, this loss function counts negative line errors—instances in which the line passes above a point—as being indicative of a good fit, cancelling out positive line errors. Look at the line in Figure 3-2. Imagine that we drew another line at the same slope but five units higher. Such a line would clearly be a poor fit to the data—it would pass above the entire cloud of points. But its total loss would be lower than the line we have drawn because every line error would be five units less.[4] Thus, minimizing the total loss resulting from the loss function $f(e_i) = e_i$ favors lines that pass above all the points by as much as possible.

We could fix this problem by abandoning the idea of minimizing the total loss $\sum_{i=1}^n e_i$ and instead trying to get $\sum_{i=1}^n e_i$ to be as close to zero as possible. This change of strategy will disallow lines that pass over all the points, but it will cause a new problem. It turns out that there are many ways to choose a and b such that the positive and negative line errors cancel each other and the sum of the line errors is exactly zero.[5] Minimizing the line errors themselves cannot give a unique line of best fit. The real problem here is that by choosing a loss function $f(e_i)$ that can take on negative values, one implicitly says that some types of errors can cancel the effects of other types of errors. If possible, we actually want to make no errors at all, of either sign, so we should choose a loss function that cannot take negative values.

At this point, many expositions of linear regression go something like this: Because we need to choose a loss function that cannot take negative values, we will minimize the sum of the squares of the line errors, $\sum_{i=1}^n e_i^2$. An immediate objection is that setting $f(e_i) = e_i^2$ is not the only way to disallow negative values of $f(e_i)$. We could, for example, choose the loss function to be $f(e_i) = |e_i|$, the absolute value of the line errors. There are many functions of the line errors that cannot be negative, so why choose the squares?

The objection is perfectly valid, but we will, like other sources, mainly rely on minimization of the sum of the squared line errors, the so-called "method of least squares." There are two reasons for this. The first but less compelling reason is that minimizing the squared errors is mathematically easy compared with minimizing many other possible loss functions. Computers make this first reason unimportant in practice—it is now easy to find lines that minimize total losses resulting from any of a virtual infinity of possible choices of loss function—but it has mattered historically. A second reason is that, as we will see in later chapters, the sum of squared errors has special meaning if we make certain assumptions about the way in which the data were generated. Though we will rely primarily on the sum of squared line errors, we will also return to the sum of the absolute values of the line errors in the exercises.

---

[4] Recall that if the line passes above a point in the dataset, the corresponding line error has a negative sign, so, even though most of the line errors have larger absolute values when the line is shifted upward, they are smaller— that is, more negative—numbers.

[5] The sum of the line errors will be zero for any line that goes through the point $(\bar{x}, \bar{y})$, where $\bar{x}$ is the mean of x, $\bar{x} = \sum_{i=1}^n x_i/n$, and $\bar{y}$ is the mean of y. To see this, notice that the equation $\sum_{i=1}^n e_i = \sum_{i=1}^n y_i - (a - bx_i) = 0$ reduces to $\sum_{i=1}^n y_i = na + b\sum_{i=1}^n x_i$, which itself reduces to $\bar{y} = a + b\bar{x}$. This means that any line that satisfies $\bar{y} = a + b\bar{x}$ has 0 as the sum of its line errors. There are infinitely many lines that satisfy this criterion by passing through the point $(\bar{x}, \bar{y})$.

3.2 Derivation: finding the least-squares line

How, then, should we minimize the sum of the squared line errors? We will see in a moment that we can use calculus. But first, allow me a brief aside to address readers who feel less than perfectly comfortable with math. We are about to embark on a somewhat involved mathematical derivation. There will be a few of these derivations before the book is over, with more among the exercises. Some students from nonmathematical backgrounds will bristle at a derivation like this, and it is reasonable that they do—it is a lot of effort, and the payoff is not always obvious at first. So why go through the work? The philosophy of this book is that it helps users of statistics immeasurably to have a sense of the ways in which statistical methods are constructed. Without such a sense, there is too much of a temptation to view statistics as "magic" and to avoid thinking about methods critically. Knowing how methods are built gives one a keener sense for the assumptions they entail. The goal is not that you should memorize a detailed procedure; read instead to understand the objective of the derivation and to see that it is achievable by a series of concrete, justified steps.

Recall from calculus (or from Appendix A) that in many cases, we can find candidate minima of a function by taking the derivative and finding the values of the arguments for which the derivative is equal to zero. We want values of a and b that minimize the sum of the squared line errors. Thus, we should write the sum of the squared line errors as a function of a and b, then take the derivative of that function and find the values of a and b for which the derivative is equal to zero. We have already written the line errors in terms of a and b (Equation 3.2):

$$e_i = y_i - \widetilde{y}_i = y_i - (a + bx_i).$$

To get the loss function, $f(e_i) = e_i^2$, in terms of a and b, we square the expression for e_i in terms of a and b:

$$f(a,b) = e_i^2 = (y_i - a - bx_i)^2.$$

The function to minimize is the *total* loss, the sum of the loss function computed on all the line errors. We will write the total loss as $g(a,b)$:

$$g(a,b) = \sum_{i=1}^{n} (y_i - a - bx_i)^2. \tag{3.4}$$

In words, we take each line error—the vertical distance between each point and the line—square it, and then take the sum of those squared line errors. The sum that results is the total loss to be minimized.

Equation 3.4 presents two possibly unfamiliar issues when finding the derivative. First, it is a function of two variables—a and b—rather than just one.[6] For our purposes, we can just consider one variable at a time. That is, we can minimize the function with respect to a and separately minimize it with respect to b, producing two equations to solve for the two unknown values. The second new issue is that the total loss has a summation over n, where n is the number of pairs of observations we have. This is also no problem—we can take the derivative of the sum by just summing the derivatives of all the terms in the sum (Equation A.4). Thus, we can treat this function in essentially the way we would treat any other.

[6] The observed x_i and y_i values need not be treated as variables. We can change the values of a and b we use in the line, but we cannot change the observed data. Thus, we minimize with respect to a and b, and we treat the other numbers as fixed coefficients.

The value of the intercept, a, that minimizes the sum of the squared line errors appears in an exercise below. Here, we will find the value of b, the slope, that minimizes the sum of the squared line errors. To start, expand the squared term that appears in the sum:

$$g(a,b) = \sum_{i=1}^{n} (y_i^2 - 2ay_i - 2bx_iy_i + a^2 + 2abx_i + b^2x_i^2).$$

To find the value of the slope, b, that minimizes the total loss—i.e., that minimizes the expression in equation 3.4—we will take the derivative of the total loss with respect to b. The total loss is a polynomial in terms of b, which means that we can apply Equation A.2 to find its derivative. Equation A.2 says that to differentiate a polynomial, such as b^k, where k is a constant, one should take the value of the exponent "out in front and down by one," so the derivative is kb^{k-1}. There are three terms in the total loss in which b does not appear: y_i^2, $-2ay_i$, and a^2. These can all be thought of as being multiplied by $b^0 = 1$. Applying the out-in-front-and-down-by-one rule, the derivatives of these terms with respect to b are all multiplied by the exponent of b—in this case, 0—and so they drop out of the equation. In two terms where just b appears, $-2bx_iy_i$ and $2abx_i$, the b is removed in the derivative, but the term is otherwise unchanged. (Think of b as b^1, so a "1" comes out front, and b becomes $b^0 = 1$.) Where b^2 appears, in the term $b^2x_i^2$, the 2 comes out in front and down by one, changing b^2 into $2b$. Combining these moves gives the derivative:[7]

$$\frac{\partial}{\partial b} g(a,b) = \sum_{i=1}^{n} (-2x_iy_i + 2ax_i + 2bx_i^2).$$

To find the value of b that minimizes $g(a,b)$, which we can call $\widetilde{b}$, we find the values of b for which the derivative is equal to zero. That is, we solve the following equation for $\widetilde{b}$:

$$\sum_{i=1}^{n} (-2x_iy_i + 2ax_i + 2\widetilde{b}x_i^2) = 0.$$

I'd like to take this one step at a time. Notice that every term has a 2 in it, so we can factor it out. Because the 2 does not depend on i, the index of the summation, we can take it out of the sum:

$$2\sum_{i=1}^{n} (-x_iy_i + ax_i + \widetilde{b}x_i^2) = 0.$$

If twice the sum on the left is equal to zero, then the sum itself is equal to zero, so we can just drop the 2:

$$\sum_{i=1}^{n} (-x_iy_i + ax_i + \widetilde{b}x_i^2) = 0.$$

We can sum each term separately and end up with the same overall sum. That is, it's allowed to write:

$$\sum_{i=1}^{n} (-x_iy_i) + \sum_{i=1}^{n} ax_i + \sum_{i=1}^{n} \widetilde{b}x_i^2 = 0.$$

---

[7] A quick note on the form of this expression: We now write $\frac{\partial}{\partial b} g(a,b)$ rather than $g'(a,b)$ to make clear that we differentiated with respect to b and not a. The curly ∂ that appears in the differential operator indicates that we are working with a function of more than one variable.

Because a and $\widetilde{b}$ do not depend on i, we can put them out in front of the sums in which they appear:

$$\sum_{i=1}^{n}(-x_iy_i) + a\sum_{i=1}^{n}x_i + \widetilde{b}\sum_{i=1}^{n}x_i^2 = 0.$$

To get $\widetilde{b}$ on one side, subtract the first two terms on the left from both sides of the equation:

$$\widetilde{b}\sum_{i=1}^{n}x_i^2 = \sum_{i=1}^{n}x_iy_i - a\sum_{i=1}^{n}x_i.$$

Now divide both sides by the sum on the left to get $\widetilde{b}$ on its own:

$$\widetilde{b} = \frac{\sum_{i=1}^{n}x_iy_i - a\sum_{i=1}^{n}x_i}{\sum_{i=1}^{n}x_i^2}. \tag{3.5}$$

This is an expression for $\widetilde{b}$, the slope of the line that will minimize the sum of the squared line errors.[8] The expression is not completely satisfying because it has a, the intercept, in it, and we do not know how to find a. To get an expression that does not have a in it, we have to minimize the sum of the squared line errors with respect to a and plug the resulting expression into this one. You will find the value of the intercept a that minimizes the sum of the squared line errors in an exercise below. It is

$$\widetilde{a} = \frac{\sum_{i=1}^{n}y_i - \widetilde{b}\sum_{i=1}^{n}x_i}{n}. \tag{3.6}$$

You may notice that the expression for the least-squares intercept $\widetilde{a}$ also has the least-squares slope $\widetilde{b}$ in it, but normally we leave it in this form. In an optional exercise, you will verify that plugging the expression for $\widetilde{a}$ into the expression for $\widetilde{b}$ above, solving for $\widetilde{b}$, and simplifying gives an expression for the least-squares slope $\widetilde{b}$ that does not depend on the intercept a:

$$\widetilde{b} = \frac{\sum_{i=1}^{n}x_iy_i - \frac{1}{n}\sum_{i=1}^{n}x_i\sum_{i=1}^{n}y_i}{\sum_{i=1}^{n}x_i^2 - \frac{1}{n}\left(\sum_{i=1}^{n}x_i\right)^2}. \tag{3.7}$$

This expression can be evaluated using the x_i and y_i values—that is, the data.

The expressions for $\widetilde{a}$ and $\widetilde{b}$ may look complicated, but they can be simplified. Adopting some standard notation, we write the sample means of the x_i and the y_i as

$$\overline{x} = \frac{1}{n}\sum_{i=1}^{n}x_i$$

and

$$\overline{y} = \frac{1}{n}\sum_{i=1}^{n}y_i.$$

You will have a chance below to prove that the expressions for $\widetilde{a}$ and $\widetilde{b}$ simplify to

---

[8] There is a little bit of extra work involved in confirming whether critical points for functions of multiple variables are maxima, minima, or saddle points. One needs to examine some properties a matrix of second derivatives. We will be skipping it as beyond our scope, but it is straightforward.

$$\tilde{a} = \bar{y} - \tilde{b}\bar{x}. \tag{3.8}$$

and

$$\tilde{b} = \frac{\sum_{i=1}^{n}(x_i - \bar{x})(y_i - \bar{y})}{\sum_{i=1}^{n}[(x_i - \bar{x})^2]}. \tag{3.9}$$

We will also see later that $\tilde{b}$ can be rewritten in terms of estimates of some basic quantities from probability theory.

Exercise Set 3-1[9]

(1) (a) Use R to calculate the least-squares intercept $\tilde{a}$ and the least-squares slope $\tilde{b}$ for the data in Figure 3-1. (Use either Equations 3.6 and 3.7 or Equations 3.8 and 3.9. For Equation 3.8 and 3.9, use the mean() function to compute $\bar{x}$ and $\bar{y}$.) The x values in Figure 3-1 can be accessed in R using anscombe$x1, and the y values can be accessed using anscombe$y1. (anscombe is one of R's built-in datasets.) Here is some R you will need: Given two numeric vectors x and y of the same length, you can compute a new vector, each entry of which is the product of the corresponding entries in x and y, using x*y. You can get a vector where each entry is the square of a corresponding entry in x with x^2 (or just x*x). To take the sum of all the entries in x, use sum(x). Finally, you can get the number of entries in x using length(x). Use parentheses to make sure your mathematical operations are done in the right order.

(b) Verify that the slope and intercept you obtain in part (a) are the same ones returned by the lm() function. To save the output of lm() in an object called mod.fit, use either

```
mod.fit <- lm(anscombe$y1 ~ anscombe$x1)
```

or

```
mod.fit <- lm(y1 ~ x1, data = anscombe)
```

After running one of the above commands, to see the least-squares slope and intercept, use summary(mod.fit).

(2) [Optional] In this problem, you'll derive Equation 3.7 and show that Equations 3.6 and 3.7 are equivalent to Equations 3.8 and 3.9.
(a) Verify Equation 3.7 by plugging the value of the least-squares intercept $\tilde{a}$ defined in Equation 3.6 in place of a in Equation 3.5.
(b) Prove that Equation 3.6 can be rewritten as Equation 3.8.
(c) [Hard] Prove that Equation 3.7 can be rewritten as Equation 3.9. (Hint: Work backwards, starting with Equation 3.9. Work with the numerator and the denominator separately.)

3.3 Conclusions

We have found one possible line of "best fit"—the line that minimizes the sum of squared line errors. Figure 3-3 shows the data from Figure 3-1 with the least-squares line drawn

---

[9] See the book's GitHub repository at github.com/mdedge/stfs/ for detailed solutions and R scripts.

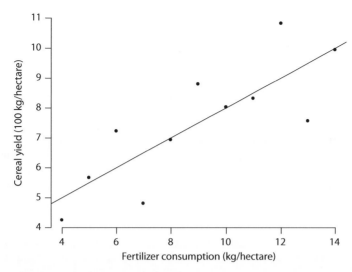

Figure 3-3 The least-squares line.

through it. As you saw in the previous exercise set, the equation for the least-squares line is $\tilde{y} = 3 + 0.5 * x$.

The act of drawing the least-squares line requires no probabilistic or statistical assumptions. We have not even introduced the concept of probability, but the least-squares line is comprehensible as a summary and a visual aid. The least-squares line is also available as a rough-and-ready tool for prediction. If we learn, for example, that a sub-Saharan African country consumes 10 kg/hectare of fertilizer, we might guess that cereal yield in that country (in 100 kg/hectare) will be approximately $3 + 0.5 * (10) = 8$. Whenever we have two paired sets of numbers, we can draw the least-squares line and interpret it in this minimal, informal way, regardless of the properties of the data.

We have already come a long way, but these interpretations are not completely satisfying. We can use the line to make predictions, but what assurances do we have that our predictions will be any good? More broadly, when might those predictions be sensible, and what types of properties might the predictions have? To go back to the beginning of Chapter 1: Suppose I claim that the line is evidence of an important relationship between the two variables being considered and you claim that there is no such relationship. How can we clarify our difference of opinion?

In considering these questions, it becomes clear that we do not want to talk solely about the data we happen to have collected, but also about the larger entity(ies)—populations or processes—from which the data are drawn. To make claims about populations, we will have to make assumptions about how the data were generated. If we write down our assumptions in mathematical form, then we can use mathematical methods to study the predictions we could make from a least-squares line, the reasonableness of the least-squares criterion, and the meaning of the intercept and slope we obtain. The branch of math that we need is probability theory, which is covered in the next two chapters.

In the following exercises, you will explore a few properties and extensions of the least-squares line.

Exercise Set 3-2

(1) (a) Verify Equation 3.6 by minimizing the sum of squared line errors (Equation 3.4) with respect to a, the intercept. Remember to apply Equations A.2–A.4 when taking the derivative.

    (b) Use the expression for $\tilde{a}$ from Equation 3.8 to show that the least-squares line always passes through the point $(\bar{x}, \bar{y})$.

(2) Suppose that for some set of data, we know that the intercept $a = 0$. For example, we might be studying the degree to which hanging a weight from a spring stretches the spring, and we might know that when no weight is added to the spring, the spring is not stretched at all. We want to find the line that minimizes the squared line errors when the intercept is constrained to be zero. What is the mathematical expression for the slope, which we can call b', that will do the job?

(3) [Optional] Imagine that instead of seeking a line of best fit of the form $y_i = a + bx_i$, we sought a line of best fit of the form $x_i = c + dy_i$. That is, rather than writing a line for the y_i in terms of the x_i, we want to write a line for the x_i in terms of the y_i. If we choose to minimize the squared line errors, what are the expressions for $\tilde{c}$ and $\tilde{d}$? Are these numbers equal to $\tilde{a}$ and $\tilde{b}$?

(4) [Optional] We chose to define the line of "best fit" as the one that minimizes the sum of the squared line errors. We could have defined it as the line that minimizes some other function of the line errors. In this problem, we consider the line that minimizes the sum of the absolute values of the line errors, sometimes called the "L1 line." The function rq() in the quantreg package can find the slope and intercept of the L1 line.

    (a) Install and load the quantreg package using the install.packages() and library() functions.

    (b) Use rq() to find the $\tilde{a}$ and $\tilde{b}$ values for the L1 line for the data from Figure 3-1, which can be accessed using anscombe$x1 and anscombe$y1. Use the default setting for the parameter tau (tau = 0.5) to find the L1 line.

    (c) Plot the least-squares line (also called the L2 line) and the L1 line on the same plot. You can do so using the following R code, which uses the abline() function to draw lines on the plot:

```
plot(anscombe$x1, anscombe$y1)
mod.fit.L2 <- lm(anscombe$y1 ~ anscombe$x1)
mod.fit.L1 <- rq(anscombe$y1 ~ anscombe$x1)
abline(mod.fit.L2, lty = 1)
abline(mod.fit.L1, lty = 2)
legend("topleft", legend = c("L2", "L1"), lty = c(1,2))
```

    What do you notice about the least-squares and L1 lines in this case?

    (d) Now try a new set of data. Consider the data in anscombe$x3 and anscombe$y3. Start by plotting the data:

```
plot(anscombe$x3, anscombe$y3)
```

    Look at the plot. Do you expect that the least-squares and L1 lines will be similar or different in this case? If they will differ, how will they differ, and why?

    (e) Now plot the least-squares and L1 lines. How do you explain what you see?

3.4 Chapter summary

One way to visualize a set of data on two variables is to plot them on a pair of axes. A line that "best fits" the data can then be drawn as a summary. In this chapter, we considered how to define a line of "best" fit—there is no sole best choice. The most commonly chosen line to summarize the data is the "least-squares" line—the line that minimizes the sum of the squared vertical distances between the points and the line. One reason for the least-squares line's popularity is convenience, but we will see later that it is also related to some key ideas in statistical estimation. We discussed the derivation of expressions for the intercept and slope of the least-squares line.

3.5 Further reading

Freedman, D. A. (2009). *Statistical Models: Theory and Practice*. Cambridge University Press.
   For a similar but more technical treatment of the least-squares line, including the more general case of multiple regression, see Chapter 2 of Freedman.

CHAPTER 4

Probability and random variables

Key terms: Bayes' theorem, Conditional probability, Continuous random variable, Cumulative distribution function, Discrete random variable, Distribution family, Event, Independence, Instance, Intersection, Parameter, Probability, Probability density function, Probability mass function, Random variable, Set, Union

> While you see a chance, take it.
>
> Steve Winwood, "While You See a Chance"

Given a set of paired observations on two variables, we can draw a line that "best fits" the data. There are many possible definitions of "best fit," but for most options we could choose, it is feasible to find a line of best fit, at least with the help of a computer.

So far, it has been less clear how we should interpret such a line. The object of the next several chapters is to provide a framework for interpretation. To start, we can distinguish between probability theory and statistical estimation and inference.

In probability theory, we think about processes that generate data, and we ask "what can we say about the data generated by such a process?" We start by coming up with a model, and then we use the model to make claims about the data we might observe. For example, suppose that we give students a quiz with ten questions, and suppose we are interested in the number of questions students answer correctly. We might make a model that assumes that a given student has a probability p of getting each question right, regardless of whether the student answered any of the other questions correctly. We can then ask questions about the scores that could result. What scores would be likely? What score would we "expect"? How much would a student's scores vary on repeated quizzes with the same properties? At this point, we do not necessarily have any information about how real students behave—we are merely asking "*If* students behaved in the way described by the model, what would the consequences be?"

In statistical estimation and inference, we work in the opposite direction. We start with data, and we ask "what can we say about the process that generated these data?" For example, we might start with a student's data on a ten-item quiz. If we assume the same model—that the student has a probability p of getting each question right, regardless of the outcomes of the other nine questions—then we could ask questions, for example, about a particular student's value of p. What is our best guess for the value of p for this student? How certain are we about our guess? Can we say whether this student's value of p is different

Statistical Thinking from Scratch: A Primer for Scientists. M. D. Edge, Oxford University Press (2019).
© M. D. Edge 2019. DOI: 10.1093/oso/9780198827627.001.0001

from another student's? These questions might not be cogent if the model doesn't match the situation well, for example if the questions vary widely in difficulty.

In this chapter and the next one, we will discuss concepts of probability, the study of data generated by specified processes. Statistical estimation and inference will come later.

We focus on the probabilistic concepts most important to statistical thinking. Some of the ideas we cover may seem distant from applications. As you read, keep in mind that all of our future work depends on the ideas in this chapter and the next. In statistics, the assumptions we start with usually take the form of a probabilistic model, which subsumes fundamental axioms that apply to all probabilistic models plus additional assumptions that reflect the situation we want to study. This leads to an idea that might be jarring—strictly speaking, the statistical ideas we introduce in later chapters are designed for application to *data that are generated by the probabilistic models we assume.* But we seldom actually *have* data that are generated by those models. Our models are mathematical idealizations; our data, in contrast, come from the real world. Thus, the credibility of our statistical claims depends on the extent to which the real world behaves like the probabilistic models we assume. Without an understanding of probabilistic models, we will not be able to assess whether our assumptions are reasonable.

This idea raises a philosophical question worth knowing about, though a full treatment is beyond our scope. The question is, "What *is* a **probability**?" Probabilities might simply be objects that exist in the minds of people who engage in a particular branch of mathematics. Such a view has some truth—probabilities are *at least* mathematical objects.[1] But probability is also used to understand the world. How, then, ought we to interpret probabilities in relation to the real world?

One major view is that probabilities are best understood as long-term frequencies. On this view, the probability of a particular event is just the frequency with which it tends to occur over many trials. The frequency view is intuitively appealing, and it describes many probabilistic situations well, like coin flips, deals from decks of cards, and spins of roulette wheels.

The frequency view becomes awkward in some situations. For example, imagine that you are told that the probability that it will rain in your town tomorrow is 25%. If you take a frequency view of probability, it is hard to make sense of this. Time and space will only offer us one instance of tomorrow's date in your town. There can be no long-term frequency of, say, rainfall in Cuernavaca, Mexico on March 25th, 2150. We only have one shot at March 25th, 2150 in Cuernavaca—it will either rain or it won't. If we want to use the frequency view of probability, we have some options. We can claim that the sentence, "The probability it will rain in my town tomorrow is 25%" is simply not a valid probability statement. If we want to save the statement, we can re-interpret it to mean something like "The probability it will rain in my town the day after days that are like today in all relevant respects is 25%." This may not be satisfactory—weather is complicated, and there may be no previous or future days that are like today in all relevant respects.

One solution to problems like these is to abandon the frequency view of probability. An alternative to the frequency view is to claim that the probability of an event measures the degree of belief in the occurrence of the event that a rational person would have. The degree-of-belief view can solve problems that come up for the frequency view in interpreting probabilities of one-shot events, but it is possible to criticize it on other grounds. The debate is ongoing.

---

[1] Of course, whether mathematical objects exist principally "in the minds of people who engage in mathematics" is a separable question.

A separable but related question is whether the universe is fundamentally probabilistic or deterministic. Put differently, if we could re-start the universe with precisely the same initial conditions, would everything turn out the same as it has this time? Some leading interpretations of one of our most successful scientific theories—quantum mechanics—suggest that at the atomic and subatomic scales, the world is fundamentally a chancy place, meaning that even with the same starting conditions, repeated instances of subatomic processes can have different outcomes. But even granting this claim of quantum mechanics, it is still not clear whether subatomic chanciness implies macroscopic chanciness. If the world—at least the pieces of it that are big enough to see—might be deterministic, what business do we have thinking of probabilities? Laplace—a noted determinist and noted probabilist—answered this question by arguing that probabilities are summaries of our own uncertainty about the world. The example of Laplace shows that it is possible take the study of probability seriously even if one thinks that probabilities would not appear in a perfect, complete understanding of nature.

It may seem troubling that we are launching into a subject whose ultimate interpretation is still unclear. We do not need to be troubled yet—our goal in this chapter and the next one is to learn the mathematics of probability, and the math is not affected by the interpretation of the outcome. That is, we start by viewing probability theory as a system of statements that follow from a set of assumptions. For now, we can assume that probabilities are simply mathematical objects that obey those assumptions—we can agree that the math is correct without agreeing on its real-world interpretation.

Probability is a vast subject, and we will cover only what is necessary to understand the statistical material in future chapters. After a brief orientation to some concepts from set theory (Box 4-1), we will start with the axioms of probability, the set of assumptions we start with when we make probabilistic deductions. We will then learn how to reason about how probabilistic events are related to each other using conditional probability, independence, and Bayes' theorem. Next, we will introduce random variables, distribution functions, and density functions. We will finish building our collection of probabilistic ideas in the next chapter.

Box 4-1 SETS AND SET NOTATION

Probability is built upon set theory. Set theory is a branch of mathematics that deals with collections of objects, or **sets**. You need to know a little set theory to understand this chapter.

A set is an unordered collection of objects. We write sets using curly brackets, like so:

$$\Omega = \{1, 2, 3, 4, 5, 6\}.$$

The objects in a set are called elements. Ω's elements are the integers ranging from 1 to 6—we might use Ω to represent the possible results of rolling a six-sided die. The order in which the elements appear does not matter. That is,

$$\Omega = \{1, 2, 3, 4, 5, 6\} = \{6, 5, 4, 3, 2, 1\}.$$

It also doesn't matter if elements appear more than once—only the unique objects in the set count:

$$\Omega = \{1, 2, 3, 4, 5, 6\} = \{1, 2, 3, 4, 5, 6, 6, 6\}.$$

(Continued)

Box 4-1 (Continued)

We usually write sets so that each element appears exactly once.

Imagine two new sets, K and L:

$$K = \{1, 3, 5\},$$

$$L = \{1, 2, 4\}.$$

K and L are both subsets of Ω—all their elements appear in Ω. K and L will let us illustrate the two most important set operations, intersection and union. The **intersection** of two sets is the set of all elements that appear in both sets. It is written with a $\cap$ symbol:

$$K \cap L = \{1\}.$$

The **union** of two sets is the set that has every element that appears in *either* of the two original sets. It is written with a $\cup$ symbol:

$$K \cup L = \{1, 2, 3, 4, 5\}.$$

In probability, we are often interested in the elements that are *not* in a particular set but that are in a larger set that contains all elements of interest. For example, if we are thinking about rolls of a six-sided die, then the set Ω contains all elements of interest, and we might want the set of all elements in Ω that are not in K. We will call this set the complement of K, which we write K^C. In our case,

$$K^C = \{2, 4, 6\}.$$

As another example,

$$L^C = \{3, 5, 6\}.$$

The set of "all possible elements" with respect to which the complement is taken is excluded from the notation—it is implicit. When we need to refer to it explicitly, we usually write it as Ω. The complement of any subset S of Ω has these properties:

$$S \cup S^C = \Omega$$

and

$$S \cap S^C = \emptyset.$$

The first property is that the union of the set and its complement are equal to Ω. The second property says that the set and its complement share nothing in common—their intersection is empty. We will follow the convention of using $\emptyset$ to denote the empty set, or the set with no elements.

4.1 [Optional section] The axioms of probability

During the past 150 years, one concern of mathematicians has been to find axioms—minimal assumptions that can be used as a logical foundation for a particular branch of mathematics. Once a field has a set of axioms, mathematicians working in that field can make claims that are deductively certain if the axioms are true. Many fields of math have more than one set of axioms. For example, geometers have a choice of starting with Euclid's axioms or starting with a different set of assumptions. Depending on which they choose, they are said to be doing "Euclidean" or "non-Euclidian" geometry. Probability has also been axiomatized in more than one way, but the most influential set of axioms is called the Kolmogorov axioms, after Andrey Kolmogorov, who proposed them in 1933. This is not the place for a formal treatment, but we will try to give a flavor of Kolmogorov's axioms.

One way to think about axioms is as ways of formalizing our most basic intuitions about objects we want to study. For example, it is hard to understand what it would mean for an event to have negative probability or probability greater than 1 (i.e., greater than 100%). We might then write axioms that outlaw these possibilities, as Kolmogorov did.

The most important object in Kolmogorov's axioms is the probability space. The probability space has three components: a set of *outcomes*, a set of *events*, and a *probability function*. For example, suppose that we are playing roulette and that for a single spin, we care about the color of the pocket in which the ball lands. Roulette wheels have pockets of three colors: black, red, and green. We can represent the set of outcomes as follows:

$$\Omega = \{B, R, G\}.$$

The curly brackets indicate that the items inside are elements of a set. Each element of the set is an outcome—B for black, R for red, and G for green.[2]

We move on to the set of events. Be cautious: the name "events" might give you the wrong idea. Outside of probability theory, it is not customary to speak of "events" like "the ball landed in a black pocket *or* it landed in a red pocket," but that is exactly what will do here. An **event** is a *subset* of the outcomes that appear in the outcome set Ω. For example, the event "the ball landed in a black pocket or it landed in a red pocket" is written $\{B, R\}$. We do not talk about events like $\{B, R\}$ when we tell stories, but we can imagine situations in which we would want to know the probability of $\{B, R\}$—for example, if you are considering placing a bet on green. Because each event is a set, the set of events is a set of sets. For example, in our round of roulette, the set of events is

$$\mathcal{F} = \{\varnothing, \{B\}, \{R\}, \{G\}, \{B, R\}, \{B, G\}, \{R, G\}, \{B, R, G\}\}.$$

$\mathcal{F}$ contains sets that hold all possible combinations of the outcomes in Ω, along with the symbol $\varnothing$. $\varnothing$ is the empty set—the set with nothing in it. In this case, $\varnothing$ represents an event like "the ball did not land in a black, red, or green pocket." It is an odd event to talk about, and we will see that it has a probability of zero.

The last component of the probability space is a probability function P. The probability function assigns probabilities to each event in $\mathcal{F}$, and it has to obey Kolmogorov's three axioms of probability. The axioms are as follows:

(i) The probability of every event in $\mathcal{F}$ is a non-negative real number. (Probabilities cannot be negative.)

(ii) The probability of the event that includes every outcome is 1. We write this axiom as $P(\Omega) = 1$. In our roulette example, we could write $P(\{B, R, G\}) = 1$, or say, "the probability that the ball will land in a black pocket, a red pocket, or a green pocket is one." (A loose English rendering is, "The probability that *some* outcome will happen is set to 1.")

---

[2] If we were interested in the outcome of many spins of the roulette wheel, then we would include outcomes representing multiple spins. Here, for simplicity, we are concerned with a single spin of the wheel.

(iii) Suppose we have two mutually exclusive events E_1 and E_2. "Mutually exclusive" means that the two events share no outcomes in common: $E_1 \cap E_2 = \emptyset$. The third axiom states that the probability of observing either E_1 or E_2 is the sum of the probabilities of E_1 and E_2. That is, if $E_1 \cap E_2 = \emptyset$, then $P(E_1 \cup E_2) = P(E_1) + P(E_2)$. In plainer English, if two events can't both happen, then the probability that either one or the other happens is equal to the sum of their individual probabilities. For example, a single ball in roulette cannot land in both a black pocket *and* a red pocket on the same trial. Thus, $P(B \cup R) = P(B) + P(R)$.

In our roulette example, the axioms of probability theory do not tell us what the probabilities are. It might be reasonable to set the probabilities of the events to be proportional to the number of pockets associated with each event, but the axioms of probability do not demand that we do so. The axioms, do, however, require that the probabilities we assign to the events have certain properties. We will pause here to consider the implications of the axioms and to practice with sets and probability spaces.

Exercise Set 4-1

(1) Suppose we know the probability of an event A, which we write as $P(A)$. What is the probability of its complement, $P(A^C)$? Use the properties of complements and the axioms of probability to justify your answer.

(2) Suppose that in the roulette example above, $P(B) = \beta$ and $P(R) = \rho$ and that the ball can land in only one color. Find the probabilities of each event in $\mathcal{F}$, and justify each of your answers using the axioms of probability and the result from Problem 1.

(3) [Optional] Show that $P(E_1 \cup E_2) = P(E_1) + P(E_2) - P(E_1 \cap E_2)$. That is, the probability that at least one of two events occurs is the sum of the probabilities that each occurs separately minus the probability that they both occur. Hint: define the set difference $E_2 \backslash E_1$ as the set of elements of E_2 that are not elements of E_1. (The complement is a special case of the set difference: $E_1^C = \Omega \backslash E_1$.)

4.2 Relationships between events: conditional probability and independence

In the previous (optional) section, we considered a single trial of a probabilistic process—a round of roulette.[3] We studied exactly one feature of the outcome of that trial—the color of the pocket in which the ball lands. Often, though, we would like to consider several features of an outcome at once, or we would like to study several trials. For example, we could simultaneously study the color and number of the pocket in which the roulette ball lands, or we could study multiple rounds of roulette. We need concepts that describe how these events are related to each other.

Imagine that we draw a card from a well-shuffled deck of playing cards.[4] We might take note of several features, including the number or face of the card, the color, or the suit. Let's

---

[3] In case you did not read the optional section on the axioms of probability, the probability rules that follow from it are: (1) Probabilities cannot be negative or greater than 1. (2) If two events A and B are mutually exclusive (they cannot both happen), then the probability that one of them happens is the sum of the probabilities that either one happens, $P(A \cup B) = P(A) + P(B)$. (3) In a valid probability setup, the probability that some event obtains is set to be equal to 1. More loosely, the probability that one of the set of possible events occurs is 1.

[4] Standard decks of playing cards have 52 cards in four suits. The suits are spades, clubs, diamonds, and hearts. Spades and clubs are black; diamonds and hearts are red. Each suit has 13 cards: ace, 2, 3, 4, 5, 6, 7, 8, 9, 10, jack, queen, and king.

call the event that the drawn card is an ace A and the event that the drawn card is black B. If each card in the deck has an equal chance of being drawn, then to get the probability of these events, we count the number of cards that correspond to each event and divide by 52, the number of cards in the deck. Thus, the probability of drawing an ace is $P(A) = 4/52 = 1/13$ (there are 4 aces in the deck), and the probability of drawing a black card is $P(B) = 26/52 = 1/2$ (there are 26 black cards in the deck). This is good, but we have said nothing about how these events are related.

Suppose that someone draws a card and hides it from you. She tells you that the card is black. What is the probability that the hidden card is *also* an ace? You know that there are 26 black cards and that two of these black cards are aces. If each black card has an equal probability of being drawn, then the probability that the black card is an ace is $2/26 = 1/13$.

We just calculated the **conditional probability** of drawing an ace given that the card is black. If the event that the drawn card is an ace is A and the event that the drawn card is black is B, then we write this conditional probability as $P(A|B)$. The condition[5] is written after the vertical bar—in this case, that the card is black. In prose, we call this the "conditional probability of A given B," the "probability of A conditional on B," or "the probability of A given B." To calculate the conditional probability for any two events A and B, we do exactly what we just did in the playing card example:

$$P(A \mid B) = \frac{P(A \cap B)}{P(B)}. \tag{4.1}$$

That is, we calculate the probability that events A and B both occur, and we divide it by the probability that B occurs. In the playing card example, $P(A \cap B) = 2/52$ because there are two black aces in the deck. In contrast, $P(B)$ is $1/2$ because half the cards in the deck are black. Taking the quotient gives $P(A|B) = 1/13$. Figure 4-1 shows a schematic.

Notice that in this example, $P(A|B) = P(A)$. That is, conditioning on the event that we draw a black card did not change the probability that the card is an ace. Knowing that the drawn card is black gives us no information about whether the card is an ace. For any events A and B, if $P(A|B) = P(A)$, then we say that A and B are **independent**. Independence is one of the probabilistic concepts we will use most. If two events labeled A and B are independent, then $P(A \cap B) = P(A)P(B)$. That is, the probability that they both occur is equal to the

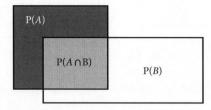

Figure 4-1 A schematic of two events, A and B, whose probabilities are represented by the areas of rectangles. $P(A)$ is the area of the dark and light gray regions; $P(B)$ is the area of the white and light gray regions, and $P(A \cap B)$ is the area of overlap—the light gray region. The conditional probability of A given B, $P(A|B)$, is the area of the light gray region divided by the entire $P(B)$ rectangle (white plus light gray), whereas $P(B|A)$ is the area of the light gray region divided by the entire $P(A)$ rectangle (light gray plus dark gray).

---

[5] People also use "condition" as a verb with the preposition "on," as in "we condition on the color of the card," which means that we compute probabilities assuming the color of the card is known.

product of the probabilities that they each occur separately. You will prove this in an exercise.

Consider a different case. Suppose that C is the event that the drawn card is a club. One-fourth of the cards are clubs, so the probability of drawing a club, $P(C)$, is 1/4. All clubs are black. What is the conditional probability of drawing a club given that the drawn card is black?

$$P(C \mid B) = \frac{P(C \cap B)}{P(B)} = \frac{13/52}{26/52} = \frac{13}{26} = \frac{1}{2}.$$

$P(C|B)$ is twice as large as $P(C)$. Because $P(C|B) \neq P(C)$, drawing a club and drawing a black card are not independent events.

Exercise Set 4-2[6]

(1) Prove that if events A and B are independent—that is, if $P(A \mid B) = P(A)$—then $P(A \cap B) = P(A)P(B)$. Prove that if $P(A \mid B) = P(A)$, then $P(B \mid A) = P(B)$.
(2) Suppose that you know $P(A \mid B)$, $P(A)$, and $P(B)$. How can you calculate $P(B \mid A)$?

4.3 Bayes' theorem

If you completed the last set of exercises, you have already proven Bayes' theorem.[7] Bayes' theorem is remarkable: rarely has such a simple, clear, and uncontroversial piece of mathematics served as the basis for so much debate. We will explore an example and the theorem itself. Some of the implications for statistics are discussed in Box 4-2.

Imagine that you are a physician, and you test a patient for hepatitis C. You know that hepatitis C is present in 0.1% of the population from which the patient comes—one person in a thousand. You also know that the test for hepatitis C is correct most of the time—specifically, if the person being tested has hepatitis C, the test will be positive 99% of the time, and if the person being tested does not have hepatitis C, the test will be negative 95% of the time. The patient's test comes back positive. On the basis of the test result and the prevalence in the population, what is the probability that she has hepatitis C?

When presented with this problem for the first time, most people intuit that the probability the tested person has hepatitis C is somewhere between 95% and 99%. This intuition is badly mistaken—it gets the conditional probability backwards. Ninety-nine percent is the probability of a positive test given hepatitis, not the probability of hepatitis given a positive test result. The datum we have is a positive test. Call the event of observing this datum D. Call the event that a person has hepatitis H. We have been given $P(D|H)$, the probability of a positive test given hepatitis, but we need $P(H|D)$, the probability of hepatitis C given a positive test. How can we get it?

For most people seeing a problem like this for the first time, the easiest way to proceed is to rephrase the problem in terms of numbers of people rather than probabilities. Figure 4-2 shows a schematic of the calculation. Say there are 100,000 people in the population, 100 of whom have hepatitis C (i.e. the frequency of hepatitis C is 0.1%). Ninety-nine of these 100 people with hepatitis C (99%) will test positive. The other 99,900 people in the population

---

[6] See the book's GitHub repository at github.com/mdedge/stfs/ for detailed solutions and R scripts.
[7] Bayes' theorem is named for the Reverend Thomas Bayes, but the story of its discovery is not straightforward (Stigler, 1983).

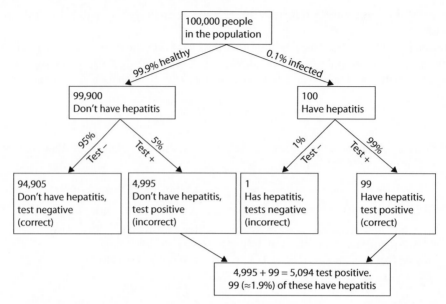

Figure 4-2 A schematic of the calculation of the probability that a person has hepatitis C given a positive test result, as described in the text.

do not have hepatitis C, but 4,995 of these people (5%) will test positive anyway. So we have 99 people *with* hepatitis C who test positive and 4,995 people without hepatitis C who test positive. The fraction of people who test positive that in fact have hepatitis C is thus $99/(99 + 4995) \approx 0.0194$, or under 2%. This is wildly out of accord with most untrained intuitions.[8]

We'll now step through the same chain of reasoning more formally, which will give the powerful abstraction that is Bayes' theorem. We need an expression for P(H|D) in terms of P(D|H). We can start by re-examining the definitions of these conditional probabilities (Equation 4.1). Remember that

$$P(D\,|\,H) = \frac{P(H \cap D)}{P(H)}$$

and that

$$P(H\,|\,D) = \frac{P(H \cap D)}{P(D)}.$$

These quantities have the same numerator, and they only differ in the denominator. Thus, we can express P(H|D) in terms of P(D|H) in just a few mathematical steps; the trick is to multiply by P(H)/P(H) = 1—multiplying by 1 does not change the value of the expression, so it is allowed— to get the third expression in this sequence:

$$P(H\,|\,D) = \frac{P(H \cap D)}{P(D)} = \frac{P(H \cap D)}{P(H)}\frac{P(H)}{P(D)} = P(D\,|\,H)\frac{P(H)}{P(D)}.$$

---

[8] Psychologists call this discordance between intuition and the correct answer as given by Bayes' theorem "base-rate neglect." It is well studied. Bad news: physicians are just as vulnerable to base-rate neglect as the rest of us (see, e.g., Hoffrage & Gigerenzer, 1998).

This is **Bayes' theorem**. For any two events A and B, Bayes' theorem holds that

$$P(A \mid B) = P(B \mid A) \frac{P(A)}{P(B)}. \tag{4.2}$$

We return to the hepatitis example. Applying Bayes' theorem, to find the probability that the tested person has hepatitis given a positive test, we need to calculate

$$P(H \mid D) = P(D \mid H) \frac{P(H)}{P(D)}.$$

We already have $P(D \mid H)$, the probability of a positive test given hepatitis. It is 99%, or 0.99. We also know that 0.1% of the people in the population have hepatitis, so $P(H) = 0.001$. The missing piece is $P(D)$, the probability of a positive test.

We were not given the probability of a positive test, $P(D)$, but we do know the probability of a positive test given that the tested person either has hepatitis, $P(D|H)$, or does not have hepatitis, $P(D|H^C)$. (Pronounce H^C as "H-complement"—remember complements from set theory, Box 4-1.) We can combine this information to get $P(D)$ with a useful trick. The patient either has or does not have hepatitis—H and H^C are mutually exclusive and collectively exhaustive. Thus, we can calculate the probability of a positive test as the sum of the probability that the patient has a positive test *and* has hepatitis plus the probability she has a positive test *and* does not have hepatitis That is,

$$P(D) = P(D \cap H) + P(D \cap H^C).$$

This is an application of a general principle called the law of total probability. Equation 4.1 implies that $P(D \cap H) = P(D|H)P(H)$. Applying this insight, we have

$$P(D) = P(D \cap H) + P(D \cap H^C) = P(D|H)P(H) + P(D|H^C)P(H^C).$$

Substituting in the values from the example gives

$$P(D) = P(D|H)P(H) + P(D|H^C)P(H^C) = 0.99 * 0.001 + 0.05 * 0.999 = 0.05094.$$

Now we have all the parts we need to calculate the conditional probability we care about—the probability that the tested person has hepatitis C given a positive test result. The computation is now straightforward:

$$P(H \mid D) = P(D|H) \frac{P(H)}{P(D)} = 0.99 * \frac{0.001}{0.05094} = 0.0194.$$

This is the same result we obtained the first time, thinking about numbers of people rather than probabilities.

Box 4-2 BAYES' THEOREM IN STATISTICS

Bayes' theorem is an essential idea in probability theory in its own right, but it is worth foreshadowing two related ways in which Bayes' theorem is at the center of important ideas in statistical theory. These ideas will be covered in more depth in the second half of the book.

First, Bayes' theorem is critical for understanding statistical hypothesis testing. Later, we will calculate p values, which can be interpreted roughly as the probability of observing data "like" those observed given that some hypothesis is true. (Here "like" means something

(Continued)

Box 4-2 (Continued)

along the lines of "as discordant with or more discordant with the hypothesis." More on this is coming in Chapter 7.) *p* values are often misinterpreted as the probability that the *hypothesis* is true given the observed data. This error is similar to the one that most people make when first presented with the hepatitis example—the conditional probability is rearranged, and it is assumed that $P(A|B) \approx P(B|A)$. The hepatitis C example reveals how wrong this interpretation can be. $P(A|B)$ and $P(B|A)$ are related, but they need not have similar values.

Second, Bayes' theorem has lent its name to one side of a debate in statistics. The two main parties are frequentists—the majority during the twentieth century—and Bayesians. One disagreement between frequentists and Bayesians is about whether it is legitimate to assign probabilities to hypotheses.

All parties agree that one can calculate the probability of the observed data given some hypothesis. Everyone also agrees that the probability of observing the data given that some hypothesis is true is distinct from the probability that the hypothesis is true given that the data have been observed. Frequentists hold that we cannot usually assign "prior"—that is, prior to the data—probabilities to hypotheses, and thus that we cannot calculate the probability that a hypothesis is true given the data. Let us call the data *D* and the hypothesis *H*. Frequentist methods assume that expressions like $P(H)$ are either meaningless or have unknowable values, so we cannot get from $P(D|H)$ to $P(H|D)$. There are some good reasons for holding this view. In the hepatitis C case, we knew the frequency of people in the population with hepatitis C, and we could use that frequency as the probability of the hypothesis that the patient now in our office has hepatitis C. But scientific hypotheses are not exactly like patients arriving at a clinic—it is not obvious that they can be viewed as members of a population where some proportion of them are true. The frequentist would rather say that the truth of the hypothesis is unknown but fixed—that is, not random—and therefore not subject to probability statements. Thus, in the frequentist view, it is impossible to know the probability that the hypothesis is true given that the data have been observed—the probability of the data given the hypothesis is the best we can do.

In contrast, Bayesians have held that it is sensible to assign prior probabilities to claims about the world—to set $P(H)$. Bayesians differ in their methods for setting $P(H)$, but they use Bayes' theorem to, for example, get from $P(H)$ and $P(D|H)$ to a statement of the *posterior* probability—$P(H|D)$, or the probability that the hypothesis is true given the observed data. Bayesians have gained ground in the last few decades for several reasons, including that many Bayesian methods produce good results in many areas, that some scientists find the Bayesian outlook satisfying, and that the computational challenges of Bayesian inference—once formidable—have been made tractable. This book spends more time on frequentist ideas than on Bayesian ones, but our focus on frequentist methods is not dogmatic, and a Bayesian approach to statistics is sketched in Chapter 10.

4.4 Discrete random variables and distributions

Up to now, we have considered the probabilities of individual events. To study more complicated scenarios, we need a way of representing random processes that captures the probabilities of many different events. The new concept of a "random variable" will do the job for us.

Suppose we roll a six-sided die once and want to construct a probability model to represent this process. One possibility might be to assume that each of the six faces is equally likely to face up after the roll.

We can represent the process of rolling the die as a **random variable**. In the case of a single die, we can use a variable labeled X. X is random, and it can take integer values from 1 to 6. X is different from the variables you have seen before in algebra and calculus. In those fields, variables represent numbers. However, our X is *not* a number yet—it is the unrealized outcome of a random process, the realization of which will provide a number. Sometimes we will want to refer to a specific **instance** of X that has occurred—a realization of the random process represented by X—such an instance *is* a number. To keep these concepts separate, we often use a capital letter to represent a random variable and a lowercase letter to indicate a particular instance of that random variable. In other words, before we roll the die, we consider X, which might take any integer value from 1 to 6. Suppose we then roll the die and observe a "2." Now we have observed lowercase x, and $x = 2$.

All the probability information about X is contained in its *distribution*. Our X is a **discrete random variable**, meaning that the number of outcomes is countable.[9] For discrete random variables, there are two ways of representing the distribution—the **probability mass function** (also pmf, or just "mass function") and the **cumulative distribution function** (also cdf, or just "distribution function"). The probability mass function is more intuitive. We write it as

$$f_X(x) = P(X = x). \tag{4.3}$$

That is, for any value x that is a possible outcome of X, the probability mass function gives the probability that the random variable X turns out to be equal to x. For example, suppose we assume that the die roll results in a "1" with probability 1/6. Then we write that $f_X(1) = P(X = 1) = 1/6$. Take a moment to parse this notation. We write f_X—the subscript is a capital X—to indicate that this function contains information about the random variable X. However, the input to the function is a specific possible value of X. In other words, we feed in actual numbers, not random variables. (Random variables, again, are numbers-to-be rather than actual numbers.) The function's output is the probability that the random variable will equal the function's input.

The probability mass function is one way to represent the distribution of a discrete random variable. The other is the cumulative distribution function. Whereas the probability mass function gives the probability that a random variable will *equal* a particular value, the cumulative distribution function gives the probability that the random variable will be *less than or equal to* a particular value. We write it as $F_X(x) = P(X \leq x)$. Figure 4-3 shows the probability mass function and cumulative distribution function for the die-roll example, assuming that each face is equally likely to appear.

Figure 4-3 suggests that if either the cumulative distribution function or the probability mass function is known, then it is possible to deduce both functions. This is true. In our example, the probability that the die shows a number less than or equal to one is just the

---

[9] "Countable" here means that the number of outcomes is either finite, or, if it is infinite, it is not of a higher "order" of infinity than the natural numbers {0,1,2,3,4,...}. The set of all natural numbers is infinite (you can go as high as you want) but countable. Roughly, countable means that you make progress in counting the members of the set, even if you can never finish. If I start counting, "0, 1, 2, 3, 4,...," I will never finish counting natural numbers, but I will never need to go back and get natural numbers I missed between 0 and 4. In contrast, the set of all real numbers is infinite and uncountable. Take any two real numbers, such as 0 and 1, and there are an infinite number of real numbers between them. It's impossible to make progress in exhaustively enumerating all possible real numbers—one can never count all the real numbers between any two real numbers.

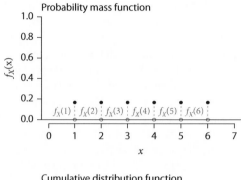

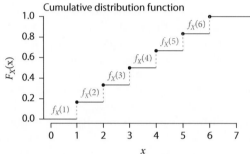

Figure 4-3 The probability mass function and cumulative distribution function for a roll of a fair six-sided die, with annotations in gray. The probability mass function on the top panel shows that each of the outcomes from one to six is equally likely. Each of the solid dots represents a probability of 1/6 of the outcome on the horizontal axis. All other outcomes have probability zero. The cumulative distribution function (bottom panel) shows that the probability that $X \leq x$ increases by jumps that correspond to the probabilities of the outcomes. The dots appear at the locations of jumps in the cumulative distribution function—in this case, the integers from 1 to 6.

probability that it shows 1 (since there are no possibilities smaller than 1), or $f_X(1)$. The probability it shows a number less than or equal to 2 is just the probability it shows 1 or 2, or $f_X(1) + f_X(2)$, and so on. You could write it like this:

$$F_x(1) = f_x(1),$$
$$F_x(2) = f_x(1) + f_x(2),$$
$$F_x(3) = f_x(1) + f_x(2) + f_x(3),$$
$$F_x(4) = f_x(1) + f_x(2) + f_x(3) + f_x(4),$$
$$F_x(5) = f_x(1) + f_x(2) + f_x(3) + f_x(4) + f_x(5),$$
$$F_x(6) = f_x(1) + f_x(2) + f_x(3) + f_x(4) + f_x(5) + f_x(6).$$

In other words, the cumulative distribution function is a series of partial sums of the probability mass function. We can write this more generally as

$$F_X(x) = P(X \leq x) = \sum_{x_i \leq x} f_X(x_i). \tag{4.4}$$

Cumulative distribution functions $F_X(x)$ *increase monotonically* in x, meaning that they never decrease as their argument increases. Further, because they output probabilities, they cannot take values larger than 1, and they either reach 1 or approach a limit of 1 as their argument increases to infinity. Let's take a moment to practice.

Exercise Set 4-3

(1) Suppose we flip a fair coin 3 times. "Fair" means that each time we flip the coin, the probability of observing a "heads" outcome is 1/2. We can model the number of heads we observe as a random variable X. X can take the values 0, 1, 2, or 3. Find the value of the probability mass function $f_X(x)$ for $x = 0$, $x = 1$, $x = 2$, and $x = 3$. (Hint: Each possible sequence of heads and tails is equally likely. Write them all out and count the number of heads that each one gives.)

(2) Suppose you summed $f_X(x_i)$ for every possible x_i. What would the sum be? If you are having trouble, try this on the probability space defined in Problem 1, and then think about the answer you obtain.

(3) In terms of F_X, what is $P(a < X \leq b)$, the probability that a random variable X falls between two numbers, a and b, with $b > a$? (Hint: Notice that $a < X \leq b$ if and only if $X \leq b$ and $X > a$.)

4.5 Continuous random variables and distributions

The probability mass function and the cumulative distribution function are adequate for describing discrete random variables—random variables that can take on a countable number of values. Suppose, though, that we want to construct a random variable that is allowed to take on any real number. We might think of random variables for values we measure with decimal numbers, like heights, weights, or times.

Suppose we want to construct a probability model for human weights. We could start by writing down a probability mass function for weights rounded to the nearest kilogram. In principle, the number of kilograms a person could weigh could take any positive integer value (ignoring physiological limits on how large or small an adult human can be). This is a countable number of possibilities, and we could write a function to assign a probability to each one. But this feels unnatural. Why are we rounding the weights to the nearest whole number? Why not allow the random variable to take any positive value?

We can allow random variables to take on any real number, but it will require extra conceptual work. If we allow the random variable to take on any real number, then the probability that the random variable will equal any *specific* number becomes zero. Consider: what is the probability that an adult human's weight will round to 70.6 kilograms? It is small. That it will round to 70.64 kilograms? Smaller still. That it will round to 70.636 kilograms? Even smaller. But consider that there are an infinite number of real numbers between 70.636 and 70.637. If we truly allow weights to take *any* real number, then we are effectively asking about the probability that a human weight will round to a decimal number with an infinite number of digits. For any real number, this probability is zero. Thus, we cannot write down a probability mass function that will meaningfully distinguish different **continuous random variables**—the probability mass function will always equal zero everywhere. One solution would be to allow only a certain degree of precision and to model random variables discretely. It is usually more convenient to change one's conceptual tools.

One cannot use probability mass functions for continuous random variables, but there is nothing to stop us from using the cumulative distribution function. It may not be coherent to ask about the probability that a human will weigh 70.0000...kilograms, but we can ask about the probability that a human will weigh 70 kilograms or less, or about the probability that a human will weigh between 70 and 71 kilograms. Thus, we are still equipped with

$$F_X(x) = P(X \leq x). \tag{4.5}$$

Take a moment to do the following exercises, which consider cumulative distribution functions for continuous random variables.

Exercise Set 4-4

(1) Suppose you have a continuous random variable X that takes values in the interval $[0, 1]$. (That is, X can be any real number between 0 and 1, including 0 or 1.) Suppose further that the probability distribution of X is "uniform." Here, by "uniform" we mean that for any two equally sized intervals in $[0, 1]$, the probability that X lands in one interval is equal to the probability that X lands in the other interval. That is, $P(a < X \le b) = P(c < X \le d)$ provided that $b - a = d - c \ge 0$ and a, b, c, and d are all in the interval $[0, 1]$.[10] Draw the cumulative distribution function $F_X(x)$ for $x \in [-1, 2]$.

(2) Continuing with X from Problem 1: How would the cumulative distribution function change if X were more likely to land in $[0.4, 0.6]$ than in any other region of length 0.2?

Box 4-3 INTEGRATION

To work with the distributions of continuous random variables, we need another calculus concept: integration. There is a crash course on integral calculus in Appendix A, and more calculus resources are mentioned at the end of that appendix.

As a brief reminder, the integral of a function $f(x)$ is written as $\int f(x)dx$. You can think of the "d" as meaning "a tiny amount of" and the $\int$ as "sum," so "$\int f(x)dx$" means something like "the sum of all the tiny pieces of $f(x)$." Integrals are closely analogous to sums. For discrete random variables, we sum over a countable set of outcomes. To perform the analogous operation for a continuous random variable, we integrate.

As discussed in Appendix A, integrals can be used to compute the area underneath the graph of a function. Specifically, if the integral (also called the indefinite integral) of a function $f(x)$ is $\int f(x)dx = F(x)$, then the area underneath the graph of $f(x)$ and above the x axis, between $x = a$ and $x = b$, is $\int_a^b f(x)dx = F(b) - F(a)$. The quantity $\int_a^b f(x)dx$ is called "the definite integral of $f(x)$ from a to b."

Remarkably, taking an integral and taking a derivative are inverse processes, meaning that if $F(x)$ is the indefinite integral of $f(x)$, then the derivative of $F(x)$ is $f(x)$. The relationship between integration and differentiation means that the rules for taking derivatives introduced in Chapter 3 imply rules for integration. In particular, for polynomials—that is, functions of the form $f(x) = ax^n$, with a and n constants (and $n \ne -1$)—the indefinite integral has the form

$$\int ax^n dx = \frac{ax^{n+1}}{n+1} + C,$$

where C is an arbitrary constant. You can memorize the phrase "up and under," which describes what happens to the exponent when a polynomial is integrated. (The exponent's value is increased by one—it goes "up"—and the new value of the exponent divides the original expression—it comes "under" in a fraction.) Further, the integral of a function times

(Continued)

---

[10] Remember from Problem 3 of Exercise Set 4-3 that $P(a < X \le b) = F_X(b) - F_X(a)$.

Box 4-3 (Continued)

a constant is equal to that constant times the integral of the original function. That is, if $g(x) = af(x)$ and a is a constant, then

$$\int g(x)dx = \int af(x)dx = a\int f(x)dx. \tag{A.6}$$

(A.6 is the equation number from Appendix A.) Finally, the integral of a sum of two functions is the sum of the integrals of the individual functions. So, if $h(x) = f(x) + g(x)$, then

$$\int h(x)dx = \int f(x) + g(x)dx = \int f(x)dx + \int g(x)dx. \tag{A.7}$$

4.6 Probability density functions

The cumulative distribution function completely specifies all the information we have about a random variable, so, in one sense, we have no problem—we can answer questions about continuous random variables using the cumulative distribution function. There is a more intuitive way to proceed, though. We can define a function that will behave analogously to the probability mass function and be defined for continuous variables. Recall that when we had a discrete random variable, the cumulative distribution function was given by partial sums of the probability mass function (Equation 4.4, reproduced here):

$$F_X(x) = \sum_{x_i \leq x} f_X(x_i).$$

We appeal to this relationship to define the probability density function, also called the pdf or just "density." The only trouble is that we cannot take a partial sum of a continuous random variable—there are uncountably many possible values of X to sum over, and we would never get anywhere. However, recall that integrals are like sums—they "sum" the area under a curve (Box 4-3). Though we cannot take a partial sum of a continuous function, we *can* take an integral. We define the **probability density function** of a continuous random variable, which we will also denote f_X, using this relationship:

$$F_X(x) = \int_{-\infty}^{x} f_X(u)du. \tag{4.6}$$

Take a moment to notice how similar this expression is to Equation 4.4 (reproduced directly above)—we just replace the partial sum with a partial integral. Remember that an integral can be interpreted as an area under a curve. This means that the cumulative distribution function computed at x, which is equal to the probability that $X \leq x$, is also equal to the area under the probability density function between $-\infty$ and x. Figure 4-4 depicts this relationship.

Because integration and differentiation are inverse procedures (Equation A.5), Equation 4.6 implies

$$f_X(x) = F_X'(x). \tag{4.7}$$

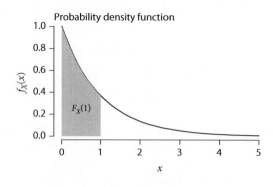

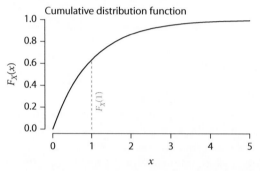

Figure 4-4 A probability density function (upper panel) and cumulative distribution function (lower panel). These functions are for what is called an "exponential" random variable with rate 1 (See Table 4-2). The random variable takes values in $[0, \infty)$. The region under the density function between 0 and 1 is shaded, and its area is $1 - e^{-1} \approx 0.632$. (Here, e is Euler's number, ≈ 2.718) The density is 0 for all $x < 0$. As a result, $\int_{-\infty}^{1} f_X(u)du = 1 - e^{-1}$, meaning that $F_X(1) = 1 - e^{1}$. The lower panel shows the cumulative distribution, with a gray, dashed vertical line at $F_X(1)$. The length of the vertical line is $1 - e^{-1}$, the same as the area of the shaded region.

That is, the probability density function is the derivative of the cumulative distribution function. This means that the probability density at a chosen point is the instantaneous slope of the cumulative distribution function at that point.

In many ways, continuous density functions are analogous to discrete probability mass functions—we will see a few of these analogies later. Plots of probability density functions can be interpreted similarly to plots of probability mass functions: random variables are more likely to take values in regions where their density or mass functions are high than in regions where their density or mass functions are low, assuming that the regions being compared are of similar width. In the following exercises, you will start to become comfortable with probability density functions.

Exercise Set 4-5

(1) Suppose that $f_X(x)$ is a probability density function. What is the total area under $f_X(x)$, or $\int_{-\infty}^{\infty} f_X(x)dx$?

(2) [Optional] Suppose there is a function $f(x)$ that takes the following values:

$$f(x) = 0 \text{ if } < x < 0 \text{ or } x > 1/10, \text{ and}$$

$$f(x) = 10 \text{ if } 0 \leq x \leq 1/10.$$

Could $f(x)$ be a probability density function for some continuous random variable? (Consider the axioms of probability from Section 4.1.) How is this situation different from that of a probability mass function?

4.7 Families of distributions

There are two main requirements for mass or density functions: First, the function must be non-negative. Second, a function of a discrete variable can be a probability mass function if the sum of all the values it takes is 1. Similarly, a function of a continuous variable can be a probability density function if the total area under its graph—that is, its integral from minus infinity to plus infinity—is 1.

Sometimes, when we want to study a phenomenon using a probabilistic model, we need to start from the beginning, defining a distribution that meets the requirements of our situation. In principle, this will work fine—all we need to do is write down a distribution, density, or mass function that describes our process, and we can start asking questions about our phenomenon. But most of the time, we work with distributions that have already been derived and studied.

For example, suppose that we are interested in the outcome of a fair die roll. We can model this process with a random variable, X. In this case, we want X to have the mass function shown in Figure 4-3. That is, we set the probability of observing each face to be 1/6. We are in a good situation: with a little math, we can learn a lot about the die roll, and with some programming, we can also simulate the die roll. However, we can save ourselves some work if we know that the distribution we have just written for X belongs to a family of distributions. X's distribution belongs to the discrete uniform family. If we know this, then we can take advantage of facts already known about discrete uniform distributions. The way to start spotting these things is to learn something about the dozen or so most commonly used distributions. You can then add to your list of known distributions throughout your life.

Not everyone uses the term **distribution family**, but the metaphor is apt. Distributions in the same family share a certain resemblance—they are often similarly shaped, and summaries of their behavior can be computed from the same functions. Distributions in the same family differ from each other because they have different **parameters**. You can think of parameters as being like the first and middle name of the distribution, whereas the family is the surname. The distribution we wrote down for our fair die roll is the Discrete Uniform(1,6) distribution. The parameters are written in parentheses after the family name. Discrete uniform distributions have two parameters: a minimum and a maximum possible value. All the intervening integers are possible outcomes, and all possible outcomes are equally likely.

Tables 4-1 and 4-2 provide a list of some of the most important discrete and continuous distribution families along with their parameters. In this book, the normal distribution is especially important.

For each of the distributions in Tables 4-1 and 4-2, you can think of the distribution family as capturing the general type of situation under study and of the parameters as capturing the specifics of the situation under study. For example, the exponential

Table 4-1 Some useful discrete probability distribution families[a]

Family	Parameters	Probability mass function	Description
Bernoulli	p	$P(X = k) = p^k(1-p)^{1-k}$ for $k \in \{0, 1\}$	A single trial with probability of success p. Example: Probability of heads on a single toss of a fair coin is Bernoulli($\frac{1}{2}$)
Binomial	n, p	$P(X = k) = \binom{n}{k} p^k(1-p)^{n-k}$ for $k \in \{0, 1, 2, \ldots, n\}$	Number of successes out of n independent trials with probability of success p on each trial. Example: Number of heads out of 10 flips of a fair coin is Binomial(10, $\frac{1}{2}$).
Geometric	p	$P(X = k) = (1-p)^k p$ for $k \in \{0, 1, 2, \ldots, \infty\}$	Number of failures before first success from independent trials with probability of success p on each trial. Example: Number of tails before first head when flipping a fair coin is Geometric($\frac{1}{2}$).
Poisson	λ	$P(X = k) = \frac{\lambda^k e^{-\lambda}}{k!}$ for $k \in \{0, 1, 2, \ldots, \infty\}$	Roughly, the number of times an event will happen in a large amount of space or time. Examples: the number of goals in a hockey game, the number of men killed by horse-kicks per year in the Prussian cavalry, or the number of mutations on a chromosome.
Discrete Uniform	a, b	$P(X = k) = \frac{1}{b-a+1}$ for $k \in \{a, a+1, a+2, \ldots, b\}$	An integer between a and b (inclusive), with each integer having an equal chance of being chosen. Example: The outcome of a single roll of a fair die is Discrete Uniform(1, 6).

[a] The $\binom{n}{k}$ in the probability mass function of the binomial distribution is pronounced "n choose k" and is the number of distinct subsets of size k that can be drawn from a set of n distinct objects. Its value is $n!/[k!(n-k)!]$, where $n! = n * (n-1) * (n-2) * \ldots * 3 * 2 * 1$. By a special definition, $0! = 1$. $n!$ is pronounced "n factorial." Factorial notation also appears in the probability mass function of the Poisson distribution.

Table 4-2 Some useful continuous probability distribution families

Family	Parameters	Probability density function	Description
Uniform	a, b	$f_X(x) = \frac{1}{b-a}$ for $x \in [a, b]$, 0 elsewhere	A real number drawn from a range running from a to b. Every number in the range has the same probability density.
Exponential	λ	$f_X(x) = \lambda e^{-\lambda x}$ for $x \in [0, \infty)$, 0 elsewhere	The waiting time to an event that happens with average rate λ per unit time. A continuous analog of the geometric distribution.
Normal	μ, σ^2	$f_X(x) = \frac{1}{\sigma\sqrt{2\pi}} e^{-\frac{(x-\mu)^2}{2\sigma^2}}$	Also called the "Gaussian" or the bell curve. We will have more to say about how it arises in Chapter 5.
χ^2	k	$f_X(x) = \frac{1}{2^{\frac{k}{2}}\Gamma(k/2)} x^{\frac{k}{2}-1} e^{-x/2}$ for $x \in [0, \infty)$, 0 elsewhere. Here and in the t distribution, Γ denotes the gamma function. If you've never seen it before, don't worry.	Take k independent draws from a normal distribution with $\mu = 0$ and $\sigma^2 = 1$. Square the k draws. The sum of the squared draws is χ^2-distributed with k "degrees of freedom." The χ^2 distribution is important in inferential statistics. (χ is not an "x"; it is a Greek "chi," pronounced "kai.")
t	k	$f_X(x) = \frac{\Gamma\left(\frac{k+1}{2}\right)}{\sqrt{k\pi}\,\Gamma\left(\frac{k}{2}\right)}\left(1 + \frac{x^2}{k}\right)^{-\frac{k+1}{2}}$	Draw a value from a normal distribution with $\mu = 0$ and $\sigma^2 = 1$ and call it Y. Independently, draw a value from a χ^2 distribution with k degrees of freedom and call it Z. The quantity $Y/\sqrt{Z/k}$ has a t distribution. The t distribution is important in inferential statistics.

distribution is often used to model waiting times to some event, like a failure of a lightbulb or a radioactive emission. The properties of the lightbulb or of the radioactive material we want to study are captured by the parameter of the distribution we use. In particular, the parameter λ is set to be the average rate at which events happen. If the lightbulbs we study burn out in one year on average, we would set $\lambda = 1$ to model the waiting times in years, or if we have a mole of uranium–238 that emits 3,000,000 alpha particles per second, then we would set $\lambda = 3,000,000$ to study the distribution of waiting times between emissions in seconds.

In the following problems, you will interrogate some probability distributions and use R to simulate random draws from probability distributions. In the next chapter, you will learn how to summarize the behavior of random variables.

Exercise Set 4-6

(1) What is the probability of drawing (i) a "0", (ii) a "1", and (iii) a "2" from a Poisson distribution with parameter $\lambda = 5$?

(2) You repeatedly flip a fair coin. What is the probability that you first see "heads" on your 6th flip? (Hint: find a suitable distribution family in Table 4-1.)

(3) You can use R to interrogate the density and distribution functions of distributions from known families. Consider a Poisson distribution with parameter $\lambda = 5$. If we want to know the value of the probability mass function for $x = 2$, $f_X(2)$, we use the dpois() function:

```
> dpois(2, lambda = 5)
[1] 0.08422434
```

To get the value of the cumulative distribution function $F_X(2)$, we use ppois():

```
> ppois(2, lambda = 5)
[1] 0.124652
```

Finally, we might want to know the *inverse* of the cumulative distribution function. That is, suppose we are interested in a probability p between 0 and 1. We might want to find the value of q that solves the equation $F_X(q) = p$. That is, we might want to find the number q such that the probability that the random variable is less than or equal to q is p. This sounds abstract, but q is just the pth percentile of the distribution. We can get these numbers using qpois():

```
> qpois(0.124652, lambda = 5)
[1] 2
```

The inverse of the cumulative distribution is also called the quantile function, and the argument to the quantile function is called a quantile.

For the normal distribution, the density functions are calculated by dnorm(), cumulative distribution functions are calculated by pnorm(), and quantile functions are calculated by qnorm(). Please do the following in R using the standard normal distribution—the normal distribution with a mean (μ) of 0 and a standard deviation (σ) of 1.

(a) Plot the probability density function for $x \in [-3, 3]$. (Hint: use seq() to generate x values).

(b) Plot the cumulative distribution function for $x \in [-3, 3]$.

(c) What value of x is at the 97.5th percentile of the standard normal distribution? (The 97.5th percentile is a quantile.)

(4) For much of the rest of the book, we will explore questions about probability and statistics using simulated random numbers. The random numbers we'll use in our

simulations are actually "pseudorandom"—they come from an algorithm designed to produce numbers that share many properties with random numbers. Methods for testing the random-like properties of pseudorandom numbers are beyond our scope. R generates pseudorandom numbers using the Mersenne Twister, a state-of-the-art algorithm.

In this question, you will familiarize yourself with generation of random numbers in R. For many distribution families, R provides a function that generates random numbers. To generate these numbers, you need to provide the function name, which specifies the distribution family, the number of independent draws from the distribution you want to generate, and the parameters you want to use. For example, if you want to generate 10 independent draws from the Poisson distribution with parameter $\lambda = 5$, you use the command `rpois(n=10, lambda=5)` or just `rpois(10, 5)`. To generate simulated values from the normal distribution, use `rnorm()`.

(a) Generate a vector of 1,000 independent draws from a standard normal distribution. (The standard normal distribution has $\mu = 0$ and $\sigma^2 = 1$.) Save the draws and plot them as a histogram. Hint: use `help(rnorm)` to look up the syntax for `rnorm()`.

(b) [Optional] Generate and save a vector of 1,000 independent draws from a continuous uniform distribution with a minimum of 0 and a maximum of 1. (Use the `runif()` function.) Now feed those values to the `qnorm()` function for a standard normal distribution and save the result. Plot them as a histogram. What do you notice? How do you think this works?

4.8 Chapter summary

In this chapter, we learned the rules of probability. Probabilities are non-negative, they sum to one, and the probability that either of two mutually exclusive events occurs is the sum of the probability of the two events. Two events are said to be independent if the probability that they both occur is the product of the probabilities that each event occurs. We used Bayes' theorem to update probabilities on the basis of new information, and we learned that the conditional probabilities $P(A|B)$ and $P(B|A)$ are not the same. Finally, we learned how to describe the distributions of random variables. For discrete random variables, we rely on probability mass functions, and for continuous random variables, we rely on probability density functions.

4.9 Further reading

There are many introductory probability texts. Here are three recommended ones. All of these require a little more math than is used here, but they cover every topic in this chapter and the next (plus more) in greater depth.

Blitzstein, J. K., & Hwang, J. (2014). *Introduction to Probability*. CRC Press, Boca Raton, FL.

Hoel, P. G., Port, S. C., & Stone, C. J. (1971) *Introduction to Probability Theory*. Houghton Mifflin, Boston, MA.

Ross, S. M. (2002). *A First Course in Probability*. Pearson, Upper Saddle River, NJ.

For more on the philosophical underpinnings and implications of probability, see

Diaconis, P., & Skyrms, B. (2017). *Ten Great Ideas about Chance*. Princeton University Press.

A fascinating book of essays with forays into many areas that neighbor probability theory, including math, physics, philosophy, history, and even psychology. It is a little technical in places, but should be readable after this chapter and the next.

Hájek, A. (2011). Interpretations of Probability. *Stanford Encyclopedia of Philosophy.* https://plato.stanford.edu/entries/probability-interpret/

An introduction to the philosophical problem of interpreting probability, including a nice account of Kolmogorov's axioms.

CHAPTER 5

Properties of random variables

Key terms: Central limit theorem (CLT), Conditional distribution, Correlation, Covariance, Dependent variable, Disturbance, Expectation (or Expected value), Homoscedasticity, Independent variable, Joint distribution, Law of large numbers, Linearity of expectation, Normal distribution, Standard deviation, Variance

In the last chapter, we learned the rules of probability, and we learned how to specify probability distributions. Here, we continue our study of probability by exploring the properties of random variables. We already know how to specify a random variable's distribution using a probability mass function, probability density function, or cumulative distribution function. But we need ways to describe the behavior of a random variable other than pointing to its distribution. What is the average value the random variable will take? How varied will the outcomes of the process be? How will the random variable be related to other random variables? We will explore ideas that respond to these questions: expectation, variance, and correlation. We will also discuss two theorems that describe the behavior of the means of large samples, the law of large numbers and the central limit theorem. Finally, we will build a probabilistic model for simple linear regression.

5.1 Expected values and the law of large numbers

Often, we want to summarize the information contained in a probability distribution. Most of the distributions we care about involve the assignment of probabilities or probability densities to a large—possibly infinite—set of numbers. It is useful to summarize the distribution with a small set of numbers. The most important summaries of distributions are answers to two questions: "Where is the distribution centered?" and "How spread out is the distribution?" The first question is about *location* (see Box 5-1), and the second is about *dispersal*. This section covers the expectation, or expected value, which is one way of describing location, and the next section covers the variance, which is one way of describing dispersal.

The **expectation** is analogous to the familiar concept of the average or the mean. Specifically, the expectation of a discrete random variable is an average of all the values that a random variable can take, weighted by the probability with which they occur. Suppose X is a discrete random variable with mass function $f_X(x)$. Further, suppose that the values that X can take—that is, the values of x that have $f_X(x) > 0$—are contained in the

Statistical Thinking from Scratch: A Primer for Scientists. M. D. Edge, Oxford University Press (2019).
© M. D. Edge 2019. DOI: 10.1093/oso/9780198827627.001.0001

Box 5-1 MEASURES OF LOCATION

Many readers of this book are likely already familiar with three measures of location, or central tendency, for samples of data: the sample mean, the sample median, and the sample mode. The sample mean is the sum of all the observations in the sample divided by the number of observations in the sample, a sample median is a number than which half of the observations in the sample are smaller, and the sample mode is the observation that occurs most frequently in the sample.

There are concepts analogous to the sample mean, median, and mode that apply to probability distributions. The analogue of the sample mean is the expectation, the subject of this section. It is a weighted average of the values that the random variable can take, where the weight is the probability mass or density associated with each value. The analog of a sample median is a distribution median, a point at which the cumulative distribution function equals one half. The analog of the sample mode is the distribution mode, the maximum of the mass or density function.

The sample mean is a popular measure of central tendency, but it is vulnerable to distortion by extreme values. For example, suppose a random sample of 1,000 Americans reports their net worth. If one of the richest 100 Americans happens to be included in the sample, then the average net worth calculated from the sample will not be representative of the typical income—the rich American's net worth is likely to be much greater than that of the rest of the sample combined. The median, in contrast, will not change if the income of the richest member of the sample is increased—it is robust to outliers. Why, then, is the sample mean more popular?

Many reasons can be given, but one reason that means are important in statistics is that they share a close affinity with expectations of random variables. We will see that expectations have many helpful properties, and these properties are not always shared by the distribution median or mode.

set $\{x_1, x_2, ..., x_k\}$. Then the expectation of X is

$$E(X) = \sum_{i=1}^{k} x_i P(X = x_i) = \sum_{i=1}^{k} x_i f_X(x_i). \tag{5.1}$$

The expectation is the sum obtained after multiplying each value the random variable can take by the probability that the random variable takes that value. For example, if Y represents a six-sided die taking the values $\{1, 2, 3, 4, 5, 6\}$ with equal probability, then

$$E(Y) = \sum_{i=1}^{K} y_i f_Y(y_i) = 1\left(\frac{1}{6}\right) + 2\left(\frac{1}{6}\right) + 3\left(\frac{1}{6}\right) + 4\left(\frac{1}{6}\right) + 5\left(\frac{1}{6}\right) + 6\left(\frac{1}{6}\right) = \frac{21}{6} = \frac{7}{2}.$$

One intuitive way to understand the expected value is to notice that it is the value the sample mean would take if we had a sample that exactly reflected the underlying probability distribution. For example, suppose we threw a fair 6-sided die 60 times and observed the 6 faces exactly 10 times each—in this case, the sample frequencies for each outcome (10/60) would exactly match the probabilities (1/6). The sample mean would be the sum of all the die rolls divided by the total number of rolls, 60, or

$$\frac{10*1 + 10*2 + 10*3 + 10*4 + 10*5 + 10*6}{60} = 1\left(\frac{10}{60}\right) + 2\left(\frac{10}{60}\right) + 3\left(\frac{10}{60}\right) + 4\left(\frac{10}{60}\right) + 5\left(\frac{10}{60}\right) + 6\left(\frac{10}{60}\right)$$

$$= 1\left(\frac{1}{6}\right) + 2\left(\frac{1}{6}\right) + 3\left(\frac{1}{6}\right) + 4\left(\frac{1}{6}\right) + 5\left(\frac{1}{6}\right) + 6\left(\frac{1}{6}\right) = \frac{7}{2},$$

which is exactly equal to the expectation.

Similarly, if X is a continuous random variable with probability density function $f_X(x)$, then the expectation of X is

$$E(X) = \int_{-\infty}^{\infty} x f_X(x) dx. \tag{5.2}$$

This is analogous to the discrete case, but instead of summing over a mass function, we integrate over a density function. A picture will give you an idea of what is meant when we say that the expectation is a measurement of location. Figure 5-1 shows density functions of two normal distributions that are alike except that their expectations differ.

The name "expectation" is slightly misleading—when we observe a random variable, we do not necessarily "expect" to observe the expected value. In fact, it is sometimes not even possible to observe the expected value—if we roll a fair 6-sided die, the expected value is 3.5, but 3.5 is not one of the possible outcomes. Rather, the expected value can be thought of as a long-term average. If a random variable is observed repeatedly, then the average of the observations will get closer to the expectation. This idea is formalized by the **law of large numbers**. One version of the law of large numbers is stated as follows:

Weak[1] law of large numbers: Suppose that $X_1, X_2, X_3, ..., X_n$ are independent random variables and that all the X_i have the same distribution. (We say the X_i are "i.i.d."—

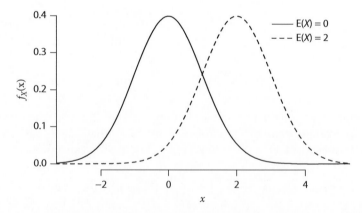

Figure 5-1 Plots of the density functions of two normal distributions with different expected values. The density functions have the same shape, but their locations are different.

---

[1] There is also a "strong" law of large numbers. The differences between forms of the law of large numbers come down to different senses in which random variables can be said to "converge" to constant values. The "weak" law presented here is easier to prove and will work just fine for us.

independent and identically distributed.) Assume that $E(X_1) = E(X_2) = \ldots = E(X_n) = \mu$ and that μ is finite.[2] Define $\overline{X}_n$ as the mean of the observations,

$$\overline{X}_n = \frac{1}{n}(X_1 + X_2 + X_3 + \ldots + X_n).$$

The weak law of large numbers holds that as n grows to approach infinity, $\overline{X}_n$ "converges in probability" to μ. This means that for any positive constant δ,

$$\lim_{n \to \infty} P(|\overline{X}_n - \mu| > \delta) = 0. \tag{5.3}$$

In words, the probability that the sample mean will be more than *any* fixed distance from the expectation—we get to pick the distance, and we can choose it to be as small as we want —approaches zero for large samples. Loosely, the sample mean will get ever closer to the expectation as the sample size increases. But notice: the law of large numbers is a statement about what happens as the sample size approaches infinity. How do we know whether the sample size is large enough that the law of large numbers has "kicked in" and that the sample mean is close to its expectation? We will introduce some ideas relevant to this question over the next few chapters. We focus mostly on the normal distribution, but there is an important caveat: for distributions with heavier tails than the normal—that is, distributions for which the probability of extreme events is higher than for the normal distribution, like the Pareto distribution—the rate of convergence can be slower, even if the law of large numbers applies.

We will delay proving the law of large numbers until the next section, where a proof will be an optional exercise.

One of the most useful facts about expectations is that they are linear. This means that the expectation of any linear function of random variables is equal to the same linear function applied to the expectations of the random variables.[3] For example, suppose that X and Y are random variables and that you want to know the expectation of $aX + bY + c$, where a, b, and c are constants (i.e., a, b, and c are not random). If you know the expected values of X and Y, then you can calculate the desired expectation as

$$E(aX + bY + c) = aE(X) + bE(Y) + c. \tag{5.4}$$

There are actually three facts here. First, $E(aX) = aE(X)$—the expectation of a random variable times a constant is the constant times the expectation of the random variable.[4] This follows from Equation A.6 and the definition in Equation 5.2. A second fact is that the expectation of a constant is just that constant: $E(c) = c$. (Think of the constant as a discrete random variable that takes the value c with probability 1.) Third, the expectation of a sum of random variables is equal to the sum of the expectations of those random variables—that is, $E(X + Y) = E(X) + E(Y)$. To prove this last part, we need the concept of a joint distribution of two random variables, which will come in a later section.

---

[2] In this book, we will only consider probability distributions with finite expectations, but it is possible for the expectation to be undefined. The Cauchy distribution, which looks much like a normal distribution to the naked eye, is one distribution for which the expectation is undefined.

[3] A linear function of a variable x is one of the form $f(x) = a + bx$, where a and b are constants. Similarly, for a function of two variables, any function $f(x,y) = a + bx + cy$ with constants a, b, and c is linear. In a linear function, you can't raise the arguments to powers, raise constants to powers of the arguments, take the sine of the arguments, etc. If there are two or more arguments, linear functions cannot include products of the arguments, so $f(x,y) = xy$ is not linear.

[4] It is crucial that a is a constant and not a random variable. For random variables X and Y, the expectation of the product $E(XY)$ depends on the expectations of X and Y separately and also on their *covariance*, a quantity discussed later in the chapter.

Linearity of expectation is amazingly handy in probability calculations—you will see it many times in the exercises. Linearity does not generally hold for distribution medians or modes, which is one reason why the expectation is a go-to indicator of location in probability. The privileged status of the expected value in probability is, in turn, one explanation for the importance of the sample mean in statistics.

One last fact about expected values. To calculate the expectation of a *function* of a continuous random variable X, we write

$$E[g(X)] = \int_{-\infty}^{\infty} g(x)f_X(x)dx, \tag{5.5a}$$

or, if X if is discrete,

$$E[g(X)] = \sum_{i=1}^{k} g(x_i)f_X(x_i). \tag{5.5b}$$

This is analogous to what we did when calculating $E(X)$—notice that if we set $g(x) = x$, then Equation 5.5a reduces to Equation 5.2, and Equation 5.5b reduces to Equation 5.1. Equation 5.5 is sometimes called the law of the unconscious statistician, perhaps because near-thoughtless applications can solve apparently difficult problems.

Exercise Set 5-1[5]

(1) See Tables 4-1 and 4-2 for relevant mass and density functions for the following questions.
   (a) What is the expected value of a Bernoulli random variable with parameter p?
   (b) Use part (a) and the linearity of expectation (Equation 5.4) to get the expected value of a binomial random variable with parameters n and p.
   (c) [Optional] What is the expected value of a discrete uniform random variable with parameters a and b? (Hint: For integers a and b with $b > a$, the sum of all the integers from a to b, including a and b, is $\sum_{k=a}^{b} k = (a + b)(b - a + 1)/2$.)
   (d) [Optional] What is the expected value of a continuous uniform random variable with parameters a and b? (Hint: $b^2 - a^2 = (b - a)(b + a)$.)
(2) In this exercise, you will explore the law of large numbers by simulation. The law of large numbers states that if one samples n independent observations from any distribution with expectation μ, then as n becomes large, the mean of the sample gets closer and closer to μ. The following lines of code draw n.samps samples of a requested size (samp.size) from a standard normal distribution and plots a histogram of the sample means:

```
samp.size <- 20
n.samps <- 1000
samps <- rnorm(samp.size * n.samps, mean = 0, sd = 1)
samp.mat <- matrix(samps, ncol = n.samps)
samp.means <- colMeans(samp.mat)
hist(samp.means)
```

   (a) What do you expect the histogram to look like when n = 1? How will it change as n increases? Test your predictions, running the code above while trying the values 1, 5, 20, 100, and 1,000 for samp.size.

---

[5] See the book's GitHub repository at github.com/mdedge/stfs/ for detailed solutions and R scripts.

(b) Modify the code you used in part (a) to use the exponential distribution instead of the normal distribution. Set the "rate" parameter of `rexp()` to be 1. What are your predictions? Test them again.

5.2 Variance and standard deviation

The expected value is a measurement of the location of a distribution. Roughly, it indicates where a random variable's distribution is centered. The **variance**, in contrast, is a measurement of dispersal. It indicates how spread out a random variable's distribution is. The most intuitive way to define the variance of a random variable X is

$$\text{Var}(X) = \text{E}\Big([X - \text{E}(X)]^2\Big). \tag{5.6}$$

Take a moment to parse this. Start inside the brackets and work outward: $X - \text{E}(X)$ is the difference between a random variable's value and its expectation. This seems to have something to do with dispersal. What if we just took the expectation of this quantity, $\text{E}[X - \text{E}(X)]$? It won't work—by the linearity of expectation (Equation 5.4), this is $\text{E}(X) - \text{E}\Big(\text{E}(X)\Big) = \text{E}(X) - \text{E}(X) = 0$, for any distribution for which $\text{E}(X)$ is defined. $X - \text{E}(X)$ seems to tell us something about the spread of the distribution of X, but it changes sign depending on whether X is larger or smaller than its expectation, and the sign changes cancel each other out.

We find ourselves in a situation similar to the one we were in when defining the line of best fit: We have to consider a function of $X - \text{E}(X)$ that is constrained to be non-negative rather than $X - \text{E}(X)$ itself. The most straightforward way to do that would be to use the quantity $|X - \text{E}(X)|$, the absolute value of the difference between the random variable X and its expectation. That quantity is called the mean absolute deviation (MAD), and it is often useful. Another option is to take the expectation of the squared difference $[X - \text{E}(X)]^2$. The expectation of this squared difference is the variance (Equation 5.6). Like the absolute difference, the squared difference cannot take negative values. The squared difference has an obvious drawback: it is in the wrong units. If we use a distribution to model height in meters, we are measuring the spread of the distribution in squared meters, which is bizarre. Still, two mathematical advantages of the variance often outweigh the inconvenience of working in squared units. First, the variance of a probability distribution is much easier to compute mathematically than an analogous quantity using absolute deviations would be. Second, and more importantly, the variances of linear functions of random variables— including sums, differences, and means—are beautifully behaved (see Equations 5.8 and 5.9 below), whereas the analogous quantities for absolute deviations can be a hassle. Figure 5-2 shows density functions of three normal distributions with the same expectation but differing variances.

Some of the nicest properties of the variance are:

(i) The variance is given by the expression

$$\text{Var}(X) = \text{E}(X^2) - [\text{E}(X)]^2. \tag{5.7}$$

To compute the variance, you need to know only the expectation of the random variable and the expectation of the square of the random variable. It is usually easier to compute the variance of a random variable in this form than in the form shown in Equation 5.6.

(ii) If we know the variance of a random variable X, then we can calculate the variance of a linear function of X as

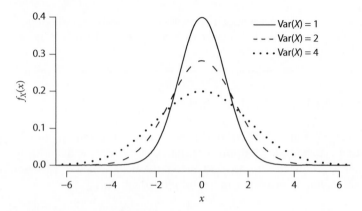

Figure 5-2 Plots of the density functions of three normal distributions with different variances but the same expected values. The density functions are centered in the same place, but those with larger variances are more spread out.

$$\mathrm{Var}(a + cX) = c^2\mathrm{Var}(X), \tag{5.8}$$

where a and c are constants. The constant a disappears—adding a constant to a random variable does not affect the variance. The variance of a random variable multiplied by a constant is the same as the square of the constant times the variance of the random variable.

(iii) This is a big one. If X and Y are independent random variables, then

$$\mathrm{Var}(X + Y) = \mathrm{Var}(X) + \mathrm{Var}(Y). \tag{5.9}$$

The variance of the sum of two independent random variables is the sum of the variances of those two random variables. This is wonderful for statistics, where we are nearly always concerned with the properties of samples containing multiple observations. You will prove Equations 5.7–5.9 in the exercises.

As mentioned before, the most obvious problem with the variance is that it is in the wrong units. The usual fix is to take the square root of the variance, which is called the **standard deviation**:

$$\mathrm{SD}(X) = \sqrt{\mathrm{Var}(X)}. \tag{5.10}$$

Though the standard deviation is less intuitive than the mean absolute deviation—$\mathrm{E}(|X - \mathrm{E}(X)|)$—the useful properties of the variance justify its use. The standard deviation is usually a somewhat larger (and never smaller) than the expected absolute deviation, and it is more sensitive to large deviations.

Exercise Set 5-2

(1) Prove Equation 5.7. (Hint: Remember the linearity of expectation (Equation 5.4), and remember that $\mathrm{E}(X)$ is a constant.)

(2) [Optional] Prove Equation 5.8 in the following two steps.
  (a) Prove that $\mathrm{Var}(X + a) = \mathrm{Var}(X)$ if X is a random variable and a is a constant. (Hint: It's easier to start with the definition in Equation 5.6 than with the identity in Equation 5.7 for this proof.)
  (b) Prove that $\mathrm{Var}(cX) = c^2\mathrm{Var}(X)$. (Hint: This time, start with Equation 5.7.)

(3) [Optional] In this question, you will prove a version of the claim that if X and Y are independent random variables, then $\text{Var}(X + Y) = \text{Var}(X) + \text{Var}(Y)$. This claim holds for both discrete and continuous random variables, but it is easier to prove in the discrete case.

(a) Assume that X and Y are discrete random variables. We will need the concept of the *joint probability mass function* of X and Y, $f_{X,Y}(x, y) = P(X = x \cap Y = y)$. In words, the joint probability mass function of X and Y gives the probability that X equals x *and* Y equals y. If X and Y are independent, what is $f_{X,Y}(x, y)$ in terms of $f_X(x)$ and $f_Y(y)$?

(b) Express $\text{Var}(X + Y)$ in terms of $\text{Var}(X)$, $\text{Var}(Y)$, and the quantity $\text{Cov}(X, Y) = E(XY) - E(X)E(Y)$. $\text{Cov}(X, Y)$ is the "covariance" of X and Y, and we will see it again in the next section.

(c) The expectation of the product of X and Y is

$$E(XY) = \sum_x \sum_y xy f_{X,Y}(x, y).$$

That is, it is the sum over all possible values of X and all possible values of Y of the product of x and y times the probability that $X = x$ and $Y = y$. This is the same as the other expectations you have learned, but now we sum over two variables instead of just one. Prove that when X and Y are independent, $E(XY) = E(X)E(Y)$. What does this imply about $\text{Var}(X + Y)$ when X and Y are independent? (Hint: Remember that if a term inside a sum does not depend on the index of summation, you can pull it out of the sum.)

(d) What is $\text{Var}(X - Y)$ when X and Y are independent?

(4) (a) What is the variance of a Bernoulli random variable with parameter p?

(b) Use part (a) and the result of Problem 3 to get the variance of a binomial random variable with parameters n and p.

(5) Suppose that $X_1, X_2, ..., X_n$ are independent random variables, each with (finite) variance σ^2. What is the variance of $\frac{1}{n}(X_1 + X_2 + ... + X_n)$, the mean of the X_i? What is the standard deviation of the mean of the X_i? (The standard deviation of a sample mean is often called a "standard error"—see Chapter 7 for more.)

(6) [Optional] In this exercise, we will prove one version of the law of large numbers—the "weak" law of large numbers.

(a) We start by proving Markov's inequality. Markov's inequality states that for a non-negative random variable X,

$$P(X \geq c) \leq \frac{E(X)}{c},$$

where c is a positive constant. We will use Markov's inequality to prove Chebyshev's inequality, and we will use Chebyshev's inequality to prove the weak law of large numbers. Prove Markov's inequality by defining a random variable Z that equals 1 if $X \geq c$ and equals 0 otherwise. Then notice that $Z \leq X/c$ and take the expectations of both sides, remembering that Z is a Bernoulli random variable.

(b) Chebyshev's inequality says that for a random variable Y with expectation μ,

$$P(|Y - \mu| \geq d) \leq \frac{\text{Var}(Y)}{d^2},$$

where d is a positive constant. Prove Chebyshev's inequality by replacing X in Markov's inequality with $(Y - \mu)^2$.

(c) The weak law of large numbers makes a statement about the mean of independent and identically distributed random variables. Repeating from the previous section: Suppose that $X_1, X_2, X_3, ..., X_n$ is a set of independent random variables and that all the X_i have the same distribution. Assume that $E(X_1) = E(X_2) = ... = E(X_n) = \mu$ and that $\mu < \infty$. Define X_n as

$$\overline{X}_n = \frac{1}{n}(X_1 + X_2 + X_3 + \ldots + X_n).$$

Then, for any positive constant δ, $\lim_{n\to\infty} P(|\overline{X}_n - \mu| > \delta) = 0$. Use Chebyshev's inequality and the results of Problem 5 to prove the weak law of large numbers. You can use the additional assumption that $\text{Var}(X_1) = \text{Var}(X_2) = \ldots = \text{Var}(X_n) = \sigma^2$ and that $\sigma^2 < \infty$. The weak law of large numbers holds even when the variance of the individual observations is not finite (as long as the expectation is finite), but the proof is easier if we assume finite variance.

5.3 Joint distributions, covariance, and correlation

We now have tools for measuring the location and spread of the distribution of a random variable. Recall, though, that we began our study of probability with the goal of better understanding the best-fit line. The best-fit line is a way of summarizing the relationship between two variables. We thus need a probabilistic measurement of the strength of the relationship between two random variables. The measurements we will use are the covariance and correlation. The **covariance** is a measurement of the extent to which two variables depart from independence (in a specific way), and the **correlation** is the covariance rescaled so that it takes values between -1 and 1.

So far, we have only considered distributions of individual random variables. If we want to talk about the relationship between two random variables, we need the notion of a **joint distribution** of two random variables. The extension is straightforward. Before, we said that the cumulative distribution function of a random variable X is (Equation 4.4)

$$F_X(x) = P(X \le x).$$

We now say that the joint cumulative distribution function of two random variables X and Y is

$$F_{X,Y}(x, y) = P(X \le x \cap Y \le y). \tag{5.11}$$

Remember that the $\cap$ symbol means "intersection," which you can translate to English as "and." In words: the joint cumulative distribution function of two random variables evaluated at (x, y) is the probability that the first random variable will be smaller than x *and* the second random variable will be smaller than y.

The joint cumulative distribution function completely defines the joint distribution of the random variables, but we will prefer to work with mass and density functions rather than with cumulative distribution functions. For discrete random variables, the joint probability mass function is the probability that the first random variable equals a specified value *and* the second random variable equals another specified value:

$$f_{X,Y}(x, y) = P(X = x \cap Y = y). \tag{5.12}$$

We can recover the mass function of either X or Y on its own—called the "marginal" probability mass function of X or Y—by summing over the joint probability mass function. The mass function $f_X(x)$ is the probability that X takes the value x. This is the probability that $X = x$ *and* $Y = y$—that is, the joint probability mass function—summed over all the possible values that Y can take. In notation,

$$f_X(x) = P(X = x) = \sum_y P(X = x \cap Y = y) = \sum_y f_{X,Y}(x, y). \tag{5.13}$$

Here, the sums over y indicate that we sum the joint mass function over all values that Y can take.

For continuous random variables, the joint probability density function can be defined in terms of the joint cumulative distribution function, as before. The joint density $f_{X,Y}(x, y)$ is defined so that

$$\int_{-\infty}^{z} \int_{-\infty}^{w} f_{X,Y}(x, y)dx\, dy = P(X \le w \cap Y \le z) = F_{X,Y}(w, z).$$

That is, the joint density is constructed so that definite integrals of the joint density yield the joint cumulative distribution (Equation 5.11). We have a double integral, but the principle is the same as a single integral, only now we find volumes under a surface rather than areas under a curve.

A property analogous to Equation 5.13 holds for continuous random variables; we just have to substitute appropriate integrals for the sums. Namely, if X and Y are jointly distributed continuous random variables, then we can integrate the joint density function with respect to one random variable to get the marginal density of the other random variable:

$$\int_{-\infty}^{\infty} f_{X,Y}(x, y)dy = f_X(x). \tag{5.14}$$

Figure 5-3 shows a joint probability density function.

With the concept of a joint distribution in hand, we can define the covariance as an extension of the variance. Namely,

$$\text{Cov}(X, Y) = \text{E}\Big([X - \text{E}(X)][Y - \text{E}(Y)]\Big) = \text{E}(XY) - \text{E}(X)\text{E}(Y). \tag{5.15}$$

The proof that these two expressions are equivalent is entirely parallel to the proof you used for the identity in Equation 5.7 (Exercise Set 5-2, Problem 1). $\text{E}(XY)$ is the expectation of the product of X and Y. Notice that the order doesn't matter: $\text{Cov}(X, Y)$ is the same as $\text{Cov}(Y, X)$. To see why we say the covariance is an extension of the variance, notice that

$$\text{Cov}(X, X) = \text{E}(XX) - \text{E}(X)\text{E}(X) = \text{E}(X^2) - [\text{E}(X)]^2 = \text{Var}(X).$$

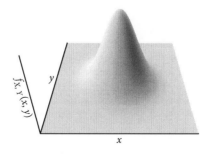

Figure 5-3 A density function of two jointly normally distributed random variables. The height of the surface gives the probability density, x goes from left to right, and y goes back into the page. The total volume under the surface is equal to 1.

In the last problem set (Exercise Set 5-2, Problem 3, Part c), you showed that if X and Y are independent discrete random variables, then

$$\text{Cov}(X, Y) = 0.$$

This is also true for continuous random variables: independence implies zero covariance. The converse does not hold: if $\text{Cov}(X, Y) = 0$, then X and Y are not necessarily independent. Nonetheless, we sometimes speak of the covariance as measuring the degree of departure from independence, even though it is possible for dependent random variables to have zero covariance.

It is worth pausing to understand the sense in which the covariance measures the degree of departure from independence. A heuristic illustrates how the covariance works. Imagine that X and Y are random variables that each have expected values of zero. Thus, $E(X)$ $E(Y) = 0$, and $\text{Cov}(X, Y) = E(XY)$.[6] Now imagine a coordinate system as in Figure 5-4. The left panel of Figure 5-4 shows the same density as in Figure 5-3, but now it is shown as a contour plot divided into quadrants. The lines dividing the plot into quadrants are the $x = 0$ and $y = 0$ lines, which are also the $x = E(X)$ and $y = E(Y)$ lines. In the upper right quadrant, both x and y are positive, and so the product xy is positive. Similarly, in the lower left quadrant, both x and y are negative, and the product xy is again positive. In the upper left and lower right quadrants, the product xy is negative.

One way to understand the expectation $E(XY)$ is to consider how a joint distribution's mass or density is divided into these four quadrants. Speaking loosely, if most of a joint distribution's mass or density is in the upper right and lower left quadrants, then $E(XY)$ is likely to be positive. If most of the mass or density is in the upper left and lower right

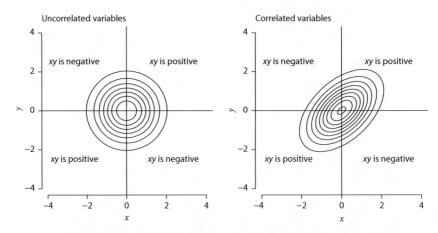

Figure 5-4 Two joint distributions, one with zero covariance and another with positive covariance. The left panel provides another view of the same joint distribution from Figure 5-3. This is a contour plot: rather than showing the surface in three dimensions, the contour lines show sets of coordinates in x and y that all have the same value of $f_{X,Y}(x, y)$, just as contour lines on a topographic map show sets of locations that are at the same elevation. The regions where the product xy is either positive or negative are labeled. In the left joint density, $\text{Cov}(X, Y) = 0$. The right joint density reflects a distribution where $\text{Cov}(X, Y)$ is positive. (Here, $\text{Cov}(X, Y) = 1/2$.) Most of the density is in the top-right and bottom-left quadrants.

---

[6] You will prove in the exercises that subtracting constants from two random variables does not affect the covariance between them, so it is always possible to rescale random variables so that this heuristic applies.

quadrants, then $E(XY)$ is likely to be negative. Thus, you can interpret a positive covariance between X and Y to mean something like "if X takes a value larger than its expectation, then Y is likely to take a value larger than its expectation, and if X takes a value smaller than its expectation, then Y is likely to take a value smaller than its expectation." The joint distribution on the left of Figure 5-4 is symmetric around both the $x = E(X)$ and $y = E(Y)$ lines. Its density is divided equally into the four quadrants, and in this joint distribution, $Cov(X, Y) = 0$. The right panel of Figure 5-4 shows another joint density where $Cov(X, Y)$ is positive. This joint distribution has more density in the upper right and lower left quadrants than in the other two quadrants.

Now we understand the type of departure from independence to which the covariance is sensitive.[7] There is a problem, however. The covariance is sensitive both to the linear dependence between two random variables and to the scaling of the variables. Consider the joint distribution from the right side of Figure 5-4. In that joint distribution, $Cov(X, Y) = 1/2$. Now suppose that we define another random variable Z, and we say that $Z = 2X$. Intuitively, we might assert that Y and Z ought to be exactly as closely related as Y and X, since Z is a simple linear function of X. If we were to draw the joint density of Z and Y, it would look exactly like the right panel of Figure 5-4, except the labeling on the horizontal axis would change: we would see $-8, -4, 0, 4, 8$ instead of $-4, -2, 0, 2, 4$. However, the covariance is different:

$$Cov(Z, Y) = Cov(2X, Y) = E(2XY) - E(2X)E(Y).$$

By the linearity of expectation (Equation 5.4), the "2"s can come out of the expectations, and then we can factor them out:

$$Cov(Z, Y) = 2E(XY) - 2E(X)E(Y) = 2[E(XY) - E(X)E(Y)] = 2Cov(X, Y) = 1.$$

Thus, the covariance of Z and Y is twice as large as the covariance of X and Y, even though Z and Y are no more closely related than X and Y are. The covariance, then, is not a pure measure of the degree of linear dependence between two variables, which we will sometimes need.

The fix is to scale the covariance. If we know the variances of X and Y, then we can derive the maximum value that $Cov(X, Y)$ can possibly achieve. This maximum possible value turns out to be $\sqrt{Var(X)Var(Y)} = SD(X)SD(Y)$. Dividing the covariance by this value fixes the problem pointed out in the last paragraph. We call this scaled version of the covariance the correlation:

$$Cor(X, Y) = \rho_{X,Y} = \frac{Cov(X, Y)}{\sqrt{Var(X)Var(Y)}}. \tag{5.16}$$

The correlation of two random variables X and Y is often written as $\rho_{X,Y}$, and its value is always between -1 and 1. Larger absolute values of the correlation indicate tighter associations between the variables. The sign of the correlation indicates direction; if positive, then larger-than-expected values of X are associated with larger-than-expected values of Y, and if negative, then larger-than-expected values of X are associated with smaller-than-expected values of Y.

---

[7] The recently developed concepts of "distance covariance" and "distance correlation" capture all types of dependence, not just linear dependence, in the sense that the distance covariance is zero if and only if two random variables are independent. See Székely & Rizzo (2017) for a review.

Exercise Set 5-3

(1) Consider two jointly distributed discrete random variables, X and Y. Their joint probability mass function is positive for the following three sets of (x, y) coordinates and zero everywhere else:

$$f_{X,Y}(-1, 1) = f_{X,Y}(1, 1) = f_{X,Y}(0, -2) = \frac{1}{3}.$$

(a) What is $\text{Cov}(X, Y)$ (Equation 5.15)?

(b) Are X and Y independent? (Remember to check whether $P(A \cap B) = P(A)P(B)$ for pairs of events representing possible values of X and Y.)

(2) Suppose that the covariance and correlation of two jointly distributed random variables X and Y are given by $\text{Cov}(X, Y) = \gamma$ and $\text{Cor}(X, Y) = \rho$. What are the correlation and covariance of Z and Y if $Z = a + bX$, where a and b are constants?

(3) [Optional, requires comfort with sums] Prove that if X and Y are discrete random variables with joint mass function $f_{X,Y}(x, y)$, then $E(X + Y) = E(X) + E(Y)$. This is one implication of the linearity of expectation. Defining the possible values of X as $\{x_1, ..., x_k\}$ and the possible values of Y as $\{y_1, ..., y_m\}$, the expectation of the sum is defined as

$$E(X + Y) = \sum_{i=1}^{k} \sum_{j=1}^{m} (x_i + y_j) f_{X,Y}(x_i, y_j).$$

5.4 [Optional section] Conditional distribution, expectation, and variance

In the previous section, we used joint distributions to specify the relationship between two random variables. Another way to understand the relationship between two random variables is to consider the **conditional distribution** of one random variable given the value of the other random variable.

For discrete random variables X and Y, the *conditional* probability mass function is defined as

$$f_{X|Y}(x|Y = y) = P(X = x | Y = y) = \frac{P(X = x \cap Y = y)}{P(Y = y)} = \frac{f_{X,Y}(x, y)}{f_Y(y)}, \tag{5.17}$$

where $f_{X,Y}(x, y)$ is the joint probability mass function of X and Y (Equation 5.12), and $f_Y(y)$ is the marginal probability mass function of Y (Equation 5.13). In words, the conditional probability mass function is the probability mass function for X when Y is known to have taken a specific value, y. Notice that the definition of the conditional probability mass function has a form parallel to the definition of conditional probability in Equation 4.1.

For continuous random variables, the conditional probability density function is defined analogously. If X and Y are continuous random variables, then the conditional density function $f_{X|Y}(x|Y = y)$ is

$$f_{X|Y}(x|Y = y) = \frac{f_{X,Y}(x, y)}{f_Y(y)}, \tag{5.18}$$

where $f_{X,Y}(x, y)$ is the joint density function of X and Y and $f_Y(y)$ is the marginal density function of Y (Equation 5.14).

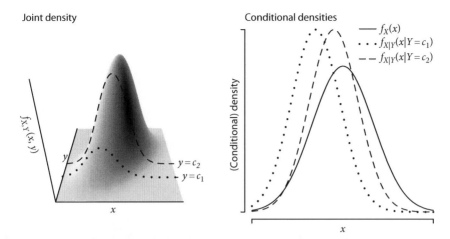

Figure 5-5 A view of a joint density and some conditional densities derived from it. The shaded surface on the left is a joint density function of two random variables, X and Y. The dotted and dashed lines each trace the surface along a line parallel to the x axis. These lines specify the shape of the distribution of X conditional on Y taking the constant values c_1 or c_2. Here, X and Y are jointly normally distributed with correlation 0.6. In the right panel, density functions for X derived from the joint density on the left are shown. The solid line shows the marginal density of X, the density of X when the value of Y is unknown (see Equation 5.14). The dotted line shows the conditional density of X when $Y = c_1$, and the dashed line shows the conditional density of X when $Y = c_2$. The area underneath each of the three density functions is 1. Though the left panel shows that the joint density is lower overall when $Y = c_1$ than when $Y = c_2$, dividing by the marginal density of Y makes the areas under the conditional densities all equal to 1 (Equation 5.18).

Figure 5-5 shows the relationship of the conditional probability density to the joint probability density.

Conditional probability density functions can be summarized in the same ways as other probability density functions. For example, one can derive a conditional expectation from a conditional probability density. The conditional expectation $E(X|Y = y)$ is the expectation of X given that $Y = y$. For continuous random variables, it is

$$E(X|Y = y) = \int_{-\infty}^{\infty} x f_{X|Y}(x|Y = y)dx. \tag{5.19}$$

If X and Y are discrete, then we replace the integral with an analogous sum. This definition is analogous to the one given for the expected value in Equation 5.2. Figure 5-5 suggests— and it is true in this case—that the conditional expectation of X given $Y = y$ is different for distinct values of y.

Similarly, the conditional distribution of X has a variance. The conditional variance of X given $Y = y$ is

$$\text{Var}(X|Y = y) = E(X^2|Y = y) - [E(X|Y = y)]^2. \tag{5.20}$$

The conditional variance is defined analogously to the variance (Equation 5.6), and it has properties analogous to the variance (Equations 5.7–5.9). (Notice that this expression is exactly like Equation 5.7, but with the condition that $Y = y$ added to both terms.) In principle, the conditional variance of X given $Y = y$ can vary for different values of y, though for the joint distribution on the left side of Figure 5-5, the conditional variance is the same for all values of Y.

One way to assess the degree of relationship between two random variables X and Y is to compare the conditional variance X given Y with the variance of X obtained without conditioning on Y. If the conditional variance is much smaller than the variance, then knowing the value of Y decreases one's uncertainty about X substantially, and X and Y can be said to be closely related. In an exercise at the end of the chapter, you will show that if the relationship between X and Y is linear, then the comparison of the conditional and unconditional variance is closely related to the correlation coefficient.

5.5 The central limit theorem

The **central limit theorem**—the last probabilistic tool we need to acquire before building a model to study simple linear regression—is like the law of large numbers, in that it concerns the behavior of means of very large samples of independent observations. It is a useful and beautiful result, but we also ought to be cautious in applying it to actual data—like any other theorem, it is not guaranteed to hold unless its assumptions are met.

Often in statistics, we study the distribution of a mean of a sample of size n. Suppose that we sample observations $X_1, X_2, X_3, ..., X_n$ independently from a distribution with expectation $E(X) = \mu$ and finite variance $Var(X) = \sigma^2$. The mean of the observations is a random variable: $\sum_{i=1}^{n} X_i/n = \overline{X}_n$. What is the behavior of the sample mean $\overline{X}_n$ for large samples? By the law of large numbers, the sample mean gets closer to μ as the sample gets large. One way of showing this, which we adopted in Exercise Set 5-2, is to show that $E(\overline{X}_n) = \mu$ and that $Var(\overline{X}_n) = \sigma^2/n$, meaning that the variance of the sample mean decreases as the sample size n increases.

Thus, we know two facts about the distribution of the sample mean $\overline{X}_n$: its expectation and its variance. We want to know more—what is the actual shape of the distribution? This is where the central limit theorem comes in. Here is one version of the theorem—we won't prove it, but you'll have a chance to convince yourself of it by simulation in an exercise:

> **Central limit theorem** Suppose that $X_1, X_2, X_3, ..., X_n$ are independent random variables and that all the X_i have the same distribution. Assume that $E(X_1) = E(X_2) = ... = E(X_n) = \mu$, that $Var(X_1) = Var(X_2) = ... = Var(X_n) = \sigma^2$, and that both μ and σ^2 are finite.[8] Define $\overline{X}_n$ as
>
> $$\overline{X}_n = \frac{1}{n}(X_1 + X_2 + X_3 + ... + X_n).$$
>
> As n approaches infinity, the distribution of $\overline{X}_n$ converges to a normal distribution[9] with expectation μ and variance σ^2/n.

---

[8] If you are used to thinking of "variance" as "sample variance," then the requirement of finite variance may be confusing—sample variances are always finite. "Infinite variance" in this case means that the value of the integral $\int_{-\infty}^{\infty} x^2 f_X(x)dx$ diverges to infinity. This integral can diverge if X has a "heavy-tailed" distribution. One example with wide application in natural and social science is the Pareto distribution, which has undefined variance for certain parameter values.

[9] This statement is a little loose—we haven't said what it means for a distribution to converge to another distribution. We will leave it informal, but one way to say what is happening is that as the number of independent samples considered increases, the cumulative distribution function of the sample mean becomes more and more like a cumulative distribution function that is part of the normal distribution family.

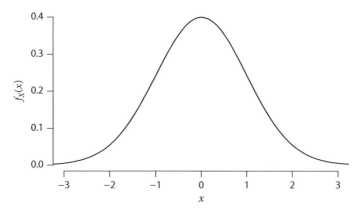

Figure 5-6 The probability density function of the Normal$(0, 1)$ distribution. The normal distribution with expectation 0 and variance 1 is also called the "standard" normal distribution.

Recall that **normal distributions** (also called Gaussian distributions) are a family of continuous probability distributions (Table 4-2). The density function of a normal distribution with expectation 0 and variance 1—also called the "standard" normal distribution—is shown in Figure 5-6. Normal distributions are symmetric, and most of the probability density is in the middle of the distribution, with decreasing density as one gets farther from the center. Normal random variables have probability density functions described by (Table 4-2):

$$f_X(x) = \frac{1}{\sigma\sqrt{2\pi}} e^{-\frac{(x-\mu)^2}{2\sigma^2}}.$$

The constants μ and σ^2 are parameters that control the location and spread of the normal distribution. If $\bar{X} \sim \text{Normal}(\mu, \sigma^2)$—the tilde means "distributed as"—then

$$E(X) = \mu \tag{5.21}$$

and

$$\text{Var}(X) = \sigma^2. \tag{5.22}$$

We won't prove Equations 5.21 and 5.22, because they require integrals that are beyond our scope, but the way to get them is conceptually familiar—we just apply the definitions of expectation and variance from Equations 5.2 and 5.6 to the normal probability density function.

Because of the central limit theorem, we expect (approximately) normal distributions to arise often. In particular, if samples are drawn from any distribution with finite variance, we expect means of large samples to be roughly normal—though how large is "large" depends on the distribution from which the original observations are drawn.[10,11] Further, we expect

---

[10] The central limit theorem only guarantees an increasingly good approximation to normality. By Cramér's theorem, if the sum or mean of n independent random variables is exactly normally distributed, then the individual random variables must *all* be normally distributed. In words, the central limit theorem guarantees that the distribution of the sample mean approaches a normal distribution, but Cramér's theorem guarantees that the distribution will never reach exact normality unless the random variables being averaged themselves have a normal distribution.

[11] The n necessary for convergence can also depend on the part of the distribution of $\bar{X}_n$ that is of interest. The tails of the distribution of $\bar{X}_n$—that is, extreme, rare events involving $\bar{X}_n$—may converge to approximate

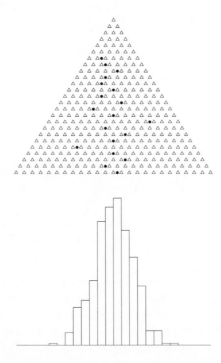

Figure 5-7 A still from an animation of a bean machine produced by the `quincunx()` function in R's `animation` package (Xie, 2013). Each black circle represents a ball falling toward the bottom of the bean machine. Each time a ball falls down one row, it hits a peg, represented by a triangle. Each time a ball hits a peg, it has a 50% chance of bouncing left and a 50% chance of bouncing right. The histogram at the bottom shows the resting places of the balls after falling through the bean machine. See the book's cover for another depiction of a bean machine.

even individual observations of processes that involve the summation of many small, independent influences to be roughly normally distributed.

You can develop intuition for how the central limit theorem works by thinking about a device called a "bean machine," also called a "Galton board" or "quincunx."[12] Figure 5-7 shows a representation of a bean machine, which is a triangular array of pegs mounted on a board. To use a bean machine, one drops a ball in at the top, at which point it hits a peg. As it hits the peg, the ball may bounce either left or right on its way down to the next row of pegs. In the next row, it will hit another peg and bounce either left or right. Then it will drop to another row of pegs, and so on until it falls into one of several bins at the bottom of the bean machine. When one drops many balls into the bean machine and looks at how they are distributed in the bins at the bottom, the shape closely resembles the density function of a normal distribution.

How can we understand the emergence of an approximately normal distribution from the bean machine? There are at least two ways. First, we can appeal to the central limit theorem itself. Imagine that the bin all the way to the left of the bean machine is labeled 0,

normality more slowly than the center of the distribution, as you will see in an optional exercise. Caution is required in applying the central limit theorem if rare events are of interest.

[12] Imagine the meeting where someone said, "No, we need a catchier name for that—like 'quincunx'."

and that the numbers of the bins increase as we move to the right, up to the far right bin, which is labeled k, where k is the number of rows of pegs on the bean machine. We can now think of each row of the bean machine as representing a Bernoulli random variable (Table 4-1). In any row, if the ball bounces left, then it is as if the random variable for that row is 0. If the ball bounces right, then it is as if the random variable for that row is 1. The bin into which the ball falls is governed by the sum of the random variables that represent each row. If the random variables are independent, then the central limit theorem says that as the number of rows in the bean machine increases, the distribution of the balls' final positions should approach a normal distribution.[13]

This explanation clarifies the connection between the bean machine and the central limit theorem, but it provides little insight into *why* the central limit theorem holds. Why is it that when we sum independent random variables, most of the sums end up having intermediate values, with only a few sums taking on extreme values? Look again at the bean machine in Figure 5-7. It has 25 rows of pegs. Imagine that a ball falls in the bin that is farthest to the left. There is only one way for this to happen—the ball must fall to the left at every single row on the way down. In contrast, there are 25 ways for a ball to fall in the bin second from the left. The ball has to fall right at one peg and fall left at the other 24, but it doesn't matter *which* peg is the one at which the ball falls right. Now think of the third bin from the left. To get there, the ball has to fall right twice and fall left 23 times, and there are 300 ways to pick the two rows at which the ball falls right. In the middle, there are over 5 million routes a ball can take through a 25-row bean machine to get to the 12th or 13th bins. Thus, as we move toward the center, there are more possible routes that a ball can take in order to land in a particular bin. As we move right from the center, the number of possible routes the ball can take decreases again, until at the far right bin, the ball can only take the path of falling right in every single row. This is one way to think about how the central limit theorem works: when many independent random variables are being summed or averaged, there are many ways to achieve intermediate values, but there are only a few ways to achieve extreme values.

It is worth remembering that the central limit theorem is about a situation that, strictly speaking, no-one has ever encountered—the behavior of a sample mean as the size of the sample approaches infinity. All invocations of the central limit theorem in practice are approximations and extrapolations—there may be some slippage between theory and practice. Often, theoretical approximations are good enough, but one needs to think carefully about individual cases. As you will see in an exercise below, the details matter, particularly when tail events—that is, outcomes far removed from the center of a variable's distribution—are important.

Exercise Set 5-4

(1) Watch a bean machine in action using the `animation` package in R. To do this, install the `animation` package using `install.packages("animation")`. (Remember that you only have to *install* a package once.) Then, load the package:

```
library(animation)
```

---

[13] We introduced the central limit theorem as a statement about means of random variables, but because the sample mean is just the sum divided by a constant, if the mean is approximately normal, then the sum must be approximately normal, too.

Use the `quincunx()` command to view the animation. You can change the number of balls you'll drop through the bean machine, the number of layers of pegs on the bean machine, and the number of balls that are dropped per second:

```
nball <- 500 #change the number of balls
nlayer <- 25 #change the number of rows of pegs on the board
rate <- 10 #change the speed at which the balls fall
ani.options(nmax = nball + nlayer - 2, interval = 1/rate)
quincunx(balls = nball, layers = nlayer)
```

You can try it with different numbers of balls and layers.

(2) In this problem, you will explore the distribution of the mean of samples from a beta distribution. Beta distributions have two parameters that control the shape of the distribution, and for certain values of those parameters, the beta distribution can depart sharply from normality. Nevertheless, the central limit theorem guarantees that for large sample sizes, the means of samples from the beta distribution will approximate normality. You can use the R function `dosm.beta.hist()`, available in `stfspack` (see Exercise Set 2-2, Problem 3), to take the means of samples of a specified size from the beta distribution. The parameter `n` controls how many observations are included in each sample (the sample size), and the parameter `nsim` controls how many samples you plot. Ten thousand samples will give you a good idea of what the density function looks like.

To see what the beta distribution looks like for a given set of shape parameters, set the sample size to 1. For example,

```
dosm.beta.hist(1, 10000, shape1 = 1, shape2 =1)
```

will give you a histogram of 10,000 observations from a beta distribution with parameters 1 and 1. If you increase the sample size, then the distribution of the sample mean gets closer to normality. Try this—starting with samples of size 1 and increasing the sample size—with the following sets of parameter values: (1, 1), (0.2, 0.2), (2, 0.5), (0.5, 2), (3, 3). Feel free to try other parameter sets—it's fun. What do you notice?

(3) [Optional, hard, but read the solution if you skip] Simulate many samples of size $1,000$ from a Pareto distribution with shape parameter 4. (You may want to look up the Pareto distribution.) Pareto distributions with shape parameter 4 have finite variance, and so the central limit theorem applies to them. Nonetheless, the sample sizes necessary to achieve convergence to the normal are very large, especially in the tails. Compare the distribution of the means of the simulated samples with the appropriate normal distribution. Experiment with changing the size of the samples and the shape parameter (keeping it greater than 2 so that the central limit theorem applies). Use the `stfspack` function `rpareto()` to simulate data from a Pareto distribution.

To see what's going on, you need to compare the proportion of observations in the tails (i.e. beyond some number of standard deviations from the expectation, like 2, 3, 4, or 5) to what is expected under a normal distribution. Use the `stfspack` function `compare.tail.to.normal()` to do it, as in:

```
compare.tail.to.normal(x, k, mu, sigma)
```

Here, `x` is a vector of data, `k` is a number of standard deviations, and `mu` and `sigma` are the expectation and standard deviation of the normal distribution to which we should compare. The output is the ratio of the proportion of extreme observations in the sample (i.e. observations more than `k` standard deviations from the sample mean) divided by the probability of observations beyond `k` standard deviations from the expectation of a normal distribution. If the quotient is close to 1, then the data's tails are "like" those of a normal distribution. If the quotient is greater than 1, then there are more extreme values in the data than would be expected under a normal distribution. The expectation and standard deviation of means from Pareto samples of size n with parameters a and b can be computed as

```
expec.par <- a * b/(a-1)
sd.mean <- sqrt(a * b^2 / ((a-1)^2 * (a-2)) / n)
```

as long as a is greater than 2 and b is positive.

5.6 A probabilistic model for simple linear regression

Now that we have gathered some probabilistic concepts, we can return to the situation that motivated us—that of understanding a linear relationship between two variables. We now build a probabilistic model of the situation suggested by the best-fit line in Chapter 3. We consider two random variables, X and Y. Connecting to the running example, Y will represent cereal yield, and X will represent fertilizer application. We will think of the value of Y as being partially determined by X and partially determined by random factors separable from X, which we lump together in a "disturbance" term.[14]

Assume that Y is a linear function of X plus a random disturbance. The disturbance is a random variable by which Y differs from the linear function of X—some people use the word "error" instead of "disturbance." To model one observation of X and Y, we write

$$Y = \alpha + \beta X + \epsilon. \tag{5.23}$$

On the right side, X and ϵ are random variables, whereas α and β are fixed coefficients. Terminology: X is the "**independent variable**," Y is the "**dependent variable**," α is the "intercept," β is the "slope," and ϵ is the "**disturbance**." In words, the model constructs Y by adding a random disturbance ϵ to a linear function of X with intercept α and slope β. This is just a model; there is no guarantee that it is a true description of any actual empirical situation. Nonetheless, we can use the probability we have learned to understand what the consequences would be if the model were true.

To say more about Y and its relationship to X, we need to make assumptions about X and ϵ. We start with few assumptions and add increasingly stringent assumptions as we proceed. At each stage, we will be able to say new things about Y and its relationship with X. Box 5-2 lists the assumptions we will make about the model and the claims we will be able to prove under each one.[15] Figure 5-8 illustrates some of the assumptions made here (and in future chapters).

You can treat most of the proofs in the remainder of this section as optional on a first reading. (Some of them depend on optional section 5.4.) The non-optional parts are understanding Figure 5-8, Assumptions A1 and A2, and Claims 5.1.1 and 5.2.3. You should also attempt the second problem of the exercise set. Let's move on with the proofs.

---

[14] This formulation suggests that X partially *causes* Y. Researchers in different fields vary in the degree to which they frame their uses of regression as potentially addressing questions of cause and effect. More on this in the Postlude, Section Post.2.1.

[15] In addition to the following assumptions, we assume throughout that X and ϵ have finite variance.

Box 5-2 ASSUMPTIONS ABOUT THE LINEAR MODEL AND CLAIMS THAT FOLLOW

(1) Assumption

$E(X) = \mu_X$, and $E(\epsilon) = 0$. In words: X has a known expectation, and the disturbance term has an expectation of 0.

Claim 5.1.1

$E(Y) = \alpha + \beta\mu_X$. In words, this means that the expectation of Y falls on the line defined by intercept α and slope β where $x = E(X)$. Remember that the least-squares line had an analogous property—it always passed through the point $(\bar{x}, \bar{y})$, where $\bar{x}$ is the sample mean of the x values and $\bar{y}$ is the sample mean of the y values.

(2) Assumption

$E(\epsilon|X = x) = 0$ for all x. In words, we extend the first assumption so that the disturbance term's expectation is 0 for all values of x, not just on average. This is also called the "linearity" assumption.

Claim 5.2.1

$E(Y|X = x) = \alpha + \beta x$. In words, for any level of x, the conditional expectation of Y is obtained directly from the line with intercept α and slope β. That is, the conditional expectation is "linear"—described by a line.

Lemma 5.2.2

X and ϵ are uncorrelated, or $E(X\epsilon) = 0$. A lemma is a statement that is proven in the service of showing that another statement is true. We will use this lemma to prove Claim 5.2.3.

Claim 5.2.3

$Cov(X, Y) = \beta\sigma_X^2$. The covariance of X and Y is the product of the slope β and the variance of X, σ_X^2.

Claim 5.2.4.

$\rho_{X,Y} = \beta\frac{\sigma_X}{\sigma_Y}$. The correlation of X and Y is a simple function of the slope β, the standard deviation of X, and the standard deviation of Y.

(3) Assumption

The variance of the disturbance is constant, regardless of the value of X: $Var(\epsilon|X = x) = \sigma_\epsilon^2$. This is also called the "**homoscedasticity**" assumption.

Claim 5.3.1

$Var(Y|X = x) = \sigma_\epsilon^2$. The variance of Y conditional on X is constant.

(4) Assumption

X and ϵ are independent. That is, the independent variable and the error term are independent. This can be seen as an extension of the second assumption, which guarantees that X and ϵ are uncorrelated, and the third assumption, which says that the variance of ϵ does not depend on X. This assumption goes further by guaranteeing that the whole distribution of ϵ is always the same, regardless of the value X takes.

Claim 5.4.1

$Var(Y) = \beta^2\sigma_X^2 + \sigma_\epsilon^2$. The variance of Y is the sum of two terms, one of which measures the systematic variation in Y that is "explained" by variation in X, and the other of which measures variation due to the disturbance term. This result is the basis of descriptions of the square of the correlation coefficient as the "proportion of variance in Y explained by X," or coefficient of determination, as you will show in an exercise.

Assumption $E(X) = \mu_X$, *and* $E(\epsilon) = 0$.

We start by assuming that X has a known expectation: $E(X) = \mu_X$. We also assume that the expectation of the disturbance is $E(\epsilon) = 0$.

Claim 5.1.1 $E(Y) = \alpha + \beta\mu_X$.

Proof of Claim 5.1.1 Using the assumption and the linearity of expectation (Equation 5.4), the expectation of Y is

$$E(Y) = E(\alpha + \beta X + \epsilon) = \alpha + \beta E(X) + E(\epsilon) = \alpha + \beta\mu_X. \tag{5.24}$$

The expectation of Y is found by plugging the expectation of X into a linear function of X. ∎

Assumption $E(\epsilon|X = x) = 0$ *for all* x.

We now add a much stronger assumption—we assume that the conditional expectation of the disturbance given that $X = x$ is zero for *all* possible values of x. We will refer to this assumption again, so we'll give it a name:

Assumption A1 $E(\epsilon|X = x) = 0$ *for all* x. *This is also called the "linearity" assumption.*

Claim 5.2.1 $E(Y|X = x) = \alpha + \beta x$.

Proof of Claim 5.2.1 When $X = x$, we find the value of Y by replacing the random variable X in Equation 5.23 with the instance x:

$$(Y|X = x) = \alpha + \beta x + \epsilon. \tag{5.25}$$

After conditioning on X, the only random term in this expression is ϵ. The linearity of expectation (Equation 5.4) and Assumption A1 give

$$E(Y|X = x) = E[\alpha + \beta x + (\epsilon|X = x)] = \alpha + \beta x + E(\epsilon|X = x) = \alpha + \beta x. \tag{5.26}$$

Assumption A1 does the work—it ensures that the conditional expectation of Y, $E(Y|X = x)$, is described by a linear function of x with slope β. Because Assumption A1 enforces linearity, it is also called the linearity assumption. ∎

We built our model to ask questions about the relationship between X and Y, and by applying the linearity assumption, we can calculate two indices of the strength of that relationship—the covariance and the correlation. To do so, we first need to show that if $E(\epsilon|X = x) = 0$, then $E(X\epsilon) = 0$, and then we calculate the covariance of X and Y from the definition in Equation 5.15. Lemma 5.2.2 is a bit harder than the other proofs in this section; you might skip it if you are willing to accept that $E(X\epsilon) = 0$ under the linearity assumption.

Lemma 5.2.2 X and ϵ are uncorrelated, or $E(X\epsilon) = 0$.

Proof of Lemma 5.2.2 We start by noting that, by Assumption A1,

$$E(\epsilon|X = x) = \int_{-\infty}^{\infty} z f_{\epsilon|X}(z|x)dz = 0, \tag{5.27}$$

where $f_{\epsilon|X}(z|x)$ is the conditional density of the disturbance ϵ given the independent variable X. The term in the middle comes from applying Equation 5.19, the definition of the conditional density. The quantity we want is $E(X\epsilon)$, which is

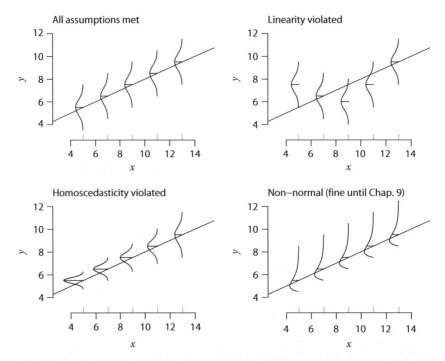

Figure 5-8 Schematics of the assumptions of the various versions of the linear model developed in this section. In each case, a "true" regression line is shown, along with density functions (rotated $90°$) representing the distribution of Y conditional on X. In the top left panel, all assumptions are met. The disturbances have expectation 0 and constant variance, regardless of the value of X. In the top right panel, the "linearity" assumption is violated, meaning that the expectation of the disturbances varies depending on X, and thus the expectation of Y conditional on X is not well described by a line. The linearity assumption is needed for all the claims made here beyond Claim 5.1. In the bottom-left corner, the "homoscedasticity" assumption is violated; the variance of the disturbances varies with X. The homoscedasticity assumption is needed for Claims 5.3 and 5.4, but, as we will see in Chapter 8, it is possible to do linear regression analysis without invoking it. In the bottom-right corner, the disturbances obey linearity and are homoscedastic but are not normally distributed—they are asymmetric, or skewed. We do not require normal disturbances for any of the claims in this chapter, but we do for the specific methods introduced in Chapter 9 (and Chapter 10).

$$E(X\epsilon) = \int_{-\infty}^{\infty} \int_{-\infty}^{\infty} xz f_{X,\epsilon}(x,z)\,dz\,dx.$$

This expression is an application of Equation 5.5, the law of the unconscious statistician. Faced with a function of two variables instead of one, we have to integrate over both of them to get the expectation. The important thing is to notice that, by Equation 5.18, the joint density $f_{X,\epsilon}(x,z) = f_{\epsilon|X}(z|x)f_X(x)$, giving

$$E(X\epsilon) = \int_{-\infty}^{\infty} \int_{-\infty}^{\infty} xz f_{\epsilon|X}(z|x)f_X(x)\,dz\,dx.$$

Notice that x and $f_X(x)$ do not depend on z. Applying Equation A.6, we pull them outside the interior integral, giving

$$E(X\epsilon) = \int_{-\infty}^{\infty} xf_X(x) \int_{-\infty}^{\infty} zf_{\epsilon|X}(z|x)dz\,dx.$$

We're home. Applying Equation 5.27, we make the replacement $\int_{-\infty}^{\infty} zf_{\epsilon|X}(z|x)dz = 0$, giving

$$E(X\epsilon) = 0 \int_{-\infty}^{\infty} xf_X(x)dx = 0. \tag{5.28}$$

By Equations 5.15 and 5.16, Equation 5.28 shows that, under the linearity assumption, the independent variable X and the disturbance ϵ are uncorrelated because $E(X\epsilon) = 0 = E(X)E(\epsilon)$. ∎

We now use the lemma to prove Claim 5.2.3:

Claim 5.2.3 $\mathrm{Cov}(X, Y) = \beta\sigma_X^2$.

Proof of Claim 5.2.3 To identify the covariance of X and Y, recall from Equation 5.15 that

$$\mathrm{Cov}(X, Y) = E(XY) - E(X)E(Y)$$

We substitute the expression in Equation 5.23 for Y, μ_X for $E(X)$, and the expression in Equation 5.24 for $E(Y)$. These steps give

$$\mathrm{Cov}(X, Y) = E[X(\alpha + \beta X + \epsilon)] - \mu_X(\alpha + \beta\mu_X).$$

Expanding both products gives

$$\mathrm{Cov}(X, Y) = E[X\alpha + \beta X^2 + X\epsilon] - \mu_X\alpha - \beta\mu_X^2.$$

The linearity of expectation (Equation 5.4) yields

$$\mathrm{Cov}(X, Y) = \alpha E(X) + \beta E(X^2) + E(X\epsilon) - \mu_X\alpha - \beta\mu_X^2.$$

Because $E(X) = \mu_X$, the $\alpha E(X)$ and the negative $\mu_X\alpha$ cancel, and, by Equation 5.28 (Lemma 5.2.2), $E(X\epsilon) = 0$, so

$$\mathrm{Cov}(X, Y) = \beta E(X^2) - \beta\mu_X^2 = \beta[E(X^2) - \mu_X^2].$$

The quantity in brackets, $E(X^2) - \mu_X^2$, is $\mathrm{Var}(X)$ (Equation 5.7). $\mathrm{Var}(X) = \sigma_X^2$, so

$$\mathrm{Cov}(X, Y) = \beta\sigma_X^2. \tag{5.29}$$

The covariance of X and Y is the variance of X times the slope. ∎

The correlation of X and Y follows quickly from the covariance.

Claim 5.2.4

$$\rho_{X,Y} = \beta\frac{\sigma_X}{\sigma_Y}.$$

Proof of Claim 5.2.4 The correlation of X and Y, $\rho_{X,Y}$, is the covariance of X and Y divided by the square root of the product of the variances of X and Y (Equation 5.16). If we define $\sigma_Y^2 = \mathrm{Var}(Y)$, then

$$\rho_{X,Y} = \frac{\mathrm{Cov}(X, Y)}{\sqrt{\mathrm{Var}(X)\mathrm{Var}(Y)}} = \frac{\beta\sigma_X^2}{\sqrt{\sigma_X^2\sigma_Y^2}} = \frac{\beta\sigma_X^2}{\sigma_X\sigma_Y} = \beta\frac{\sigma_X}{\sigma_Y}. \tag{5.30}$$

The first step follows from replacing $\text{Cov}(X, Y)$ with the expression in Equation 5.29. ∎

There is one term in Equation 5.30 for which we do not have an expression: σ_Y. We now make further assumptions in order to study the variance of Y.

Assumption $\text{Var}(\varepsilon) = \sigma_\varepsilon^2$ *and* $\text{Var}(\varepsilon|X = x) = \sigma_\varepsilon^2$.

Relying on assumptions about the expectation of the disturbance ε, we showed that $\text{E}(Y|X = x)$ is given by a linear function of x (Equation 5.26) and that $\text{E}(Y)$ is found by replacing x with μ_X in that same equation (Equation 5.24). We also found expressions for the covariance (5.29) and correlation (5.30) of X and Y.

We now add assumptions about the variance of the disturbance to prove a claim about the conditional variance of Y. We make the following assumption:

Assumption A2 *The variance of the disturbance is constant, regardless of the value of X:* $\text{Var}(\varepsilon|X = x) = \sigma_\varepsilon^2$. *This is also called the "homoscedasticity" assumption.*

Claim 5.3.1 $\text{Var}(Y|X = x) = \sigma_\varepsilon^2$.

Proof of Claim 5.3.1 Consider the conditional variance of Y given that X is known. Conditional on the value of X, Y is a linear function of ε (Equation 5.25). Equation 5.8 gives the variance of a linear function of a random variable, which, by Assumption A2, is

$$\text{Var}(Y|X = x) = \text{Var}[\alpha + \beta x + (\varepsilon|X = x)] = \text{Var}(\varepsilon|X = x) = \sigma_\varepsilon^2. \tag{5.31}$$

In principle—if our assumptions had been different—the conditional variance of Y might have varied with X, but our assumption that $\text{Var}(\varepsilon|X = x) = \sigma_\varepsilon^2$ ensures that the conditional variance of Y is constant, regardless of the value X takes.

Assumption X *and* ε *are independent.*

We add one more assumption before completing the chapter—the assumption that the independent variable X and the disturbance ε are independent. We have already assumed that the value of X does not affect the conditional expectation or variance of ε. To assume that X and ε are independent is to go further—we now assume that the value of X does not affect the conditional distribution of ε in any way.

Claim 5.4.1 $\text{Var}(Y) = \beta^2 \sigma_X^2 + \sigma_\varepsilon^2$.

Proof of Claim 5.4.1 If X and ε are independent, then Equation 5.9 applies—the variance of the sum of two independent random variables is the sum of their variances. Therefore the variance of Y can be found by applying Equations 5.8 and 5.9:

$$\text{Var}(Y) = \text{Var}(\alpha + \beta X + \varepsilon) = \beta^2 \text{Var}(X) + \text{Var}(\varepsilon) = \beta^2 \sigma_X^2 + \sigma_\varepsilon^2. \tag{5.32}$$

The variance of Y is thus neatly decomposed into a component due to the relationship between X and Y, namely, $\beta^2 \sigma_X^2$, and a component due to variation of the disturbance term, namely, σ_ε^2.[16] ∎

---

[16] It is possible to prove Equation 5.32 without assuming that X and the disturbance ε are independent and to prove an analog of Equation 5.31 without assuming that $\text{Var}(\varepsilon|X = x) = \sigma_\varepsilon^2$. To do so, one needs a result—the law of total variance—that is slightly beyond our scope.

We started by writing a simple model for one random variable in terms of another two random variables and two coefficients (Equation 5.23). Using the model and some assumptions about its components, we deduced the expectation and variance of Y, the conditional expectation and variance of Y given X, and the covariance and correlation of X and Y. This is exactly the kind of thing we wanted to be able to do when we started studying probability —we wanted to be able to make some assumptions about a process that generates data and deduce properties that the generated data would have.

In the rest of this book, we will use this framework to study this question in reverse. Assuming that we have data, what statements can we make about processes that might have generated the data?

In the following exercises, you will study an important interpretation of the correlation coefficient, and you will have a chance to simulate data under a model like the one we just analyzed.

Exercise Set 5-5

(1) [Optional] Under the model developed in the last section—including all assumptions—write the square of the correlation coefficient (Equation 5.30) in terms of the variance of Y (Equation 5.32) and the conditional variance of Y given X (Equation 5.31).

(2) In this exercise, you will simulate a model like the one developed in the last section. To simulate data, you will use the function `sim.lm()`, which is shown here:[17]

```
sim.lm <- function(n, a, b, sigma.disturb = 1, mu.x = 8, sigma.x = 2, rdis-
turb = rnorm, rx = rnorm, het.coef = 0){
    x <- sort(rx(n, mu.x, sigma.x))
    disturbs <- rdisturb(n, 0, sapply(sigma.disturb + scale(x) * het.
coef, max, 0))
    y <- a + b * x + disturbs
    cbind(x,y)
}
```

`sim.lm()` simulates one instance of a model similar to the one we just developed. The slope and intercept are specified by a and b, the variance of the disturbances is `sigma.disturb`, and the mean and standard deviation of X are `mu.x` and `sigma.x`. By default, `sim.lm()` adds three assumptions to the list we accumulated in the last section. First, it assumes that there are many random variables with the properties of X and Y rather than just one of each—in particular, there are n of each. We refer to the ith pair of random variables as X_i and Y_i. The separate pairs are independent and identically distributed—that is, for $i{\neq}j$, X_i and X_j are independent and identically distributed, and ϵ_i and ϵ_j are independent and identically distributed. Second, our new model assumes (by default) that the X_i are normally distributed. (Normality of X can be relaxed by changing the `rx` argument.) Third, we assume (by default) that the disturbances ϵ_i are normally distributed and homoscedastic. (Normality of disturbances can be relaxed with the `rdisturb` argument, and homoscedasticity with the `het.coef` argument.) These assumptions do not change the conclusions we developed in equations 5.23-5.30, which are still true for all X_i, Y_i pairs that meet the necessary conditions.

---

[17] Also available in the book's R package, `stfspack` (see Exercise Set 2-2, Problem 3 for installation instructions).

Once `sim.lm()` is defined, you can plot the outcome of one simulation on a scatterplot. For example, if I wanted to simulate the model with 50 pairs of observations, a=0, b=1, and all the other parameters set to the default, then I could use

```
> sim_0_1 <- sim.lm(n = 50, a = 0, b = 1)
> plot(sim_0_1[,1], sim_0_1[,2])
```

Try plotting the output of `sim.lm()`. Try plotting several results of `sim.lm()` run with the same parameters, and also try varying the values of the parameters. In particular, try varying a, b, sigma.disturb, and `sigma.x`. What do you see in each case? We will use `sim.lm()` to test many of the ideas in the next few chapters.

5.7 Chapter summary

We summarized the behavior of random variables using the concepts of expectation, variance, and covariance. The expectation is a measurement of the location of a random variable's distribution. The variance and its square root, the standard deviation, are measurements of the spread of a random variable's distribution. Covariance and correlation are measurements of the extent of linear relationship between two random variables.

We also presented two important theorems that describe the distribution of means of samples from a distribution. As the sample size becomes larger, the distribution of the sample mean becomes bunched more tightly around the expectation—this is the law of large numbers—and the distribution of the sample mean approaches the shape of a normal distribution—this is the central limit theorem.

Finally, we wrote a model describing a linear relationship between two random variables, and we analyzed the properties of those two random variables. So far, the most important assumptions we have made are

Assumption A1: Linearity $\mathrm{E}(\varepsilon|X = x) = 0$ *for all x.*

Assumption A2: Homoscedasticity *The variance of the disturbance is constant, regardless of the value of X:* $\mathrm{Var}(\varepsilon|X = x) = \sigma_\varepsilon^2$.

Working with this model for simple linear regression is exactly what we wanted to be able to do—to make assumptions about how data might be generated and to use those assumptions to predict properties of the resulting data. In the next several chapters, we will consider the inverse problem—that of using observed data to make statements about the process that might have generated them.

5.8 Further reading

The probability textbooks listed at the end of Chapter 4 cover all the topics here (except the probabilistic model for simple linear regression).

Lyon, A. (2013). Why are normal distributions normal? *The British Journal for the Philosophy of Science*, 65, 621–649.

This article scrutinizes scientists' invocations of the central limit theorem, which are sometimes loose and informal.

Taleb, N. N. (2008). The fourth quadrant: a map of the limits of statistics. *Edge*. https://www.edge.
org/conversation/the-fourth-quadrant-a-map-of-the-limits-of-statistics

A popular essay about the limits of guarantees like the law of large numbers and the central limit
theorem. Convergence can be slow, and it can be impossible to tell whether convergence has
occurred for tail probabilities. If one is concerned with a tail (i.e. extreme) event, like a high-
magnitude earthquake, there may be no data available to inform about its probability—by
definition, these events are rare. We may only have model predictions built on the basis of data
closer to the center of the distribution. But as you saw in Problem 3 of Exercise Set 5-4, it is
possible for processes to look as though they obey one distribution—such as a normal—most of
the time, but to have much higher probabilities of extreme events than would be predicted under
the look-alike distribution.

Interlude

We pause between two pursuits. For the past two chapters, we have considered probability theory—the study of the random outcomes of specified processes. We turn next to statistical estimation and inference—the attempt to make claims about unknown processes given samples of data.[1] In probability, we start with a process and make claims about data that might result from that process. In statistics, we reverse direction, starting with data and making claims about processes that might have generated the data.

The turn to statistics is thus a return to data. Figure Int-1 is a reproduction of Figure 3-3 showing the running example, data on grain yield and fertilizer use in sub-Saharan Africa. Figure Int-1 includes the least-squares line that we identified in Chapter 3.

We ended Chapter 3 equipped with the least-squares line, but our interpretations of the line were minimal. If someone were to ask what the line represents, we could reply with the definition: "It is the line that minimizes the squares of the vertical distances between the points and the line." That is a nice justification for a graphical data summary, but it is thin gruel on its own. Does the line *mean* anything?

In the following chapters, we will invest the least-squares line with meanings. To give the line meaning, we make assumptions. Here is one way to classify approaches to data analysis into four categories, according to the assumptions they entail.[2]

Level 1: Exploratory data analysis We conducted exploratory data analyses in Chapters 1 and 2. The researcher conducting exploratory data analyses does not make probabilistic assumptions about the data. She makes plots, ranging from basic to elaborate, and she calculates summary statistics—means, medians, interquartile ranges. She checks the data for errors. (Are any human heights negative? That is a mistake.) Looking at data is a critical part of any data analysis plan—thus, all data analyses should involve attention to level 1; they *may* include one or more of the other levels.

Level 2: Nonparametric or semiparametric data analysis In contrast to exploratory data analyses, nonparametric and semiparametric data analyses invoke probabilistic models. A nonparametric model is a model that cannot be described by a finite set of parameters— we say it is "infinite-dimensional." (Practically, people use nonparametric methods when they cannot assume that some probability distribution in their model belongs to a known distribution family or another distribution they can write down.) A semiparametric model has a parametric part—a part described by parameters—and a nonparametric

---

[1] Estimation and inference are by no means the only goals of statisticians or data analysts. For example, making predictions about new data is a major topic. Nonetheless, this book focuses on estimation and inference as central pursuits. (Prediction is discussed briefly in the Postlude.)

[2] A note for interpreting the levels: "parametric" is the adjective form of "parameter."

Statistical Thinking from Scratch: A Primer for Scientists. M. D. Edge, Oxford University Press (2019).
© M. D. Edge 2019. DOI: 10.1093/oso/9780198827627.001.0001

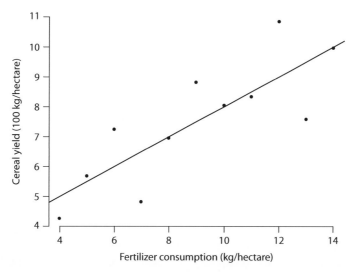

Figure Int-1 Scatterplot of fertilizer consumption and cereal yield, including the least-squares line.

part, which is not described by parameters.[3] We have already seen an example. Consider the model from the end of Chapter 5 (Equation 5.23):

$$Y = \alpha + \beta X + \epsilon.$$

The "parametric" in "semiparametric" indicates that the model has parameters. In this case, the parameters include the intercept α and the slope β. In semiparametric data analysis, the parameters are targets of estimation and inference. The "semi" in "semi-parametric" indicates that there is another part of the model that is not described by parameters. In our case, the part of the model that is not fully described by parameters is the distribution of the disturbance, ϵ, about which we may not want to make many assumptions. In Chapter 8, we consider what we can do when we do not assume that we know the probability distribution family of the disturbance.[4]

Level 3: Parametric data analysis In parametric data analysis, we retain assumptions made in level 2 and add assumptions about the probability distribution families of random variables in the model. For example, in simple linear regression, we begin a fully para-metric analysis by assuming that the disturbance is drawn from a particular distribution family, such as a normal distribution.

Level 4: Bayesian data analysis In Bayesian data analysis, we retain the assumptions of level 3 and add the assumption that the model's parameters have *prior* distributions. That is, we assign probability distributions to the parameters that summarize our uncertainty about the parameter's value. Bayesian data analysts use Bayes' theorem (Equation 4.2) to

---

[3] Recall from Chapter 3 that parameters are numbers that govern the specifics of a probability distribution. For example, the normal distribution has two parameters, μ and σ, that control the location and spread of the distribution. The binomial distribution has two parameters, n and p, where p is a probability of success on each trial and n is the number of trials. The distribution of Y in Equation 5.23 is governed in part by α and β, so α and β are parameters.

[4] Though this is a regression model, and it is semiparametric, statisticians usually mean something different from this when they say "semiparametric regression." Typically, semiparametric regression means that the deterministic (conditional the independent variables) part of the regression has a nonparametric part, and not just that the disturbance distribution may be nonparametric.

combine prior distributions and data into posterior distributions, which can summarize one's uncertainty about the parameter after observing the data.[5]

In the following chapters, we will consider semiparametric, parametric, and Bayesian approaches to estimation and inference for simple linear regression. The least-squares line will appear repeatedly, but each time we see it—that is, at each of levels 2–4—it will have a different meaning. The line remains; our assumptions change.

In Chapters 6 and 7, we cover estimation and inference in the abstract. In Chapter 6, we consider properties we would like estimators to have. Chapter 6 focuses on *point* estimation, the attempt to estimate an unknown quantity with a single number. Chapter 7 considers *interval* estimation—the attempt to define an interval that (we hope) contains an unknown quantity of interest—along with the related pursuit of hypothesis testing. Chapters 8, 9, and 10 outline semiparametric, parametric, and Bayesian approaches to estimation and inference, in that order. Throughout these chapters, we consider methods for checking assumptions. Because one's assumptions influence the conclusions one makes, we would like to know whether the assumptions are true. It is not often possible to confirm *all* of a model's assumptions—this should make you nervous—but we can often check the plausibility of some of them.

In the Postlude, we consider some extensions of simple linear regression.

---

[5] Three clarifications about this scheme: (1) It is not an exhaustive catalog of approaches to data analysis—it is merely a way to organize the approaches to data analysis in this book. (2) The nested structure of the levels does not always apply. There are, for example, nonparametric Bayesian methods. (3) The ordering of the levels does not imply that one level is "better" than any other. The four types of data analysis have distinct goals.

CHAPTER 6

Properties of point estimators

> **Key terms**: Bias, Consistency, Decision theory, Efficiency, Estimation (Estimand, Estimate, Estimator), Mean squared error (MSE), Outlier, Point estimation, Risk, Robustness, Sampling distribution, Unbiased

Estimation is the attempt to discern the values of numbers that describe a data-generating process—or, more properly, a process by which we *assume* the data were generated (see Box 6-1). That is, we postulate that data were produced by a particular process—that's the model—and we use the data to estimate properties of the process. The model might be minimal—we might only say that the data are independent samples from an unknown distribution, and we might want to estimate the expected value of that distribution. Or the model might be complex—it might feature several probability density functions with many parameters each, and we might want to estimate the values of all the parameters.

We make **estimates** by applying **estimators** to data. An estimator is a procedure—more precisely, a function—that can be applied to data. One might estimate the expected value of a distribution using the sample mean. The procedure "take the sample mean" is an estimator, and the specific sample mean the data analyst obtains is an estimate.[1] In principle, one can design many estimators for any one quantity. How do we choose? This chapter introduces some of the criteria that statisticians use.

Statisticians weigh many criteria when choosing estimators, many of which can be understood as responses to straightforward questions. First, if the data are actually generated by the assumed model, does the estimator produce estimates that are correct on average? That is, is the estimator accurate? Second, if we apply the same estimator to many datasets produced by the same process, will the estimates vary widely, or will they fall in a narrow range? That is, is the estimator precise? Figure 6-1 illustrates the distinction between accuracy and precision.

There are subtler questions to ask, too. Does the estimator improve in either accuracy or precision when the number of available data increases? We hope so. Suppose that we can quantify the costs of different types of errors—perhaps overestimates are catastrophic whereas underestimates are tolerable. Can we calibrate an estimator to minimize the expected costliness of error? In practice, we will often find that we have to make trade-offs between desirable properties—for example, deciding between two estimators when one is less precise but more accurate than the other. We may have to use our judgment in such cases, and we will sometimes have to find a principled way to combine different sources of information in making a decision. Finally, what if the model is wrong in some

---

[1] Another way to say this is that *estimators* are random variables—they are functions of other random variables, which represent the data—and that *estimates* are instances of *estimators*.

Statistical Thinking from Scratch: A Primer for Scientists. M. D. Edge, Oxford University Press (2019).
© M. D. Edge 2019. DOI: 10.1093/oso/9780198827627.001.0001

Box 6-1 PROCESSES, POPULATIONS, AND SAMPLES

In the remainder of the book, we will often invoke assumptions like, "the data are independently sampled from a normal distribution with unknown parameters," or "the data are generated according to the linear model from the end of Chapter 5." We will then discuss procedures that can lead to knowledge about the process that generated the data.

Statistical methods are also often discussed as an attempt to learn something about a *population* on the basis of a sample. For example, one might want to know the proportion of people in a population who support a certain candidate in an election on the basis of a sample from that population. In this case, the data-generating process is drawing samples from a population.

Referring to a data-generating process instead of a population captures important goals that don't fit neatly in the population framework. For example, we might believe that the measured time for a cannonball to fall from a tower of height h is $t = \sqrt{2h/g} + \epsilon$, where g is a gravitational constant and ϵ is a measurement error. We might drop the ball many times, perhaps varying the height, and use the measured times to estimate g. In this case, there is no "population" of drops from the tower; there are merely outputs of a data-generating process that we assume is described by a physical model.

Whether we are talking about samples from a population or from some other type of data-generating process, the trustworthiness of the conclusions reached will hinge on whether the data are representative of the process we care about and accurately measured. For example, if we are curious about who will win a city-wide election, and if all the people we interview are from a single wealthy neighborhood, then the sample will likely not reflect the entity we want to learn about—the city's population of voters. A sample of randomly selected people from across the city will be much better. Inferences from a sample of voters will also fail if the interviewed people provide inaccurate information, reporting that they support a candidate for whom they will not in fact vote. And if we want to learn about gravitational force, but all the objects we drop are susceptible to non-negligible air resistance (for example, pieces of paper), then the data will reflect not just the process we care about—gravitation—they will reflect gravitation plus air resistance, and we will get gravity badly wrong if we ignore the air resistance.

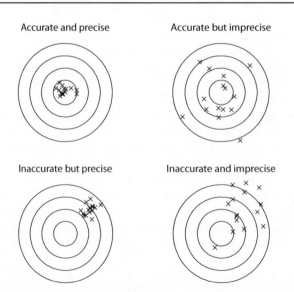

Accurate and precise · Accurate but imprecise · Inaccurate but precise · Inaccurate and imprecise

Figure 6-1 Accuracy and precision. The illustrations in top row illustrate accuracy—in the top row, the marks on the target are roughly centered on the bullseye. The illustrations in the left column illustrate precision—the marks on the target are tightly clustered.

respect or the data are contaminated? Can we design estimators that will still provide useful information?

We consider properties of estimators that address these questions. Below, assume that we are trying to estimate a quantity, also called the **estimand**, whose value is θ. The quantity we estimate might be a parameter of a distribution or another feature of a probability distribution, such as an expectation.

To make an estimate of θ, we need data. We represent the data as a random variable D. We can think of D as a set containing n elements, $D = \{D_1, D_2, ..., D_n\}$. Each element of D is a random variable or a set of random variables. For example, in this section, we will use an illustration in which each entry in the data set D is a draw from a normal distribution, and later we will assume that each element in D is a pair of observations: $D_i = \{X_i, Y_i\}$.

To get an estimate, we apply a function—an estimator—to the data. Estimators are often written with the same letter—often a Greek letter—that is used to indicate the estimand, with the addition of an accent on top. For example, if we estimate the quantity θ, we can call our estimator $\widehat{\theta}$, pronounced "theta-hat." If we are comparing two distinct estimators, we can write them with two different accents, such as $\widehat{\theta}$ and $\widetilde{\theta}$.

We need to be careful about notation. You will often see, for example, notation like $\widehat{\theta}$ used to refer to both the estimator—the function being applied to data—and the estimate— the result of applying the estimator to data. The intent of the writer is usually clear from the context, but in this chapter, we will use some extra conventions when we need to avoid confusion.

The data, D, are random. Thus, the result of applying the estimator to the data, $\widehat{\theta}(D)$, is also random—the estimates of θ that the estimator produces can vary from dataset to dataset. The distribution of an estimator is sometimes called a **sampling distribution**. We imagine drawing samples of data repeatedly and computing the estimator each time; the distribution that would result is the sampling distribution of the estimator. In this chapter, the properties of the random variable $\widehat{\theta}(D)$'s sampling distribution—its expectation, variance, and other qualities—are of central interest.

We can also observe an instance of the random variable D; the instance d represents an observed dataset. By applying the estimator to d, we get the estimate $\widehat{\theta}(d)$.

Finally, the size of the dataset often matters. To specify that an estimator is being applied to a dataset with n observations, we use a subscript n, for example $\widehat{\theta}_n(D)$. Notation is summarized in Table 6-1.

Table 6-1 Summary of notation for targets of estimation, estimators, and estimates

Notation	Meaning
θ	The value of the quantity being estimated is often indicated with a Greek letter, such as θ. θ might be the value of a parameter of a probability distribution or another function of a probability distribution, like an expectation.
$\widehat{\theta}$	An estimator is a function that is applied to data to produce estimates of a true quantity. Estimators are indicated by an accent—often but not always a "hat"—above the symbol used to indicate the true quantity. Depending on context, $\widehat{\theta}$ on its own can be used to represent an estimator or a specific estimate that results from applying the estimator.
$\widehat{\theta}(D)$	Estimators are applied to data, and data can be represented as a random variable, D. The value of an estimator applied to a random variable D is itself a random variable, and we can study its properties, including, for example, its expected value and variance.
$\widehat{\theta}(d)$	After we have observed data, they are no longer random. We refer to the instance of the data as d, and $\widehat{\theta}(d)$ is the result of applying an estimator to the observed data. That is, $\widehat{\theta}(d)$ is an estimate of θ. $\widehat{\theta}(d)$ is not random.
$\widehat{\theta}_n(D)$	We use the subscript n to specify that an estimator is being applied to a data set containing n observations. We will often be interested in how estimators behave when they are applied to datasets of different sizes.

Example Consider n independent and identically distributed random variables $X_1, X_2, ..., X_n$. Assume that for all i, $X_i \sim \text{Normal}(\theta, 1)$. That is, each observation is drawn from a normal distribution with an unknown first parameter θ and a second parameter $\sigma^2 = 1$. (Recall that the first parameter of a normal distribution is equal to its expectation (Equation 5.21), so $E(X_i) = \theta$ for all i Also remember that the second parameter of a normal distribution is equal to its variance, so $\text{Var}(X_i) = 1$ for all i.) One estimator of the unknown parameter θ is the sample mean,

$$\widehat{\theta}_n(D) = \frac{1}{n}\sum_{i=1}^{n} X_i. \tag{6.1}$$

To make the example more concrete, draw a sample from a normal distribution and take the mean. In R, use

```
> d <- rnorm(25,0,1)
```

to draw a sample of 25 independent observations with $\theta = 0$ and $\sigma^2 = 1$.[2] To calculate the sample mean, $\widehat{\theta}_n(d)$, use

```
> mean(d)
```

When I followed this procedure, I obtained an estimate $\widehat{\theta}_n(d) = 0.17$. You may get a larger or smaller value. The value I obtained is close to θ, which we specified to be 0. That's good. But how can we more formally assess how well the sample mean performs as an estimator of θ? We will return to this example as we define properties that assess how well an estimator works.

We will start with criteria that correspond roughly to accuracy and precision. Following that, we will consider a criterion that combines considerations of accuracy and precision, and two criteria that consider how the accuracy and precision of the estimator change with the number of observations used. Next, in an optional section, we take up decision theory, a framework that can accommodate the details of specific estimation problems. Finally, we will briefly discuss how estimators behave when model assumptions are violated.

6.1 Bias

The **bias** of an estimator is the difference between the expected value of the estimator and the true value of the quantity we want to estimate. Informally, an unbiased estimator is an accurate one. Estimators that behave like upper panels of Figure 6-1, with distributions centered on their targets, are less biased than those that behave like the lower panels, which have distributions centered off-target. More formally, the bias is

$$B(\widehat{\theta}_n) = E[\widehat{\theta}_n(D)] - \theta. \tag{6.2}$$

---

[2] Whereas we use σ^2 as the second parameter of a normal distribution, R uses σ. In this case, setting $\sigma = 1$ also sets $\sigma^2 = 1$, but remember that in R, the second parameter of the normal distribution is the standard deviation. Adjust your simulations accordingly whenever necessary.

If an estimator's bias is positive, then its expectation is an overestimate of the true value, and if the estimator's bias is negative, then its expectation is an underestimate of the true value. Other things equal, we prefer estimators that are **unbiased**—that have bias equal to 0.

Exercise Set 6-1[3]

(1) What is the bias of the estimator in Equation 6.1 (the sample mean) when applied to the model in which the data are independently and identically distributed Normal(θ, 1)?

(2) In this exercise, you will use simulation to explore the properties of an estimator. In particular, we will draw many samples of a specified size from a normal distribution, apply an estimator to each of the samples, and see what the mean value of the resulting estimates is.

We draw samples from a normal distribution using this function:[4]

```
mat.samps <- function(n, nsim = 10000, rx = rnorm, ...){
  samps <- rx(n*nsim, ...)
  matrix(sample(samps), nrow = nsim, ncol = n)
}
```

By default, the function `mat.samps()` draws `nsim` samples of size `n` from a standard normal distribution. (Changing the `rx` parameter allows you to generate data from other distributions, as you will see below.) All observations are independent. The output is a matrix: each row of the matrix is one sample of size `n`. To save 10,000 simulated samples of 25 observations each, for example,

```
> s.mat <- mat.samps(n = 25, nsim = 10000)
```

Once you have a matrix of samples, you can use several methods to calculate estimates of θ from each sample. One is a `for()` loop:

```
ests.mean <- numeric(10000)
for(i in 1:10000){
  ests.mean[i] <- mean(s.mat[i,])
}
```

The first line makes an empty numeric vector of the same length as the number of samples we have drawn, and the `for()` loop fills each entry of that vector with the mean of the corresponding row in the matrix of samples. A leaner method is to use `apply()`, which you have not yet seen unless you have read Appendix B:

```
ests.mean <- apply(s.mat, 1, mean)
```

The `for()` loop and `apply()` strategies give exactly the same results, which you can verify using the `identical()` function. The `apply()` function's first argument is a matrix to which we would like to apply a function, the second argument indicates whether we would like to apply the function to the rows or columns of the matrix—"1" means rows, whereas "2" would mean columns—and the third argument gives the function we would like to

---

[3] See the book's GitHub repository at github.com/mdedge/stfs/ for detailed solutions and R scripts.
[4] Available in the book's R package, `stfspack` (see Exercise Set 2-2, Problem 3 for installation instructions).

apply. So `apply(s.mat, 1, mean)` tells R, "Return a vector in which each entry is the mean of the corresponding row of `s.mat`."

And now, the question: Consider the sample *median*—implemented in R as the `median()` function—as an estimator of the first parameter of a normal distribution. Is it biased? If so, in which direction? Use simulations to justify your answer.

6.2 Variance

An estimator's variance is a measure of the spread of its distribution. The variance of an estimator is defined in the same way as the variance of any other random variable (Equation 5.6):

$$\text{Var}(\widehat{\theta}_n) = \text{E}\left[\left(\widehat{\theta}_n(D) - \text{E}[\widehat{\theta}_n(D)]\right)^2\right]. \tag{6.3}$$

You can also use the identity in Equation 5.7 to calculate the variance of an estimator:

$$\text{Var}(\widehat{\theta}_n) = \text{E}[\widehat{\theta}_n(D)^2] - \left(\text{E}[\widehat{\theta}_n(D)]\right)^2.$$

Whereas an estimator's bias is an index of its accuracy, an estimator's variance indicates the precision of the estimator. Conceptually, low-variance estimators produce tightly clustered estimates, like the ones in the left-side panels of Figure 6-1, whereas high-variance estimators produce dispersed estimates, like those in the right-side panels of Figure 6-1. Other things equal, we prefer estimators that have smaller variance and are thus more precise.

Exercise Set 6-2

(1) What is the variance of the estimator in Equation 6.1 (the sample mean) when applied to the model in which the data are independently and identically distributed Normal(θ, 1)? If we do not know the distribution family from which the data are drawn, what is the variance of the mean of a sample of independent observations as an estimator of the expectation $\text{E}(X_i)$? (Assume the unknown distribution has finite variance σ^2.)

(2) Using simulation, we convinced ourselves that the sample median is an unbiased estimator of the first parameter of a normal distribution (Exercise Set 6-1, Problem 2). Use similar simulations to determine whether the sample median has larger or smaller variance than the sample mean when the data are drawn from a normal distribution. (Assume that for large samples, the sample variance—calculated in R using `var()`—can be trusted as an indicator of the true variance.)

6.3 Mean squared error

Though the bias and the variance of an estimator inform us about an estimator's accuracy and precision, respectively, statisticians need to consider *both* accuracy and precision at once. An estimator that is unbiased but varies wildly is of little use. Similarly, a precise estimator that is wildly inaccurate is useless. (Imagine the estimator $\widehat{\theta} = 42$, which estimates the parameter value to be 42 regardless of the data. It has zero variance but will usually be inaccurate.) One way to compare estimators on the basis of both their accuracy and precision is to use a metric that combines bias and variance into one measure. The **mean squared error** is one such metric.

The mean squared error is the expected squared difference between the value of an estimator and the true value it seeks to estimate:

$$\mathrm{MSE}(\widehat{\theta}_n) = \mathrm{E}\left[\left(\widehat{\theta}_n(D) - \theta\right)^2\right]. \tag{6.4}$$

The formula for the mean squared error is similar to the formula for the variance of the estimator. Whereas the variance of the estimator measures how widely the estimator varies with respect to its expectation, the mean squared error measures how widely the estimator varies with respect to the estimand itself. For an unbiased estimator, the expectation of the estimator is equal to the estimand, and so the mean squared error is equal to the variance.

The mean squared error provides one answer to an important question about an estimator—how far off do we expect the estimate to be? Other things equal, we prefer estimators with lower mean squared error—estimators that we expect to be closer to the true value we seek to estimate.

You will prove in an exercise that the mean squared error is equal to

$$\mathrm{MSE}(\widehat{\theta}_n) = \mathrm{B}(\widehat{\theta}_n)^2 + \mathrm{Var}(\widehat{\theta}_n), \tag{6.5}$$

where the bias and variance are as defined in Equations 6.2 and 6.3. That is, the mean squared error of an estimator is equal to the square of the estimator's bias plus the estimator's variance. Thus, in Figure 6-1, the top panels illustrate estimators with lower mean squared error than those in the corresponding bottom panels (because squared bias is lower in the top panels), and the left-side panels illustrate estimators with lower mean squared error than those to their right (because variance is higher in the right-side panels).

Exercise Set 6-3

(1) [Optional] Prove Equation 6.5.
(2) When estimating the first parameter of a normal distribution, which has lower mean squared error: the sample mean or the sample median?

6.4 Consistency

Consistency is another criterion that combines considerations of accuracy and precision. In particular, a **consistent** estimator is one that, loosely speaking, gets closer to the estimand as the number of data used to make the estimate increases. More formally, a consistent estimator is one that *converges in probability* to the true value it aims to estimate. We saw convergence in probability once before, when we considered the law of large numbers (Equation 5.3). In this context, the requirement of convergence in probability means that for any positive number δ,

$$\lim_{n\to\infty}\mathrm{P}\left(|\widehat{\theta}_n(D) - \theta| > \delta\right) = 0. \tag{6.6}$$

That is, the probability that the estimator is wrong by more than δ decreases toward zero as the number of data used increases to infinity, no matter how small δ is. With an infinitely large sample, consistent estimators get the right answer—a nice feature, if a little abstract.

Using an approach similar to the one we took to prove the weak law of large numbers (Exercise Set 5-2, Problem 6), it is possible to show that an estimator $\widehat{\theta}_n$ is consistent if

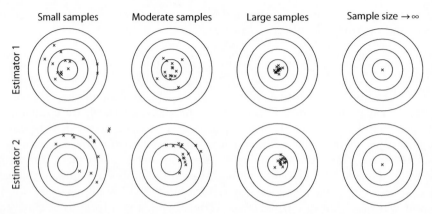

Figure 6-2 Conceptual schematic of the behavior of two hypothetical consistent estimators as the sample size increases. Estimator 1 (upper row) is unbiased at all sample sizes, and its precision increases as the sample size increases (plots from left to right). Because it is consistent, it converges to the estimand as the sample size approaches infinity. Estimator 2 (lower row) is substantially biased when applied to small and moderate sample sizes. Nonetheless, because it converges to the estimand as the sample size approaches infinity, it is consistent. One lesson is that consistency is a large-sample property, and consistent estimators might behave badly in small samples.

$$\lim_{n\to\infty}\mathrm{MSE}(\widehat{\theta}_n) = 0 \qquad\qquad (6.7)$$

That is, if the mean squared error of an estimator decreases to zero as the number of data used increases, then the estimator is consistent. Often, the easiest route to proving that an estimator is consistent—that it satisfies Equation 6.6—is to prove that it satisfies Equation 6.7.[5]

One way to understand consistency is to consider its relationship with bias. Considering Equations 6.5 and 6.7 together, we see that an unbiased estimator is consistent if its variance approaches zero as the sample size increases. In practice, the variance of many sensible estimators does decrease to zero as the sample size increases, so many unbiased estimators are also consistent. However, *biased* estimators can also be consistent, so long as their bias decreases to zero as the sample size approaches infinity. Figure 6-2 shows a schematic that exemplifies this relationship.

Exercise Set 6-4

(1) Is the sample mean consistent as an estimator of the first parameter of a normal distribution? Is it a consistent estimator of the expectation of a distribution from an unspecified family (as long as that distribution has a finite variance)?

(2) [Optional] Use simulations to convince yourself that the sample median is a consistent estimator of the first parameter of a normal distribution.

(3) Consider data $X_1, X_2, ..., X_n$ drawn independently from a Normal$(\theta, 1)$ distribution. For each of the following proposed estimators of θ, answer both these questions: (i) is the estimator unbiased? (ii) is the estimator consistent?
   (a) The sample mean, $\overline{X} = \frac{1}{n}\sum_{i=1}^{n}X_i$.
   (b) A shifted sample mean, $\frac{1}{n}\sum_{i=1}^{n}X_i + 1$.

---

[5] If an estimator satisfies Equation 6.7, then it satisfies Equation 6.6, but an estimator might still satisfy Equation 6.6 without satisfying Equation 6.7.

  (c) The first observation, X_1.
  (d) A "shrunk" sample mean, $\frac{1}{n+1}\sum_{i=1}^{n}X_i$
(4) [Optional] Prove that if Equation 6.7 holds for an estimator $\widehat{\theta}_n$, then Equation 6.6 also holds. You may use Markov's inequality and Chebyshev's inequality, both of which are given in Exercise Set 5-2, Problem 6. (For this problem, I find it easier to start with Markov's inequality than with Chebyshev's inequality.)

6.5 Efficiency

Consistency is an admirable quantity in an estimator. Suppose we have a model that correctly describes a process we care about. If we also have consistent estimators for the model's parameters, then we have a clear path to success: keep collecting data. As we gather more data, the estimators will converge to the true values of the parameters.

But consistency is not everything. It nearly always costs something—money, time, effort, or all three—to collect data. So we want estimators to do more than converge to the true values: we want them to converge to the true values as quickly as possible. Efficiency captures this desideratum: a more **efficient** estimator achieves a given level of performance with fewer data than it would take a less efficient estimator to achieve the same level of performance. Figure 6-3 is a conceptual illustration of the behavior of two estimators that differ in efficiency.

One way to define the relative efficiency of two estimators $\widehat{\theta}_n$ and $\widetilde{\theta}_n$ is as

$$RE(\widehat{\theta}_n, \widetilde{\theta}_n) = \frac{MSE(\widetilde{\theta}_n)}{MSE(\widehat{\theta}_n)}. \tag{6.8}$$

A small mean squared error is good—if $MSE(\widehat{\theta}_n) < MSE(\widetilde{\theta}_n)$, then $RE(\widehat{\theta}_n, \widetilde{\theta}_n) > 1$, and we say that $\widehat{\theta}_n$ is more efficient than $\widetilde{\theta}_n$. In principle, the relative efficiency of two estimators can depend on the sample size. For example, one estimator might be more efficient when the

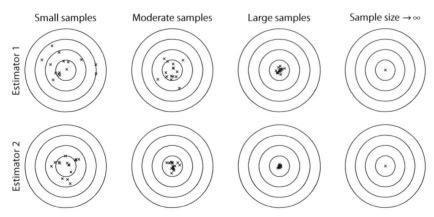

Figure 6-3 Conceptual schematic of the behavior of two hypothetical estimators that differ in efficiency as the sample size increases. Both estimators are consistent; they converge to the true value of the estimand as the sample size approaches infinity. However, estimator 2 (lower row) approaches convergence faster than does estimator 1 (upper row)—estimator 2 is more precise given the same amount of data. Thus, we say that estimator 2 is more efficient.

sample is small but less efficient when the sample is large. The asymptotic relative efficiency is the limit of the relative efficiency as the sample size increases to infinity:

$$\text{ARE}(\widehat{\theta}_n, \widetilde{\theta}_n) = \lim_{n \to \infty} \frac{\text{MSE}(\widetilde{\theta}_n)}{\text{MSE}(\widehat{\theta}_n)}. \tag{6.9}$$

Other things equal, we prefer more efficient estimators over less efficient estimators.

Exercise Set 6-5

(1) In this problem, you will explore the relative efficiency of the mean and median for normal samples of different sizes. Consider a Normal(0, 1) distribution.

   (a) Estimate the relative efficiency of the mean and median for normal samples of size 5. To estimate the mean squared error, use 10,000 normal samples of size 5. To estimate the mean squared error of the sample median, calculate the median of each of your 10,000 samples, then compute the squared difference between each median and the true value of the first parameter. The mean of these squared differences is an estimate of the mean squared error of the sample median. Do the same thing for the sample mean, and estimate the relative efficiency of the sample mean and median by taking the quotient of the estimated mean squared errors.

   (b) [Optional] Repeat part (a) for sample sizes of 2, 10, 20, 50, 100, 200, and 500. Do you have a guess for the asymptotic relative efficiency of the sample mean vs. the sample median for normally distributed data?

(2) [Optional, but read the solution if you skip] Repeat Problem 1, but for the Laplace distribution, which has heavier tails—that is, higher probabilities associated with extreme events—than the normal distribution. To draw samples from a Laplace distribution with expectation 0 and standard deviation 1, use the `rlaplace()` function[6] defined here:

```
rlaplace <- function(n, mean = 0, sd = 1) {
  exp1 <- rexp(2*n, 1)
  x <- exp1[1:n] - exp1[(n+1):(2*n)]
  x * sd/sqrt(2) + mean
}
```

You can organize Laplace-distributed random data into a matrix by setting the `rx` parameter of `mat.samps` to `rlaplace`, as in `mat.samps(25, 10000, rx = rlaplace)`.

6.6 [Optional section] Statistical decision theory and risk

Some of the ways in which we assess the performance of estimators may remind you of a problem we faced in Chapter 3. There, we needed to minimize a function of the vertical distances between data points and a line. We found ourselves, for several reasons, emphasizing the sum of the squares of these differences.

You may have noticed another prioritization of squared errors in this chapter, especially as we discussed the mean squared error. Mean squared error is a useful measurement of an estimator's overall performance, but why should we care about the mean of the squared errors that an estimator makes? Why not, for example, consider the mean of the absolute differences between the value of an estimator and the true quantity it estimates? **Statistical**

---

[6] Available in the book's R package, `stfspack` (see Exercise Set 2-2, Problem 3 for installation instructions).

decision theory—a vast field we will barely sketch—provides a framework for assessing how well estimators perform with respect to any metric, whether squared errors or some entirely different criterion.

Mean squared error implicitly imposes a loss function on the estimation errors we make. We saw loss functions in a different context in Chapter 3, when we were trying to find a line that "best" fits a set of points. There, the loss function provided a way of measuring the fit between a line and the points through which it is drawn (Equation 3.3). In general, a loss function associates a decision with a loss or cost. In our context, the loss function measures the "loss" or "cost" induced by an estimator that misses its target.

In the estimation context, the arguments of a loss function are the value of an estimate and the true value being estimated. We will write loss functions as $\lambda(\theta, \hat{\theta}(d))$, where $\hat{\theta}(d)$ is an estimate and θ is a true value being estimated. For example, the squared error loss function is $\lambda(\theta, \hat{\theta}) = (\hat{\theta} - \theta)^2$. It will help to work with some examples before moving on.

Exercise Set 6-6

(1) For each of the following, draw a plot of the loss function in the requested range. In each plot, put $\lambda(\theta, \hat{\theta})$ on the vertical axis. The horizontal axis will represent $\hat{\theta}$, and θ corresponds to a point on the horizontal axis.

   (a) Imagine that you are in charge of a wheat farm. Because your farm is remote and you cannot transport your goods, you can only sell your annual harvest to one buyer, the local baker. The baker will buy an unknown quantity of wheat, θ bushels. You earn a profit of $1/bushel for all wheat that the baker buys. It also costs you $1 to grow a bushel of wheat, and if you grow more than the baker wants, then that money is lost. If you grow exactly θ bushels of wheat, then your profit is maximized. If you grow more or less than that, then your profit is less than the maximum. $\lambda(\theta, \hat{\theta})$ is the amount by which your profit is decreased from the maximum, $\hat{\theta}$ is the amount of wheat (in bushels) that you estimate the baker will want to buy (and that you will thus plan to grow), and θ is the amount of wheat the baker actually wants to buy. Plot $\lambda(\theta, \hat{\theta})$ assuming that $\theta = 1000$ for values of $\hat{\theta}$ from 0 to 2000. (A hand-drawn plot is fine.)

   (b) Draw another plot for the same scenario as in part (a), this time assuming that it costs $2 to grow a bushel of wheat.

   (c) You are betting on a horse race. Six horses are running, and they are labeled 1,2,3,4,5, and 6. A bet on the winning horse, θ, is worth $1. Betting on any of the other horses earns you nothing. Draw the loss function assuming that horse 3 will actually win the race.

Loss functions capture the specifics of an estimation problem. Though the squared error loss function is mathematically convenient, it may not reflect the needs of people using the estimate. Using a squared error loss function imposes some hidden assumptions. In particular, users of squared error loss functions presume that errors that are twice as large are four times as costly and that underestimates and overestimates of the same absolute size are equally costly. These presumptions might be reasonable in some settings, but they are not true in many practical situations. Three examples—fictional, but realistic in principle—appeared in Exercise Set 6-6. Table 6-2 shows some commonly used loss functions, but infinitely many other loss functions can be imagined.

How can we use a loss function to choose an estimator? One way is to define a quantity called the **risk** of an estimator. The risk of an estimator is its expected loss—"expected" in the sense of "expected value." More formally, the risk is

Table 6-2 Some common loss functions

Function	Name
$\lambda(\theta, \widehat{\theta}) = (\widehat{\theta} - \theta)^2$	Squared error loss
$\lambda(\theta, \widehat{\theta}) = |\widehat{\theta} - \theta|$	Absolute error loss
$\lambda(\theta, \widehat{\theta}) = \begin{cases} K_1(\theta - \widehat{\theta}), \widehat{\theta} \le \theta \\ K_2(\widehat{\theta} - \theta), \widehat{\theta} > \theta \end{cases}$	Linear loss. Notice that if $K_1 = K_2$, then linear loss is proportional to absolute error loss.
$\lambda(\theta, \widehat{\theta}) = |\widehat{\theta} - \theta|^P$	L^P loss. (Pronounced "LP loss.") Notice that if $P = 1$, we have absolute error loss, or "L1 loss," and if $P = 2$, we have squared error loss, or "L2 loss."
$\lambda(\theta, \widehat{\theta}) = \begin{cases} 0, \widehat{\theta} = \theta \\ 1, \widehat{\theta} \ne \theta \end{cases}$	0-1 loss. (Pronounced "zero-one" loss.)

$$R(\theta, \widehat{\theta}) = E\left(\lambda(\theta, \widehat{\theta})\right). \tag{6.10}$$

You have already seen an example of a risk. The mean squared error (Equation 6.4) is the risk for a squared error loss function. With Equation 6.10, we simply generalize the mean squared error by substituting a loss function we choose for the squared error loss.

Other things equal, we prefer an estimator with lower risk to one with higher risk. This decision is uncomplicated when we are dealing only with estimators with risk that does not depend on θ. For example, consider the sample mean as an estimator of the first parameter of a normal distribution with variance 1, and assume a squared-error loss function. Given a sample of n independent observations, the risk of the sample mean is always $1/n$ (the variance of the sample mean), regardless of the true value of θ.[7] But imagine this laughable estimator: suppose that regardless of the data I collect, I always estimate θ to be 6. Call this estimator $\breve{\theta}$. If θ actually does equal 6, then the risk of $\breve{\theta}$ is zero, which is always less than $1/n$, the risk of the sample mean. How do we use the risk to choose an estimator when estimators' risks vary with the unknown estimand? We cannot generally use the value of the estimand to decide—the reason we undertake estimation is that we do not know the value of the estimand. Some relevant concepts will come in the exercises, and a Bayesian approach to the same problem will appear in Chapter 10.

There is much more to learn, but the main lesson is this: Many estimators are designed to have good properties with respect to squared error loss. The squared error loss function is mathematically convenient, but it may not reflect your priorities. Decision-theoretic thinking is a reminder to evaluate estimators using metrics that matter for the problem at hand. If your loss function will not yield to mathematics, you can often get useful information by simulation.

Exercise Set 6-7

(1) Consider estimation of θ using 100 independent observations from a Normal$(\theta, 1)$ distribution. We again consider the sample mean and the sample median as estimators.
   (a) The risk of the sample mean under squared error loss is $1/100$. Use simulations to determine the approximate risk of the sample median. (In this case, the risk of the sample median does not depend on θ).

---

[7] See Exercise Set 6-7, Problem 2, part (a).

(b) What are the approximate risks of the sample mean and sample median under absolute error loss? (Hint: use the abs () function to compute the absolute values of all entries in a vector.)

(c) What are the approximate risks of the sample mean and sample median under L^3 loss?

(2) In this problem, we consider risk functions, which help us think about estimators for which the risk varies with the value of the estimand, θ. The risk function $R(\theta, \widehat{\theta}) = E(\lambda(\theta, \widehat{\theta}))$ is a function of θ. Consider the problem of estimating the first parameter of a normal distribution using 3 independent samples from a Normal$(\theta, 1)$ distribution.

Parts (a)–(d). Draw a pair of axes with θ on the horizontal axis, ranging from 4 to 8, and $R(\theta, \widehat{\theta})$ on the vertical axis, ranging from 0 to 4. For each of the following estimators, draw the risk function assuming a squared error loss function. That is, plot $R(\theta, \widehat{\theta})$ as a function of θ assuming that $\lambda(\theta, \widehat{\theta}) = (\widehat{\theta} - \theta)^2$. Draw all your risk functions on the same plot.

(a) The sample mean. (Remember that $n = 3$.)

(b) The first observation. That is, if the three observations that make up the data are X_1, X_2, X_3, then draw the risk function for the value of X_1 as an estimator of θ.

(c) $\breve{\theta}$, where, as defined above, $\breve{\theta} = 6$ irrespective of the data collected.

(d) $\dot{\theta}$, where $\dot{\theta} = (\widehat{\theta} + \breve{\theta})/2$, $\widehat{\theta}$ is the sample mean, and $\breve{\theta} = 6$ as defined in part (b). Hint: Remember that the risk for squared error loss is the mean squared error and use the identity in Equation 6.5.

(e) An estimator $\widehat{\theta}_n$ is said to *dominate* another estimator $\widetilde{\theta}_n$ if $R(\theta, \widehat{\theta}) \leq R(\theta, \widetilde{\theta})$ for all values of θ, and $R(\theta, \widehat{\theta}) < R(\theta, \widetilde{\theta})$ for at least one value of θ. In words, an estimator dominates if it has lower risk in some situations and never has higher risk than a competing estimator. An estimator that is not dominated by any other estimator is said to be *admissible*. Which of the estimators in parts (a)–(d) might be admissible? Which estimators are dominated by another estimator? (Note: to *prove* an estimator is admissible can be difficult. We will only consider whether our estimators are dominated by any of the other estimators displayed on the same plot, and only in the range shown.)

(f) The *minimax* estimator is the estimator that minimizes the maximum risk.[8] More formally, a minimax estimator $\widehat{\theta}$ has the property that $\max_\theta[R(\theta, \widehat{\theta})] < \max_\theta[R(\theta, \widetilde{\theta})]$ for any competing estimator $\widetilde{\theta}$. Which of the four estimators in parts (a)–(d) is a candidate minimax estimator? (As with admissibility, *proving* that an estimator is minimax can be difficult—one has to show that it is minimax with respect to all possible competing estimators.)

6.7 Robustness

All of these criteria for evaluating estimators—bias, variance, mean squared error, consistency, efficiency, and risk—assume that we know a lot about the process that generated the data. They let us ask questions like, "If I continue to observe independent random numbers from a normal distribution, will the median of those numbers converge to θ? How fast will the sample median converge to θ compared with the rate at which the sample mean converges to θ?" The answers to these questions are constructive.

---

[8] In Book 12 of the *Odyssey*, Circe urges a minimax decision. Odysseus must sail a narrow passage between two cliffs, each with a danger. On one cliff is Scylla, a six-headed, unstoppable monster who will surely kill six of Odysseus' comrades if they come into her range. On the other side is Charybdis, who sucks in water three times a day, creating a whirlpool that could consume the whole ship. The loss function says that if one tacks toward Scylla, one is sure to lose six sailors. Tacking toward Charybdis may lead to no loss but may also lead to catastrophe. Scylla's side is the minimax path—six sailors, the worst possible outcome on Scylla's side, is less than the whole ship, the worst outcome with Charybdis. Or as Circe puts it, "Ah, shun the horrid gulf! By Scylla fly. 'Tis better six to lose, than all to die."

But in many—even most—practical situations, we cannot be completely confident about the process that generated the data. We may, for example, write a model postulating a distribution family that fits the data poorly. We may assume that all the data are generated by one process when they are actually generated by two or more. When we are unsure about the processes we are studying, then we value robust estimators.

Loosely, a statistical procedure is **robust** if it continues to give sensible and approximately correct answers even when the assumptions that underlie it are incorrect in some way. We also sometimes say that a procedure is robust if it invokes relatively few assumptions. To be more precise about robustness, we have to answer the question, "robust against what?" For example, as we will see in an exercise below, the median is robust against **outliers**— occasional observations that reflect a process different from the one under study. We might also seek procedures that are robust to misspecification of distributional assumptions, like assumptions of normality.[9] But an estimator that is robust in one of these senses is not necessarily robust in the other. For example, we will see that the least-squares estimator of the linear model from the end of Chapter 5 has some degree of robustness against distributional assumptions—it is consistent under a wide range of assumptions about the distribution of the disturbance term. At the same time, it is not robust against outliers. A single outlying point can exert a large influence on the least-squares line, depending on where it is placed.

Two of the most important questions an applied statistician can ask are, "How plausible is it that my model correctly describes the phenomenon I want to understand?" and "If the model assumptions are wrong, will my procedure yield any kind of useful information?" In the following exercises, we will study robustness using simulations.

Exercise Set 6-8

(1) We again consider the problem of estimating θ using independent samples from a Normal$(\theta, 1)$ distribution. Suppose that the situation is modified slightly so that for each observation, there is a small probability that it is generated by a different distribution. A few examples: (i) You are studying beak length in a species of finch, but some of your samples may have come from a closely related species, difficult to distinguish reliably, that relies on different sources of food; (ii) You are studying memory for words in Alzheimer's disease, but some of the patients you study will turn out, on autopsy, to have a different form of dementia; (iii) You are studying radioactive emissions from samples of plutonium, but your Geiger counter is faulty and breaks down occasionally, leading to artificially low measurements of radioactivity from some samples.

For this problem, assume that we draw n independent observations. Independently for each observation, with probability $1 - \gamma$ it comes from a Normal$(\theta, 1)$ distribution, and with probability γ it comes from a "contaminating" Normal$(\Lambda, 1)$ distribution. Here is an R function called `rnorm.contam()` that can simulate data under these circumstances[10]:

```
#n is the number of observations in each sample, contam.p is
#the probability that each observation is from the non-target
#distribution, mu is the parameter to be estimated, sigma is the
#standard deviation of the target distribution, contam.mu is the
#expectation of the non-target distribution, and contam.sigma
```

---

[9] John Tukey, one of the founders of robust statistics, thought of contaminated distributions in part as a way of modeling distributions that are slightly different from the assumed one (Stigler, 2010).

[10] Available in the book's R package, `stfspack` (see Exercise Set 2-2, Problem 3 for installation instructions).

```
#is the standard deviation of the target distribution. The
#output is a vector of length n, with the outlying observations
#(if any) at the end.
rnorm.contam <- function(n, mu = 0, sigma = 1, contam.p = 0.01, contam.mu = -
5, contam.sigma = 0) {
  ncontam <- rbinom(1, n, contam.p)
  c(rnorm(n - ncontam, mu, sigma), rnorm(ncontam, contam.mu, contam.
sigma))
}
```

Use rnorm.contam() to study the sample mean and sample median as estimators of θ. You can set nsim = 1000 and n = 100. Try the following values of γ: 0.001, 0.01, 0.1, 0.2, and the following values of Λ: 3, 5, 10, 100. You can generate matrices of random values using mat.samps, setting the rx argument to rnorm.contam(), followed by the arguments to rnorm.contam(), separated by commas. For example, this produces simulated data under the first requested set of parameters:

```
dat <- mat.samps(n = 100, nsim = 1000, rx = rnorm.contam, contam.p = 0.001,
contam.mu = 3, contam.sigma = 1)
```

Observe the behavior of the sample mean and median using bias, variance, and histograms. What do you conclude?

6.8 Estimators for simple linear regression

Consider the simple linear regression framework from the end of Chapter 5, in which an outcome variable, Y, is modeled as a linear function of an independent variable, X, plus a disturbance term, ϵ. That is,

$$Y = \alpha + \beta X + \epsilon.$$

For example, in the running agricultural example, Y would represent cereal yield, and X would represent fertilizer use. We might collect data from n countries on cereal yield and fertilizer use, giving the paired observations $(X_1, Y_1), (X_2, Y_2), ..., (X_n, Y_n)$. Can we use these data to estimate the intercept and slope parameters, α and β?

We might consider using the intercept and slope of the least-squares line from Chapter 3 to estimate α and β. In many cases, this is a reasonable choice. But we also might consider other "best-fit" lines as estimators. For example, we might consider the line that minimizes the sum of the absolute line errors—that is, the sum of the vertical distances between the points and the line—rather than the sum of the squared line errors. With the probabilistic model developed in Chapter 5 and the concepts for evaluating estimators from this chapter, we can compare candidate estimators under different versions of the model, such as versions with different distributions for the disturbance term. In the next set of exercises, you will perform such comparisons for the least-squares and least-absolute-errors line.

Exercises in comparing specific candidate estimators under specific conditions are useful, and they also go some way toward answering some of our main questions about the least-squares line. Nonetheless, we will seek more general frameworks that guide us toward particular estimators—given an assumed model—rather than leaving us with an infinite set of candidates to consider. Some possible frameworks will appear in Chapters 8–10.

Exercise Set 6-9

(1) In this problem, we will explore the properties of the least-squares and least-absolute-errors lines as estimators under a version of the linear regression model from the end of Chapter 5. In particular, we will consider a version of the model in which the disturbance terms are normally distributed (and of constant variance). We will use the sim.lm() function (Exercise Set 5-5, Problem 2), along with sim.lm.ests() (defined below),[11] which calls sim.lm() many times and applies an estimator to each simulated dataset.

```
#Applies estimators of a and b to nsim data sets of size n
#pairs each generated by sim.lm(). Returns a matrix with two
#columns and nsim rows. Each row contains two estimates
#made from the same simulated dataset, a in the first
#column and b in the second.
sim.lm.ests <- function(n, nsim, a, b, sigma.disturb = 1, mu.x = 8, sigma.x
= 2, rdisturb = rnorm, rx = rnorm, het.coef = 0, estfun = lm) {
  ests <- matrix(nrow = nsim, ncol = 2)
  for(i in 1:nsim) {
    dat <- sim.lm(n, a, b, sigma.disturb, mu.x, sigma.x, rdisturb, rx, het.
coef)
    ests[i,] <- estfun(dat[,2] ~ dat[,1])$coef
  }
  ests
}
```

(a) Once sim.lm() and sim.lm.ests() are defined, you can examine mean and variance of least-squares estimates from 1,000 simulated samples of size 50, each with intercept 3 and slope 0.5, using

```
ests <- sim.lm.ests(n = 50, nsim = 1000, a = 3, b = 1/2)
colMeans(ests)
apply(ests, 2, var)
```

You can also examine histograms of the estimates; for example, hist(ests[,2]) shows the slope estimates. Simulate samples of size 10, 50, 100, and 1,000. Can you make any guesses about the bias of the least-squares estimators in the scenario being simulated? What about their consistency?

(b) An alternative estimator is the least-absolute-errors line, which minimizes the sum of the absolute values of the vertical distances between the line and the points on a scatterplot. One way to identify least-absolute-errors lines in R is to use the rq() function from the quantreg package (Koenker et al., 2017). Install (if necessary) and load the quantreg package (install.packages("quantreg") and then library(quantreg)). Now repeat part (a), this time examining the least-absolute-errors line, by changing the estfun argument of sim.lm.ests() to rq, as in

```
ests <- sim.lm.ests(n = 50, nsim = 1000, a = 3, b = 1/2, estfun = rq)
```

Do you think the least-absolute errors estimators are biased? Consistent?

(c) Which estimators are more efficient in this situation: the least-squares estimators or the least-absolute-errors estimators?

[11] Available in the book's R package, stfspack (see Exercise Set 2-2, Problem 3 for installation instructions).

(2) In Problem 1, you simulated data under the linear model from Chapter 5 with normally distributed disturbances. Another way to concretize the model from Chapter 5 is to assume that the disturbances have a Laplace distribution, which has heavier tails than a normal distribution. You can simulate data from a Laplace distribution with a specified mean and standard deviation using `rlaplace` (Exercise Set 6-5, Problem 2).

(a) Simulate and view some datasets with Laplace-distributed disturbances by setting the `rdisturb` argument of `sim.lm()` to rlaplace. For example,

```
plot(sim.lm(n = 50, a = 3, b = 0.5, rdisturb = rlaplace))
```

How would you describe the resulting scatterplots, compared with plots produced using normally distributed disturbances?

(b) Repeat Problem 1 using Laplace-distributed disturbances. (You can re-use your code from Problem 1, adding the `rdisturb = rlaplace` argument to every call of `sim.lm.ests()`.) Are the least-squares and least-absolute-errors lines biased? Consistent? Is their relative efficiency the same as in Problem 1?

(3) In this problem, you will examine the robustness of the least-squares and least-absolute errors estimators. In particular, you will explore their robustness to small amounts of data contamination. Use the `rnorm.contam()` function[12] (Exercise Set 6-8, Problem 1).

(a) Simulate and view some datasets with disturbances drawn from a distribution contaminated with outlying points. Because of the way `sim.lm()` is written, the points with disturbance terms drawn from the contaminating distribution will be the ones with the largest x values. To see one plot, use

```
plot(sim.lm(n = 100, a = 3, b = 0.5, rdist = rnorm.contam))
```

This draws a sample of size 100, and since the probability of contamination is only 1% per point, you may see no outlying points in your first simulation. Increase the sample size n and/or run the command a few times. How would you describe the resulting scatterplots?

(b) Repeat Problem 1 using disturbances that are 1% contaminated. (You can re-use your code from Problem 1, adding the `rdist = rnorm.contam` argument to every call of `sim.lm.ests()`.) Are the least-squares and least-absolute-errors lines biased? Consistent? Is their relative efficiency the same as in Problem 1?

(4) [Optional, but please check the solution even if you don't work through it] In this problem, you will explore confounding, sometimes called omitted variable bias. Imagine that we observe paired observations that we believe are described by the model

$$Y = \alpha + \beta X + \epsilon.$$

Unbeknownst to us, the model is incorrect. Y is in fact a linear function of X, ϵ, and another random variable, Z:

$$Y = \alpha + \beta X + \gamma Z + \epsilon.$$

We don't know about Z, so we use least squares to estimate α and β while ignoring Z.

(a) How do you predict that ignoring Z will affect our estimates of β? Consider separately the cases in which X and Z are positively correlated, negatively correlated, and independent. Assume that γ is positive.

(b) To test your conjectures, generate samples independently according to the model $Y = \alpha + \beta X + \gamma Z + \epsilon$. X and Z are assumed to be jointly normally distributed with $E(X) = E(Z) = 1$, $Var(X) = Var(Z) = 1$, and $Cor(X, Z) = \rho$. The disturbance ϵ is assumed to be normally distributed with $E(\epsilon) = 0$. The function we will use requires the package MASS, which can be installed and loaded with the following two lines of code:

---

[12] Available in the book's R package, `stfspack` (see Exercise Set 2-2, Problem 3 for installation instructions).

```
> install.packages("MASS")
> library(MASS)
```

The function `sim.2var()` below[13] simulates data under the model with X and Z included. The function parameter n controls the number of pairs of observations in each simulated sample, nsims is the number of simulated samples to draw, a is α, b1 is β, b2 is γ, sigma.disturb is the standard deviation of ϵ, and correl is ρ, the correlation of X and Z.

```
sim.2var <- function(n, nsim, a, b1, b2 = 0, sigma.disturb = 1, correl = 0) {
    sig <- matrix(c(1, correl, correl, 1), nrow = 2)
    ivs <- MASS::mvrnorm(n*nsim, mu = c(0,0), sig)
    x <- ivs[,1]
    z <- ivs[,2]
    disturb <- rnorm(n*nsim, 0, sigma.disturb)
    y <- a + b1 * x + b2 * z + disturb
    xmat <- matrix(x, nrow = nsim)
    ymat <- matrix(y, nrow = nsim)
    list(xmat, ymat)
}
```

The output of the function is a list, the first element of which is a matrix of x values (one simulation per row), and the second a matrix of y values. To simulate and analyze 1,000 samples of 50 pairs of observations each, with true $\alpha = 0$, $\beta = 0.3$, $\gamma = 0.4$, $SD(\epsilon) = 1$, and $\rho = 0.5$, we could use

```
#Choose parameters and simulate
n <- 50
nsims = 1000
beta <- .3
gamma <- .4
rho <- .5
dat <- sim.2var(n, nsims, 0, beta, gamma, 1, rho)
ests <- numeric(nsims)
for(i in 1:nsims) {
    ests[i] <- lm(dat[[2]][i,] ~ dat[[1]][i,])$coef[2]
}
hist(ests)
summary(ests)
```

You may not be familiar with the list syntax used inside the `for()` loop if you have not read Appendix B, but it needn't concern you if you are not interested. At the end, the ests object holds least-squares-estimated slopes from 1,000 simulations.

Use these functions to test your conjectures from part (a). Using the simulations and keeping $\alpha = 0$, posit a mathematical expression that relates $\mathrm{E}(\tilde{\beta})$ to β, γ, and ρ.

(c) How do the results of (a) and (b) affect your interpretation of the least-squares estimator? Consider two cases: (i) suppose you want to predict Y from X, and (ii) suppose you want to manipulate X to change Y. Will your estimate of β (that ignores the omitted variable Z) be useful in either case?

---

[13] Available in the book's R package, stfspack (see Exercise Set 2-2, Problem 3 for installation instructions).

6.9 Conclusion

We now have many concepts that help us evaluate candidate estimators. Other things equal, we seek estimators that are accurate and precise and that converge to the correct answer with a large enough sample. We also attend to the specifics of the practical situation in which we find ourselves, choosing estimators that help us avoid the costliest kinds of errors. At all times, we are alert to the possibility that our assumptions are incorrect, and, whenever possible, we choose procedures that will give us valid information even if our assumptions are wrong.

There are two outstanding issues. First, though we now have ways of comparing candidate estimators against each other, we have said nothing about how to come up with candidate estimators. Three frameworks for identifying estimators will be presented in Chapters 8, 9, and 10. Second, we have focused strictly on **point estimation**, the practice of identifying a single number to serve as a "best guess" for an unknown value. But statistics is about making claims under uncertainty, and one might guess that point estimates could be even more effective when paired with descriptions of the degree to which we are uncertain about them. That would be correct. The subjects of the next chapter, interval estimation and hypothesis testing, are frameworks for describing uncertainty about estimates.

6.10 Chapter summary

Point estimation is the attempt to identify a value associated with some underlying process or population using data. The unknown number that is the target of estimation is called an *estimand*. An *estimator* is a function that takes in data and produces an *estimate*. We evaluate estimators according to a number of criteria. An unbiased estimator is one whose expected value is equal to the estimand—in lay terms, it is accurate. We also value low-variance estimators, which are precise. Consistent estimators converge to the estimand as the number of data collected approaches infinity. Mean squared error is the expected squared difference between the estimator and the estimand. Efficient estimators are those that converge to the estimand relatively quickly—i.e., fewer data are necessary to get close to the right answer. A robust estimator is one that can still provide useful information even if the model is not quite right or the data are contaminated.

6.11 Further reading

The material in this chapter (and the next few) is covered in more depth and with more mathematical rigor by a number of mathematical statistics texts. Three standard references are:

Bickel, P. J., & Doksum, K. A. (2007). *Mathematical Statistics: Basic Ideas and Selected Topics*, Volume I, 2nd Edition. Pearson Prentice Hall, Upper Saddle River, NJ.
Casella, G., & Berger, R. L. (2002). *Statistical Inference*. Duxbury Press, Belmont, CA.
Hogg, R. V., McKean, J., & Craig, A. T. (2005). *Introduction to Mathematical Statistics*. Pearson Education, Harlow, Essex.

The above three books are all thorough and fairly large references. The following are two books that are more concise and cover similar material at a mathematical level between this book and the above three:

Wasserman, L. (2013). *All of Statistics: A Concise Course in Statistical Inference*, 2nd Edition. Springer, New York.

Young, G. A., & Smith, R. L. (2005). *Essentials of Statistical Inference*. Cambridge University Press.

I use the Wasserman book often. Silvey's (1975) *Statistical Inference* is a concise reference.

Finally, two relevant books that are not introductory mathematical statistics texts:

Good, P. I., & Hardin, J. W. (2012). *Common Errors in Statistics (and How to Avoid Them)*. Wiley, Hoboken, NJ.
Has an excellent chapter on estimation.

Wilcox, R. R. (2011). *Introduction to Robust Estimation and Hypothesis Testing*. Academic Press, Cambridge, MA.
An introduction to robust statistics written at a mathematical level similar to this book.

CHAPTER 7

Interval estimation and inference

> **Key terms**: Confidence interval, Coverage probability, Effect size, False discovery rate, Hypothesis test, Interval estimation, Level, Null hypothesis, *p* hacking, *p* value, Power, Power function, Publication bias, Significance (statistical), Size, Standard error, Test statistic, Type I error, Type II error, Winner's curse

When engaging in *point* estimation, statisticians use data to guess the value of an unknown quantity. Once that guess is identified, **interval estimation** and inference—the subjects of this chapter—raise their heads. Interval estimation is the attempt to answer an interlocutor's first question after being presented with a point estimate: "How sure are you about that?" Loosely speaking, whereas point estimation is about arriving at a best guess for the value of an estimand, given the data, interval estimation is about identifying a *set* of possible values of the estimand that are consistent with the data. Inference, also called hypothesis testing, is another, related way to assess the degree to which a candidate value of an estimand—or, more generally, a candidate explanation—is compatible with data.

In this chapter, we put aside the problem of finding methods for identifying interval estimates and for conducting hypothesis tests. Our goal is to understand how interval estimates and hypothesis tests are to be interpreted. This chapter focuses on frequentist approaches; some Bayesian alternatives will be introduced in Chapter 10.

7.1 Standard error

We have already encountered one answer to the question "How sure are you about that estimate?" Faced with such a question, we can supply a number describing the spread of the estimator's distribution. In the previous chapter, we considered the variance of an estimator as a criterion for comparing estimators with each other—other things equal, estimators with lower variance are usually preferred. But the variance of the estimator is also a measurement of the spread of the estimator's distribution, and thus again to the purpose.

As discussed in Chapter 5, the variance—despite its other admirable traits—is in the wrong units. The fix is, once again, to take the square root of the variance, giving the standard deviation of the estimator. If an estimator is written $\widehat{\theta}_n$, then the quantity

$$\mathrm{SE}(\widehat{\theta}_n) = \sqrt{\mathrm{Var}(\widehat{\theta}_n)} \tag{7.1}$$

is what we want, where $\mathrm{Var}(\widehat{\theta}_n)$ is defined in Equation 6.3. This is the standard deviation of the estimator, but we adopt a new name, which is somewhat confusing at first but useful

Statistical Thinking from Scratch: A Primer for Scientists. M. D. Edge, Oxford University Press (2019).
© M. D. Edge 2019. DOI: 10.1093/oso/9780198827627.001.0001

once accepted. The quantity in Equation 7.1 is called the **standard error** rather than the standard deviation. The rule is this: the standard error measures the spread of the distribution of an *estimator's* sampling distribution, whereas we reserve "standard deviation" to describe the spread in the distribution of *data*. The two terms refer to mathematically analogous objects—measures of distributional spread—but they are applied to the distributions of different quantities. This distinction will be reinforced in an exercise.

We do not yet know how to estimate the standard error in general. Bootstrapping, developed in the next chapter, provides a general approach, and another method applicable in many situations will come in Chapter 9. In the exercises, we will calculate the standard error of the expectation of a normally distributed random variable when the standard deviation is known. When the standard deviation is unknown, it can be estimated and plugged into the same procedures you will use below.

Exercise Set 7-1[1]

(1) You plan to draw a sample of n independent observations $X_1, X_2, ..., X_n$ from a Normal(θ, σ^2) distribution and estimate θ using the sample mean, $\hat{\theta}_n = \frac{1}{n}\sum_{i=1}^{n}X_i$. The parameter σ^2 is known.
   (a) What is the standard deviation of the observations?
   (b) What is the standard error of the estimator, SE$(\hat{\theta}_n)$?
   (c) Suppose that $n = 25$ and $\sigma^2 = 1$. What is SE$(\hat{\theta}_n)$?
   (d) Suppose that in the situation in part (c), you decide to use the sample median as an estimator of θ, which you call $\tilde{\theta}_n$. What, approximately, is SE$(\tilde{\theta}_n)$? (Use simulation.)
(2) Again imagine that you plan to draw a sample of n independent observations from a Normal(θ, σ^2) distribution and estimate θ using the sample mean, $\hat{\theta}_n = \frac{1}{n}\sum_{i=1}^{n}X_i$. Suppose that the standard error of $\hat{\theta}_n$ is known to be SE$(\hat{\theta}_n) = \omega$. Consider the interval $(\hat{\theta}_n - \omega, \hat{\theta}_n + \omega)$. (That is, the interval has $\hat{\theta}_n - \omega$ as its minimum value and $\hat{\theta}_n + \omega$ as its maximum value.) This interval is random because its central point, $\hat{\theta}_n$, is a random variable.
   (a) If you repeatedly drew independent samples according to these settings, what proportion of the time would the interval $(\hat{\theta}_n - \omega, \hat{\theta}_n + \omega)$ contain θ? That is, how often would it be true that $\hat{\theta}_n - \omega < \theta < \hat{\theta}_n + \omega$, or, equivalently, that $|\hat{\theta}_n - \theta| < \omega$?
   (b) What is the probability that the interval $(\hat{\theta}_n - 2\omega, \hat{\theta}_n + 2\omega)$ contains θ?

7.2 Confidence intervals

The most popular method of frequentist interval estimation is the confidence interval. If you completed the last set of exercises, then you have already encountered a confidence interval.

Confidence intervals are written (V_1, V_2), where V_1 is the lower bound of the interval and V_2 is the upper bound of the interval. V_1 and V_2 are random variables. To refer to an instance of a confidence interval that has already been calculated from data, we write (v_1, v_2).

To form a $1 - \alpha$ **confidence interval** for an estimand θ, V_1 and V_2 are chosen to ensure that

$$P(V_1 < \theta < V_2) \geq 1 - \alpha. \tag{7.2}$$

In words, Equation 7.2 specifies that the probability that θ is contained in the interval ranging from V_1 to V_2 is at least $1 - \alpha$. This is the most intuitive way to translate

---

[1] See the book's github repository at github.com/mdedge/stfs/ for detailed solutions and R scripts.

Equation 7.2 into English, but it suggests a misinterpretation. In frequentist statistics, θ is a fixed number. Fixed numbers display no random behavior—they simply are equal to themselves with probability one. Instead, it is V_1 and V_2 that are random, and the probability statement is about V_1 and V_2, not θ. Thus, a better English rendering of Equation 7.2 is that $1 - \alpha$ is the probability that the interval ranging from V_1 to V_2 contains—or *covers* —θ. The value $1 - \alpha$ is often called a **coverage probability**. Another equivalent statement of Equation 7.2 is

$$P(V_1 < \theta \cap V_2 > \theta) \geq 1 - \alpha.$$

This second statement is harder to read but easier to interpret correctly: $1 - \alpha$ bounds the probability that *both* of two random events happen—namely, that the lower bound of the confidence interval falls below θ *and* the upper bound of the confidence interval falls above θ. Figure 7-1 shows fifty different 90% confidence intervals—that is, $1 - \alpha$ confidence intervals with $\alpha = 0.1$—computed from independent samples.

In the previous set of exercises, you encountered one way of constructing a confidence interval. Suppose one has an unbiased point estimator that is normally distributed with a known standard error. Then one can construct a confidence interval by setting the lower bound some number of standard errors below the estimate and the upper bound some number of standard errors above the estimate. For a normally distributed, unbiased estimator $\hat{\theta}$ with standard error ω, one[2] possible $1 - \alpha$ confidence interval is

$$(\hat{\theta} - \omega z_{\alpha/2}, \hat{\theta} + \omega z_{\alpha/2}), \tag{7.3}$$

where $z_{\alpha/2} = \Phi^{-1}(1 - \alpha/2)$ and Φ^{-1} is the inverse of the cumulative distribution function of the Normal$(0, 1)$ distribution.[3] In words, $z_{\alpha/2}$ is chosen so that the probability that an

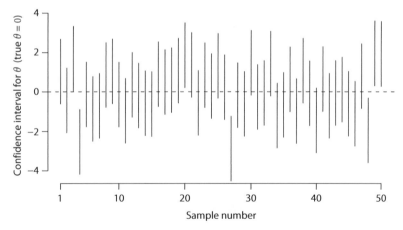

Figure 7-1 Fifty independent 90% confidence intervals for an estimand, θ. Here, $\theta = 0$, represented by the dashed horizontal line. The confidence intervals are represented by vertical lines. Approximately 90% of the vertical lines cover the true value of $\theta = 0$. (In this simulation, 43 of 50 intervals cover θ, but in the long run, the success rate would converge to 90%.)

---

[2] This confidence interval is symmetric around the estimated value of the parameter. If symmetry around the estimated value is not required, then there are many (actually, an infinite number of) confidence intervals at a given coverage level in this setting. But this one is by far the most commonly used.

[3] You can access the inverse of the cumulative distribution function for the normal distribution using the qnorm() function in R. For example, qnorm(0.95, 0, 1) reveals that to build a 90% confidence interval, one

observation drawn from a normal distribution falls within $z_{\alpha/2}$ standard deviations of the mean is $1 - \alpha$. You will learn a more general approach for constructing confidence intervals in the next chapter.

As with all statistical methods, the attractive properties of confidence intervals are only guaranteed if the assumptions that the method entails are satisfied. But even if the necessary assumptions are met, interpreting confidence intervals is a subtle game.[4] One correct interpretation is that $1 - \alpha$ is the probability that a future confidence interval—for example, one based on a sample that has not yet been collected—will contain the true value of the estimand. But what about a confidence interval that we have *already* calculated? Can we say that the probability that the estimand is inside the interval is $1 - \alpha$? We cannot. It is worth dwelling on this puzzle: Why can we make probability claims about future confidence intervals yet be unable to make probability claims about confidence intervals we have already calculated?

One correct response is to note that in frequentist statistics—of which confidence intervals are a part—estimands are treated as fixed numbers, even though they are unknown. The upper and lower bounds of a confidence interval that has already been calculated are also fixed quantities, v_1 and v_2. For a frequentist, then, the statement $P(v_1 < \theta < v_2)$ makes little sense because it involves no random variables. One either has a statement like $P(1 < 2 < 3)$, which must equal one, or a statement like $P(1 < 4 < 3)$, which must equal 0. We don't know which we have because θ is unknown, but in either case, it doesn't make sense to assign a probability other than 0 or 1.

The response is legitimate. Confidence intervals are creatures of frequentist statistics, and, without further argument, their properties cannot be extended to situations that frequentists do not consider. Nonetheless, there is something unsatisfying about the response, which seems to depend on a technicality. If you are not satisfied, you can press the case further in Box 7-1 by altering—for a moment—your interpretation of probability.

Box 7-1 A THOUGHT EXPERIMENT ABOUT CONFIDENCE INTERVALS

The probability statement $P(v_1 < \theta < v_2)$ does not make sense in frequentist statistics—it contains no random terms—and so already-computed confidence intervals do not imply probability statements about the unknown estimand. But suppose we decide to interpret probabilities as degrees of belief that a rational person might hold. If we interpret probability in this way, then we might apply probability statements to quantities that are unknown or uncertain, like the estimand θ, even if we believe that they are "really" fixed numbers. Under this framework, $P(v_1 < \theta < v_2)$ could be interpreted as the degree to which a person believes that θ falls between v_1 and v_2, and we might quantify a person's value of $P(v_1 < \theta < v_2)$ by asking her to make a wager. For example, if a person will accept an even wager on the side of θ falling between v_1 and v_2, we could say—assuming that she will not accept wagers she believes to be disadvantageous—that her value of $P(v_1 < \theta < v_2)$ is at least 1/2. This "probability" statement is not a probability in the frequentist sense, so let's call it probability-W: a probability-like quantity inferred from the acceptance of a wager.

(Continued)

---

should set $z_{\alpha/2} \approx 1.64$. Similarly, for a 95% confidence interval—perhaps the most commonly used confidence level—one sets $z_{\alpha/2} \approx 1.96$, and for a 99% confidence interval, $z_{\alpha/2} \approx 2.58$.

[4] In one study of thousands of psychologists and psychology students, over 95% endorsed at least one incorrect interpretation of confidence intervals (Hoekstra et al., 2014).

Box 7-1 (Continued)

Thus, even frequentists, though unwilling to make probability statements about esti-mands, might in principle make probability-W statements about them. Now we can ask a new version of our question about confidence intervals. Can we interpret a confidence interval that has *already* been calculated as a probability-W statement about the estimand? If so, then if we have a valid 90% confidence interval for an estimand θ, for example, then it should be reasonable to accept any wager with odds more favorable than 1:9 that θ falls within the interval. It would certainly be reasonable to take such a wager about a *future* confidence interval. Suppose I repeat a study many times, computing a 90% confidence interval for an estimand each time. Each time, I offer you a wager: if the interval contains the estimand, I give you $10; if it does not contain the estimand, you give me $50. In the long run, you will win $90 ($10 on 9 trials) for every $50 that I win, and—assuming you are not bankrupted by an early, unlucky streak—your fortune will grow without limit.

So probability-W works for future confidence intervals, but does it work for past intervals? Consider the confidence intervals shown in Figure 7-1 again. If a confidence interval is a valid probability-W statement, then you should be willing to accept the same wager—$10 winnings if the estimand is covered; $50 loss if not—for each and every interval in the sequence, for they are all valid 90% confidence intervals. This means that you should accept the wager not just for the entire set of intervals, and not just for a subset that I agree to choose randomly—either of those moves would be defensible here—but for *any* specific confidence interval that an opposing gambler could pick. Now you are in trouble.

Imagine that you and I do not know that $\theta = 0$ but that we do have access to all the 90% confidence intervals shown in the figure. But suppose I ask you about the final confidence interval from the sequence of 50 shown in the figure. Of the 50 confidence intervals drawn, it is one of the most extreme. Given the full sequence of 50 intervals shown in Figure 7-1, would you accept the same wager, paying $50 if the estimand is not covered and receiving $10 if it is? That would be a questionable move—given the sequence of intervals, it would be reasonable to guess that the estimand is somewhere in the center of the cluster they form, and the last interval is not close to the center. Further, if you are willing to take such a bet on *any* confidence interval in the sequence, then an opposing gambler with access to the whole sequence will find it easy to fleece you. For example, for the sequence shown in Figure 7-1, one could bet against both the last interval in the sequence and the third-to-last interval in the sequence. These two intervals do not overlap and so cannot both include the estimand. The challenger would be guaranteed to win at least one of the bets, and because she faces a $50 payoff for a win and only a $10 penalty for a loss, she would be guaranteed a profit.

The upshot of this thought experiment is this: by interpreting probability as a degree of belief, one can make statements like $P(v_1 < \theta < v_2)$ sensible. But if one makes this move, then $P(v_1 < \theta < v_2)$—the probability that a confidence interval that has already been calculated contains θ—does not necessarily equal $P(V_1 < \theta < V_2)$, the probability that a confidence interval that we have not yet calculated will contain θ. The critical point is that we can use outside information—in this case, the whole set of confidence intervals, but in principle, any kind of outside information—to assess whether a specific interval (v_1, v_2) is plausible as an interval estimate of θ, and this outside information should sometimes affect our beliefs. Bayesians and frequentists agree on this and differ only in their response. Frequentists address the problem by refusing to make probability claims about estimands, whereas Bayesians attempt to incorporate related evidence. This is not an argument for or against frequentist or Bayesian approaches; it is a clarification of their distinct goals.

Confidence intervals are the primary tool for frequentist interval estimation. As you will see in the exercises and the next section, confidence intervals are connected to p values, a primary tool for frequentist hypothesis testing.

Exercise Set 7-2

(1) In this problem, you will explore a 50% confidence interval with a lower bound of 2 and an upper bound of 4 for the number of days by which an intervention decreases the duration of chickenpox symptoms. Assume that the estimator of the number of days by which the intervention decreased symptoms is normally distributed, unbiased, and has a known standard error. Assume further that the confidence interval was constructed according to the method outlined in Equation 7.3.

    (a) What is $z_{\alpha/2}$ for $\alpha = 0.5$? (Use qnorm() or a table of quantiles for the standard normal distribution.) Draw a probability density function of a normal distribution with expectation 0 and variance 1 with the area under the curve from $-z_{\alpha/2}$ to $+z_{\alpha/2}$ shaded. What is the area of the shaded region?

    (b) What was the value of the instance of $\hat{\theta}$ considered in the example?

    (c) What is the standard error of $\hat{\theta}$?

    (d) Give a 95% confidence interval for the number of days by which the intervention decreases the duration of chickenpox symptoms

    (e) How many standard errors from 0 is $\hat{\theta}$?

    (f) For what value of α is the lower bound of the confidence interval on $\hat{\theta}$ equal to zero? (Hint: In R, pnorm() gives the cumulative distribution function of the Normal(0, 1) distribution, or Φ.)

    (g) Assuming that $\theta = 0$, what is the probability of obtaining an estimate $\hat{\theta}$ such that $|\hat{\theta}|$ is larger than the estimate considered in this example (part b)? Note the absolute value stipulation in the question.

7.3 Frequentist inference I: null hypotheses, test statistics, and p values

We now come to inference, a pursuit that apparently has little to do with estimation. In fact, inference and interval estimation are closely connected, but it will take some time to see how. In the meantime, we start fresh.

Studies are often conducted because a researcher would like to test a particular hypothesis. For example, one might like to test whether a medication eases symptoms, whether an evolutionary model describes data from a family of organisms, or whether an advertisement induces people to buy a product. Statistical inference—along with more general principles of study design—serves these goals.

Here is a sketch of the whole procedure. The hypothesis subjected to statistical testing is called the **null**[5] **hypothesis**, often written H_0. We test the null hypothesis by comparing the value of a statistic we compute from data—called a **test statistic**[6]—

---

[5] "Null" is sometimes taken to mean that the hypothesis asserts that some parameter is equal to 0—perhaps there is no difference between two groups or there is no association between two variables. Though null hypotheses often take this form, they do not need to. For example, one could test a null hypothesis that a difference in group means is equal to 3, or that a correlation coefficient falls between –0.2 and 0.4. Do not assume that the hypothesis to be *nullified* must be a hypothesis of *nil* difference.

[6] The test statistics we consider in this book are generally summaries of the data, but the test statistic could, in principle, be the data set itself.

with the distribution that the test statistic would have if the null hypothesis were true. The comparison of the test statistic and a null hypothesis provides a *p* **value**, or just *p*. Roughly, *p* answers the question, "Assuming that the null hypothesis is true, what is the probability that the test statistic would differ as extremely or more extremely from null expectations than it actually did?" What "more extremely" means may depend on how the hypothesis test is constructed. Small values of *p* indicate that the test statistic would be surprising if the null hypothesis were true. If the test statistic is improbable under the null hypothesis, then we have some evidence against the null hypothesis.

To construct a **hypothesis test**, we need a null hypothesis and a test statistic, and we need to know the distribution the test statistic would have if the null hypothesis were true.

Here is a contrived example. Suppose that, drawing inspiration from Bergmann's rule in ecology, I have a theory that shorter people will be more attracted to hotter climates because smaller bodies shed heat more effectively. To test this theory, I propose a study. I will measure the heights of a sample randomly drawn from the population of adult women from Arizona. If the Arizonan women are shorter than women from the United States overall, then I will take my theory to be supported.

The study is a poor test of the theory. In an exercise below, you will give some reasons why. But suppose we carry it out anyway. Assume that we know before the study starts that across the United States, heights of women are approximately normally distributed with a mean of 64 inches and a standard deviation of 4 inches.[7] That is, we assume that the heights of adult women are well described by a Normal$(64, 4^2)$ distribution, where 64 is the expected height in inches and $4^2 = 16$ is the variance in squared inches, which implies that the standard deviation is 4 inches. We sample 25 women from Arizona, measure their heights, and calculate the mean height in the sample. It is 63 inches.

The mean height of the Arizonan women in the sample, 63 inches, is less than the mean height in the United States. Should I celebrate the vindication of my theory? Not yet, for many reasons. One reason: In principle, it is possible to draw a sample with a mean height of 63 inches by chance even if the mean height of women in Arizona is 64 inches. That is, it could be that if I measured the heights of all the women in Arizona, I would find that it is exactly 64, but that I happened to draw a sample with a lower height than average by chance. What is the probability that *even if Arizonan women are just as tall as women everywhere else*, I would draw a random sample of 25 women with a mean height one inch (or more than one inch) lower than average?

To answer this question, start by assuming, in contrast to the theory, that Arizonan women are indistinguishable from women everywhere else in the United States in terms of height—their mean height is 64 inches, the standard deviation is 4 inches, and their heights follow a normal distribution. Recall that if $X_1, X_2, ..., X_n$ are independent random variables from a Normal(μ, σ^2) distribution, then the sample mean $\overline{X} = \sum_{i=1}^{n} X_i/n$ has a Normal$(\mu, \sigma^2/n)$ distribution.[8] Thus, in our case, if Arizonan women were exactly as tall as women elsewhere in the United States, then the mean of a sample of 25 women would have a Normal$(64, 4^2/25)$ distribution. The probability of drawing an observation of 63 or smaller from such a distribution can be looked up using R's pnorm() function. R uses the

---

[7] These assumptions are fabricated but approximately correct (McDowell et al., 2008).

[8] That the sample mean has expectation μ and variance σ^2/n follows from Equations 5.4 and 5.8. That the distribution is *approximately* normal follows from the central limit theorem as long as *n* is large. Showing that the distribution of the sample mean is *exactly* normal when the observations are normal requires methods beyond our scope.

standard deviation instead of the variance as the second parameter when referencing the normal distribution, so the command requires setting the standard deviation to $\sqrt{4^2/25} = 4/5$,

```
> pnorm(63, mean = 64, sd = 4/5)
[1] 0.1056498
```

The probability of observing a sample mean of 63 or less is greater than 10%, even if Arizonan women are exactly as tall as women from other parts of the United States. Thus, the result of the study provides little support for the claim that Arizonan women are shorter than average women from the United States—the data can be accommodated easily without resorting to the hypothesis that Arizonan women are shorter on average.

We have just conducted a hypothesis test. I obscured some technical components to make the reasoning easier to follow, but it is worth reviewing our procedure more thoroughly. Our null hypothesis was

H_0: mean height of women from Arizona is equal to the mean height of women from across the United States.

If the mean height in inches of women from Arizona is written θ, then the null hypothesis can be more succinctly written as

$$H_0: \theta = 64.^9$$

As a test statistic, we used the sample mean, $\bar{x} = 63$. To identify the distribution of the test statistic under the null hypothesis, we assumed that the heights of Arizonan women were normally distributed with standard deviation 4. We also assumed that the 25 women in our sample were independent, randomly selected women from the population of Arizonan adult women—that is, we assumed that every adult woman in Arizona had an equal probability of being part of our sample, irrespective of who else was included in the sample. If these assumptions hold, then the mean of a sample of 25 has a Normal$(64, 16/25)$ distribution. Finally, to arrive at p, we calculated the probability that a random variable with a Normal$(64, 16/25)$ distribution would take a value smaller than 63.

A few comments are in order. First, we conducted what is called a "one-sided" hypothesis test—that is, we considered the probability that if the null hypothesis were true, the test statistic would differ from the null expectation by more than we observed in *one direction*. Formally, we considered $p = P(\bar{X} \leq 63) = P(\bar{X} - 64 \leq -1)$. To conduct a two-sided test, we would consider the probability that the test statistic departed from the null expectation in either direction, or $P(|\bar{X} - 64| \geq 1)$. Figure 7-2 illustrates the difference. In the one-tailed test we conducted, p is equal to the area shaded in black to the left of the expectation under the null hypothesis. In a two-tailed test, p is equal to the sum of the areas of the black and gray shaded regions, or

```
> pnorm(63, mean = 64, sd = 4/5) + (1 - pnorm(65, mean = 64, sd = 4/5))
[1] 0.2112995
```

---

[9] This notation is really a shorthand. The hypothesis under test is actually the *conjunction* of this parameter value and the postulate that the heights are independent samples from a normal distribution with standard deviation 4, as discussed below.

It is customary to report two-tailed rather than one-tailed p values. Two-tailed p values are more conservative, and it is possible to cheat with one-tailed tests. For example, suppose I had obtained a sample mean of 65.5 instead of 63. If I were to stick with my one-sided research hypothesis that Arizonan women are shorter than other women from the United States, then I would compute p as

```
> pnorm(65.5, mean = 64, sd = 4/5)
[1] 0.9696036
```

This is no evidence in support of my hypothesis. But suppose I decide, after seeing the result, that really it would have been more reasonable to predict that Arizonan women are taller on average than women from other parts of the United States. If I conduct a one-sided test in the other direction, then p is

```
> 1 - pnorm(65.5, mean = 64, sd = 4/5)
[1] 0.03039636
```

This new p might suggest that my theory has scored a modest success, when in fact my theory has been jerry-rigged to fit facts I already know—this p value is invalid. Of course you, Dear Reader, would never do this. But using two-tailed tests will ease everyone's mind.[10] Better yet, preregister your hypothesis (Munafò et al., 2017), and everyone wins—you get

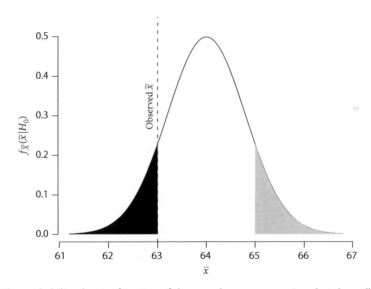

Figure 7-2 The probability density function of the sample mean assuming that the null hypothesis is true. The observed sample mean ($\bar{x} = 63$) is represented with a dashed vertical line. For a one-tailed test sensitive to the possibility that the expected height is less than 64, p is equal to the area of the black shaded region (approximately 0.106). For a two-tailed test, p is equal to the sum of the areas of both shaded regions (approximately 0.211).

---

[10] There are also good reasons for using one-tailed tests, and there is no reason to back away from one-tailed tests in such cases. For example, you may only be interested in whether a medication is more effective than the current standard treatment, not whether it is less effective.

to use a one-tailed test, and your audience knows that you didn't change your plans after seeing the data.

Second, notice that p is *not* the probability that the null hypothesis is true given the data. Instead, p is the probability of observing data "like" those we observed *assuming* that the null hypothesis is true. More precisely, let D represent the event that we observe data more extremely discordant from the expectation under the null hypothesis than we actually did. The two-sided p is $P(|\overline{X} - 64| \geq 1 | \theta = 64)$, or $P(D|H_0)$.[11] Many researchers interpret p in ways that suggest that they have reversed the order of conditioning, assuming that $P(H_0|D)$, the probability of the null hypothesis given the data, is approximately equal to $P(D|H_0)$, or p. Recalling Bayes' theorem (Equation 4.2), the actual relationship between these quantities is, supposing that all the necessary terms can be defined,

$$P(H_0|D) = \frac{P(D|H_0)P(H_0)}{P(D)}.$$

In frequentist statistics, $P(H_0)$ and $P(D)$ are undefined, making $P(H_0|D)$ unattainable. Bayesian statisticians do consider $P(H_0|D)$ by formally defining $P(H_0)$ and $P(D)$. As we saw in Section 4.3, $P(H_0|D)$ and $P(D|H_0)$ are related, but they need not have similar values. The temptation to misunderstand p as a probability statement about a hypothesis rather than about data is parallel to the temptation to misunderstand a confidence interval as a probabilistic claim about the value of an unknown quantity.

Like all procedures for estimation and inference, p values are valid only if the assumptions upon which they rely hold. Paraphrasing David Freedman (2009, Section 5.7), there are three possibilities for explaining a low p value: (i) an unlikely event occurred; or (ii) the model is wrong; or (iii) the model is right and the parameter values differ from what the null hypothesis asserts. So?

The p value is a summary of the degree of evidence against the null hypothesis. In the example, the null hypothesis was that the mean height of adult women from Arizona is the same as the mean height of adult women from other parts of the United States. But in testing this hypothesis, we assumed not just that the mean heights of Arizonan women are the same as women from other places, but also that Arizonan women's heights are normally distributed with a known standard deviation[12] and that the sample we drew was representative of Arizonan women. Our hypothesis test is actually a test of the conjunction of all these assumptions, not just of the equality of means—in other words, the failure of any of these assumptions could make low p values more probable.

Finally, we have argued that, other things being equal, smaller values of p are stronger evidence against the null hypothesis. We have not said, however, how to calibrate judgments about any particular value of p. Should we take the null hypothesis as falsified if $p < 0.1$, $p < 0.01$, or $p < 0.001$? Should we use a fixed standard for making judgments? We will consider these questions in the next section.

---

[11] Using the notation $P(D|H_0)$ to mean "the probability of D assuming the null hypothesis" may strike you as strange. Everything else we have conditioned on thus far—that is, everything that has appeared to the right of a | in a probability expression—has been a random variable. In a frequentist framework, H_0 is not random, and this is not "really" a conditional probability; it's just a probability computed under a particular set of assumptions. Some writers use $P(D; H_0)$ to describe this situation, but I will continue with the $P(D|H_0)$-style notation.

[12] We actually assumed that the *mean* of a sample of 25 Arizonan women is normally distributed, which might be approximately true even if the women's heights themselves are not quite normally distributed.

Exercise Set 7-3

(1) In the example, we considered a theory that smaller people will be attracted to hotter climates because smaller bodies shed heat more efficiently. To test this theory, I proposed testing a hypothesis, derived from the theory, that Arizonan women would be shorter than women from other parts of the United States. Give at least three reasons why the hypothesis test I proposed is a poor test of the theory. That is, give three reasons why the outcome of the hypothesis test is indecisive about the truth of the theory. (Take for granted that the assumptions we made about the distribution of women's heights are valid.)

(2) You work in a factory that makes nails. On one of the production lines, if everything works correctly, the nails made are 100 millimeters (mm) long on average. If the line breaks, the nails produced could be longer or shorter. Whether the production line is functioning correctly or not, the nail lengths are normally distributed with a standard deviation of 2 mm. Each nail length is independent of the others. To test whether the line is working, you sample four nails, measure them, and calculate the mean of the measurements.

   (a) What is the standard error of the mean of a sample of four nails?

   (b) How would you construct a 95% confidence interval for μ, the true mean of the length of the nails made by the production line?

   (c) Suppose you draw a sample of four nails and calculate the mean as $\bar{x}$. For what values of $\bar{x}$ would a two-sided test of the hypothesis $H_0 : \mu = 100$ produce $p < 0.05$?

   (d) Write R code to simulate data assuming that the null hypothesis is true. Simulate 10,000 samples of size four, and for each sample, use R to calculate p for a two-sided test of the hypothesis that $\mu = 100$. (You can use `mat.samps()` from Exercise Set 6-1, Problem 2 to simulate the a matrix of appropriate data with `mat.samps(n = 4, nsim = 10000, rx = rnorm, 100, 2)`.) Save the values of p that result. What proportion of the tests produce $p < 0.05$? How about $p < 0.1$? Plot the p values in a histogram. How are they distributed?

   (e) Modify your R code from part (d) to simulate data under the assumption that $\mu = 101$. Again conduct tests of the hypothesis that $\mu = 100$ on 10,000 samples of size four. Now what proportion of the tests produce $p < 0.05$? How about $p < 0.1$? Plot the p values in a histogram. How are they distributed?

   (f) Repeat part (e), changing the true mean from $\mu = 101$ to $\mu = 102$. What do you notice?

   (g) Repeat part (e) once more, keeping $\mu = 101$ but changing the sample size to 16. What do you notice?

7.4 Frequentist inference II: alternative hypotheses and the rejection framework

The hypothesis-testing framework covered in the last section provides one way of measuring the degree of evidence against a particular null hypothesis: the p value. Other things equal, the smaller the p value is, the stronger is the evidence against the null hypothesis. Nonetheless, the p value does not dictate a decision in light of the data.

Jerzy Neyman and Egon Pearson proposed an influential framework for such decisions. For illustration, recall the example in Problem 2 of Exercise Set 7-3, in which the problem is to monitor a production line's output of nails, which are supposed to be 100 millimeters long on average. If there is reason to believe that the production line is malfunctioning, then we might want to stop the line and inspect it. But if the production line is functioning appropriately, then we should keep it running.

Here, the null hypothesis, H_0, is that the expected nail length is 100 millimeters, which is what it would be if the production line were functioning appropriately. The Neyman–

Table 7-1a Decisions in the significance testing framework

		True state of the world	
		H_0 is true	H_a is true
Decision	Fail to reject H_0	Correct failure to reject	Type II error
	Reject H_0	Type I error	Correct rejection

Table 7-1b A confusion matrix in the significance testing framework. In the matrix, t_n, f_n, t_p, and f_p represent probabilities of true negative, false negative, true positive, and false positive events, respectively, with $t_n + f_n + t_p + f_p = 1$

		True state of the world	
		H_0 is true	H_a is true
Decision	Fail to reject H_0	t_n	f_n
	Reject H_0	f_p	t_p

Pearson rejection framework differs from the setup discussed in the previous section in that it also includes an explicit *alternative hypothesis*, H_a. In this case, the alternative hypothesis might be that the expected nail length is *not* 100 millimeters—which would suggest that the production line is malfunctioning—or it might be something more specific. Perhaps it is that the line is producing nails that are 90 millimeters long in expectation. We will call the decision to act[13] as if the null hypothesis is false "rejection" of the null hypothesis—if the data do not support such a rejection of the null hypothesis, then we have little evidence against H_0, and we say that we "fail to reject" the null hypothesis. If the null hypothesis is true, then a failure to reject is a correct decision and a rejection of the null hypothesis is called a **type I error**. In contrast, if the alternative hypothesis H_a is true, then rejecting the null hypothesis is the correct decision, and failing to reject is a **type II error**. Table 7-1a illustrates the possibilities.

Table 7-1b presents a confusion matrix, which describes the performance of a test in a given context. t_n, f_n, t_p, and f_p are the probabilities of the events shown in Table 7-1a in a particular setting.[14] t_n stands for "true negative," f_n stands for "false negative," f_p stands for "false positive," and t_p stands for "true positive." You can imagine that t_n, f_n, t_p, and f_p apply to a particular test applied in a particular research setting. For example, they might apply to

[13] In addition to the presence of alternative hypotheses, it is often claimed that a difference between the Neyman–Pearson framework and the framework from the last section—which is due mainly to Fisher—is that Fisher wanted to make inferences whereas Neyman and Pearson wanted decide how to behave. In fact, it is not hard to find quotations from Fisher, Neyman, and Pearson that support this distinction. However, Mayo (2018, pp. 173–182) convincingly argues that the distinction is overstated and that both parties were open to both behavioral and inferential interpretations in their own work.

[14] The meaning of "particular setting" may need clarification. One setting might be the testing of a particular null hypothesis, which is either true or false. If the null hypothesis is true, then it is impossible to correctly reject it or to make a Type II error. In that case, only results in the left column are possible, and $f_n = t_p = 0$. Similarly, if the null hypothesis is false, then only results in the right column are possible. On the other hand, a "particular setting" might be defined more broadly—perhaps the setting is that of hourly inspections of a manufacturing line over the course of a year. Sometimes the line will be working correctly (and thus H_0 will be true in some cases), and sometimes it will be malfunctioning. In that case, all four cells in the table might be positive. This is a dash of Bayesianism, implicitly putting a (possibly unknowable) probability distribution on the hypotheses.

the test applied to a production line manufacturing nails in Exercise Set 7-3, Problem 2. In such a setting, the "true state of the world" is whether the production line is working correctly (H_0 is true) or has a problem (H_0 is false). A rejection of or failure to reject H_0 might determine whether we decide keep the line running (fail to reject H_0) or to stop the production line and inspect it (reject H_0).

In any decision-making framework like this one, we would like to have a perfect procedure—one that rejects the null hypothesis whenever it is false and never rejects it when it is true. But such a procedure is usually impossible. The rejection approach also assumes that it is usually impossible to know the true state of the world. Specifically, we cannot know the proportion of null hypotheses that are in fact true or that are in fact false, represented here by the sums of the columns of Table 7-1b.

One thing that the researcher *can* sometimes control—assuming that it is possible to develop an appropriate statistical model for the situation—is the probability that the null hypothesis is rejected *given* that it is true, or

$$P(R|H_0) = \frac{f_p}{t_n + f_p},\qquad(7.4)$$

where R represents the event that the null hypothesis is rejected. The value of the expression in Equation 7.4 is called the "type I error rate" or the **size** of the test. The data analyst picks a number α that represents the maximum tolerable type I error rate. α is called the **level** or "significance level," and a result that leads to a rejected null hypothesis is called **statistically significant** or just "significant."

The rejection framework need not be explicitly tied to p values,[15] but one way to conduct a hypothesis test in the rejection framework amounts to calculating a p value under the null hypothesis. If p is less than the level, α, then one rejects the null hypothesis. If p is greater than or equal to α, then one fails to reject the null hypothesis. Another way to say this is that the p value is equal to the smallest possible level α that would dictate a failure to reject the null hypothesis.

Let's pause for an exercise before considering a connection between p values and confidence intervals, followed by some ways in which the significance-testing ideas presented in this section and the previous one have been misused.

Exercise Set 7-4

(1) In this problem, you will use the quantities displayed in Table 7-1b to define some metrics for assessing the performance of a test. Recall that Table 7-1b applies to a specific testing procedure and considers only two possibilities about the state of the world: the null hypothesis is true or the alternative hypothesis is true. This is of course too simple—in many cases, if the null hypothesis is false, it can be slightly wrong or grossly wrong. We will take up this complication soon.

   (a) Using the notation in Table 7-1b, what is the proportion of the time that the null hypothesis is true, $\tau = P(H_0)$?

   (b) What is the proportion of the time that the null hypothesis is false, $\varphi = P(H_0^C)$?

   (c) What is the probability that the null hypothesis is rejected given that it is false, $\pi = P(R|H_0)$? This quantity is called the test's power, and it is the subject of Section 7.7.

---

[15] In fact, it is more standard in a Neyman–Pearson framework to talk about a "rejection region"—a term introduced in the next optional section—than to talk about p values. But typically, the p value can be defined in terms of a rejection region for a Neyman–Pearson test, so there is little harm in describing Neyman–Pearson testing in terms of p values.

(d) What is the probability that the null hypothesis is true given that it is rejected, $P(H_0|R)$, also called the **false discovery rate**? Express the false discovery rate in terms of the level of the test, α (Equation 7.4), and the quantities you defined in parts (a)–(c). How does the false discovery rate change if the power increases (but other values stay the same)?

(e) What is the probability that the null hypothesis is true given that it is not rejected, $P(H_0|R^C)$, also called the negative predictive value? Express the negative predictive value in terms of the level of the test (Equation 7.4) and the quantities you defined in parts (a)–(c). How does the negative predictive value change if the power increases (but other values stay the same)?

7.5 [Optional section] Connecting hypothesis tests and confidence intervals

There is a connection between confidence intervals and hypothesis tests. Suppose we consider a "point" null hypothesis H_0, which claims that $\theta = \theta_0$, where θ_0 is a specific value that θ might take. The fact that θ_0 is a single value rather than a range of values is what makes this a "point" null hypothesis.

If one has a valid procedure for constructing $1 - \alpha$ confidence intervals, then it is possible to use the resulting confidence intervals to test the null hypothesis that $\theta = \theta_0$ at level α, simply by observing whether θ_0 falls outside the confidence interval—if it does, then the hypothesis $\theta = \theta_0$ is rejected. Similarly, if one has a procedure[16] for testing $\theta = \theta_0$ at level α, then the set of values of θ_0 that do not result in a rejection of the null hypothesis are a $1 - \alpha$ confidence interval for θ. We will state these results more precisely and prove them for a one-dimensional parameter—that is, θs that are a single number rather than a vector.[17]

> **Theorem (Part a)** Suppose that (V_1, V_2) is a $1 - \alpha$ confidence interval for θ.[18] Then a test that rejects the null hypothesis that $\theta = \theta_0$ if and only if θ_0 falls outside (V_1, V_2)—that is, if and only if $V_1 > \theta_0$ or $V_2 < \theta_0$—has level $\leq \alpha$.
>
> *Proof (Part a)* By the definition of a confidence interval, $P(V_1 \leq \theta \cap V_2 \geq \theta) \geq 1 - \alpha$. The complement of the event $V_1 \leq \theta \cap V_2 \geq \theta$ is $V_1 > \theta \cup V_2 < \theta$, and so $P(V_1 > \theta \cup V_2 < \theta) \leq \alpha$. If the null hypothesis $\theta = \theta_0$ is true, then we can substitute θ_0 for θ, giving $P(V_1 > \theta_0 \cup V_2 < \theta_0) \leq \alpha$. Thus, a test that rejects the null hypothesis that $\theta = \theta_0$ if and only if $V_1 > \theta_0 \cup V_2 < \theta_0$ will reject the null hypothesis with probability no more than α given that the null is true; in other words, it has level $\leq \alpha$. ∎

To state and prove the second part of the theorem, a few more definitions are helpful. Suppose we conduct a Neyman–Pearson hypothesis test using a test statistic T. The test's *rejection region* $R(\theta_0)$ is the set of possible values of T that would lead us to reject the null hypothesis $\theta = \theta_0$.[19] The complement of the rejection region, $R^C(\theta_0)$, is the set of possible values of T that would *not* result in a rejection of the null hypothesis $\theta = \theta_0$. $R^C(\theta_0)$ is also called the "acceptance region." Finally, a *confidence set*, like a confidence interval, is

---

[16] Specifically, the proof relies on a procedure that entails construction of a *rejection region*, defined below.

[17] The proof for a vector is similar, but requires heavier notational machinery than we have been using.

[18] V_1 and V_2 are random variables, so we are talking about a *procedure for building confidence intervals* rather than a specific instance of the procedure applied to data.

[19] The rejection region is sometimes defined in terms of the data themselves rather than a summary statistic computed from the data.

constructed using a procedure that captures the true estimand with a specified probability, but it need not be an uninterrupted segment of the number line.

Theorem (Part b) Suppose that we can conduct a level-α test of the hypothesis that $\theta = \theta_0$ for any permissible θ_0 by constructing a rejection region. Then the set of values of θ_0 that are not rejected by such a test are a $1 - \alpha$ confidence set for θ.

Proof (Part b) We conduct the test by computing a test statistic, T—if T falls in the rejection region $R(\theta_0)$, then we reject the null hypothesis that $\theta = \theta_0$; otherwise T falls in the acceptance region $R^C(\theta_0)$, and we do not reject $\theta = \theta_0$. Label the set of values of θ_0 for which T falls in the acceptance region as $C = \theta_0 : T \in R^C(\theta_0)$. The test has level α, which means that $P\left(T \in R(\theta)\right) \leq \alpha$, where θ is the true value of the estimand. $R^C(\theta)$ is the complement of $R(\theta)$, so $P\left(T \in R^C(\theta)\right) \geq 1 - \alpha$. In words, $T \in R^C(\theta_0)$ means that the null hypothesis that $\theta = \theta_0$ is not rejected, so $\theta \in C$ if and only if $T \in R^C(\theta)$. Thus, $P(\theta \in C) = P\left(T \in R^C(\theta)\right) \geq 1 - \alpha$, and C is a $1 - \alpha$ confidence set for θ. (Note that $P(\theta \in C)$ is a probability statement about C, which is random, and not θ.) ∎

7.6 NHST and the abuse of tests

The significance testing procedures covered in the previous sections were developed in the first half of the twentieth century, principally by R. A. Fisher, Jerzy Neyman, and Egon Pearson. In many disciplines, the practice of significance testing has been modified in ways that would dismay its founders. The modified procedure is often called null hypothesis significance testing (NHST). There is no single, agreed-upon definition of NHST, but many commentaries (some of which appear as recommended reading at the end of the chapter) focus on the features laid out in the next six subsections.

7.6.1 *Lack of replication*

In NHST, a rejected null hypothesis from a single study is often taken as a basis for establishing the alternative hypothesis. This would surprise Fisher, who said "[W]e may say that a phenomenon is experimentally demonstrable when we know how to conduct an experiment which will rarely fail to give us a statistically significant result" (quoted in Mayo, 2018, p. 4). Historically, academic researchers in many fields have had little incentive to replicate previous studies exactly; replications can be hard to publish and may be granted substantially less prestige than the original work. At this writing, the situation is beginning to change as researchers realize anew the importance of replication.

7.6.2 *Ossification of $\alpha = 0.05$*

In principle, the rejection framework described in Section 7.4 is flexible—one can choose the level, α, according to the specific decision being made. Increasing α will typically increase the probability of rejecting both true and false null hypotheses, which means that we will make more type I errors when the null hypothesis is true but fewer type II errors when the null hypothesis is false. This might be desirable. For example, if the cost of stopping a production line to inspect it decreases, then type I errors become more tolerable, and the user can increase the level of the test to decrease the type II error rate. But in practice, the rules for conducting hypothesis tests have ossified. In NHST, the level of the test is typically set to $\alpha = 0.05$, and no justification is given for why $\alpha = 0.05$ is a sensible level. In fact, the field of research in which the test occurs, the design of the study, and the stakes of any decisions being made on the basis of the test are all important to consider—but are often ignored—when setting α.

7.6.3 $\alpha = 0.05$ *as a gatekeeper*

Not only is the $\alpha = 0.05$ criterion inflexible; in fields under the sway of NHST, it is a gatekeeper—studies with results pass the threshold are deemed "significant," potentially published, and sometimes accepted as part of the discipline's empirical basis. Studies whose results fail to reach $\alpha = 0.05$ risk being ignored, regardless of their importance. Beyond encouraging binary thinking, this use of $\alpha = 0.05$ causes **publication bias**—a situation in which it is easier to publish significant than non-significant results in academic journals. Publication bias is also called the "file-drawer" problem (Rosenthal, 1979)—results that fail to reach significance remain in the researcher's file drawer rather than being published. If publication bias is severe enough, the published literature ceases to reflect the state of the world. The evidence against any given null hypothesis is overstated by the published record, and a record of overestimated effect sizes accumulates. This bias—along with the tendency of some NHST users to exclude relevant information beyond the p value from their reports—makes it difficult to synthesize past work.

Publication bias interacts with some of the other NHST pathologies listed. Several ways of addressing the gatekeeping function of α have been proposed, and one of the most promising is preregistration (Munafò et al., 2017). When a researcher preregisters, a study design or data analysis plan is registered with an external body before the researcher carries it out. In some cases, peer review may take place at the preregistration stage, and academic journals may make decisions about whether to accept a paper based on the plan. For many studies, preregistration is a great idea. The researcher benefits because she can get credit for her idea and useful feedback on her plans. If she interests a journal in her proposed study, she may secure a venue for publication that does not depend on the results, freeing her to be more objective in her data analysis. The research community also benefits from the transparency; one's colleagues get to see what was planned and what was actually done, which makes it easier to evaluate the evidence. Some object that preregistration limits exploration, but it does not—it merely makes clear which parts of the analysis are exploratory.

7.6.4 *Identification of scientific hypothesis with a statistical hypothesis*

In NHST, a research hypothesis is conflated with a null or alternative hypothesis. To bring back an example from earlier, we discussed a theory that smaller people are attracted to warmer climates and considered a test of the hypothesis that the mean height of women from Arizona is equal to the mean height of women from elsewhere in the United States. Testing the hypothesis is a terrible test of the theory, for reasons discussed in the solution to the first problem of Exercise Set 7-3. In NHST, the intervening steps between null hypothesis and substantive theory are often ignored, and a rejected null hypothesis is taken to establish a tenuously connected theory.

7.6.5 *Neglect of other goals, such as estimation and prediction*

In NHST, researchers may focus to an undue extent on whether null hypotheses are strictly true or not. If there is evidence that the null hypothesis is false—generally in the form of a mechanical rejection at $\alpha = 0.05$—then the NHST researcher presses forward with publication of this bare fact. But the p value alone doesn't tell us what we need to know. It doesn't even say how large the effect in question might be. (The term "significant" suggests a large or important discrepancy from the null hypothesis, but that would be an incorrect inference—statistically significant may be associated with small departures from the null hypothesis, and, in small samples, very large effects might not attain statistical significance.)

A single-minded hypothesis testing approach pushes aside other scientific goals. The importance of these other goals will vary by context. There are some situations where a hypothesis test is closely responsive to one's needs. One example would be in testing whether a new medication is more effective than the standard treatment—there are clear null and alternative hypotheses to test, and the test result summarizes the evidence. But there are other situations where estimation is much more important. For example, psychologist Paul Meehl (1990, point 6) posited a "crud factor" in "soft" psychology, which encompasses roughly clinical, social, personality, and organizational psychology. The crud factor refers to the nonzero—though possibly small—association of most variables in this domain with most other variables. Meehl illustrated the crud factor with a survey study of ~57,000 people, in which he and his colleague collected data on 45 different variables spanning family history, demographics, occupation, and preferred use of leisure time. Out of 990 possible associations among these variables, 92% were significant with $p < 0.05$, and 78% were significant with $p < 0.000001$. One of Meehl's points was that if one is working with large samples in a domain like this—one where everything is related to everything else—then tests of the null hypothesis that pairs of variables are independent provide little new information. We already know that most pairs of variables are not independent. In such a context, it is more useful to estimate the degree of dependence and to test explanations for the observed networks of association.

7.6.6 A degraded intellectual culture

NHST is at the center of a degraded intellectual culture in some disciplines. For many academic researchers, career success is predicated on research productivity, research productivity is measured by numbers of publications, and results are only publishable if they are statistically significant (see Section 7.6.3). Thus, researchers are incentivized to reach statistical significance, which is not always consistent with the goal of contributing to knowledge. In relatively rare cases, people have committed outright fraud, falsifying data to achieve significance. But there is a continuum of actions that are not so brazen—and that, in some cases, may not even be intentional[20]—that nonetheless invalidate reported p values, usually in a way that overstates the degree of evidence against the null hypothesis. These practices are called by many names, including "*p*-hacking," "researcher degrees of freedom," "data dredging," "data snooping," "data torture" (as in, "until they confess"), "selection effects," and "the garden of forking paths." Unfortunately, they are common (John, Loewenstein, & Prelec, 2012) and sometimes even encouraged.[21]

[20] Malpractice needn't be intentional, and it needn't *seem* like bad practice to the researcher. Gelman and Loken (2013) discuss the "garden of forking paths"—borrowing the title of a story by Borges—in which researchers allow details of the analysis to be dictated by data, invalidating nominal p values.

[21] In a now-infamous example, and in the midst of an article filled with otherwise good writing advice, Bem (2000) wrote, "Examine [the data] from every angle. Analyze the sexes separately. Make up new composite indexes. If a datum suggests a new hypothesis, try to find additional evidence for it elsewhere in the data. If you see dim traces of interesting patterns, try to reorganize the data to bring them into bolder relief. If there are participants you

There are many ways to *p*-hack, but a common thread is that the decision about how, whether, or when to perform or report a hypothesis test is made on the basis of the data. Here are a few of the most-observed methods, some of which you will simulate in the exercises below. (Simmons et al., 2011, also recommended at the end of the chapter, is a great reference on these problems.) To motivate the examples, suppose that we are carrying out a study of the effects of a lifestyle intervention on cardiovascular health.

(i) *Multiple testing*: One might collect data on several different dependent variables, test them all, and report only the hypothesis tests that achieve significance. For example, in a cardiovascular health study, one might collect information on many indicators of cardio-vascular health, including heart rate, heart rate variability, cholesterol levels (LDL and HDL), blood pressure (systolic and diastolic), and VO_2 max, among others. That is at least seven variables to test, assuming that the researcher doesn't come up with additional indices that combine several of the variables. Even if the intervention in fact has no effect on cardiovascular health, the probability of observing $p < 0.05$ in *at least one* of the tests— also called the familywise error rate—will be larger than 0.05. (You will simulate this in an exercise below.) If the researcher reports a result with $p < 0.05$ but fails to correct for—or worse, fails even to report—the multiple tests, then the evidence against the null hypothesis will be overstated.

(ii) *Optional stopping*: In this form of *p* hacking, data are collected and, at intervals, repeatedly tested for significance. When a given level of significance is achieved, the data collection stops, and the results are reported. If *p* values are not corrected for these multiple checks, then it is easy to achieve low *p* values even when the null hypothesis is true. As long as one is willing to collect data for long enough and check for significance every so often, then the probability that $p < .05$ at some point approaches 1.

(iii) *Data slicing*: If the planned test does not produce a significant hypothesis test, a data slicer might try to attain significance by looking at a subset of the data, perhaps leaving out an experimental condition, or participants below a certain age, or all the men. A variant of this approach is to drop individual data points on the basis of their status as outliers—some targeted dropping can make the difference between $p > 0.05$ and $p < 0.05$.

(iv) *Model selection*: In this book, we consider simple models with only one or two variables (until the postlude chapter). In most real-life research contexts, there are many variables that might be reasonable to include as part of a statistical model. For example, in a cardiovascular health study, it might be reasonable to fit a model that includes not just the target intervention as an independent variable, but also age, gender, and socioeconomic status. Or just age. Or just gender. Or just gender and age. There are many possibilities, and they may differ in terms of whether the intervention appears to be significant. If many models are tried and the most favorable is chosen, then reporting just one as if it was pre-planned may exaggerate the evidence.

To be clear, these methods for *p* hacking are not harmful in and of themselves. Many of them are key parts of research—for example, optional stopping is important in clinical trials, where one wants to offer placebo to as few patients as possible before giving an effective treatment the green light. And exploration of data may lead to genuine unexpected discoveries. The harmful thing is reporting data-dependent analysis choices as if they were planned in advance.

---

don't like, or trials, observers, or interviewers who gave you anomalous results, drop them (temporarily). Go on a fishing expedition for something—anything—interesting." Such practices are almost sure to reveal some *p* values less than 0.05 in any moderately complex dataset, even if that dataset is pure statistical noise.

There are three non-exclusive ways to deal with data-dependent steps in one's analysis. First, in some cases, it is possible to adjust one's statistical procedures in a way that accounts for the data-dependent steps. For example, there are large literatures on correcting for multiple testing (Benjamini & Hochberg, 1995; Shaffer, 1995), optional stopping (Whitehead, 1997), and model selection (Taylor & Tibshirani, 2015). A second approach—always handy—is to replicate one's study. If full replication is not possible, a good second choice is to hold some of the data aside during the data-dependent steps of analysis. When the analysis is complete in the first subset, one can check whether the observed patterns replicate in the held-aside subset. Finally, whether or not correction or replication is possible, be transparent about the data-dependent steps in your analysis, describing them fully and, if possible, making data and analysis code accessible. If you are a researcher, there may be incentives pushing you away from best practices, but it is your choice whether to follow those incentives.

7.6.7 Evaluating significance tests in light of NHST

Each of the highlighted features of NHST can lead to major problems, and NHST is at the center of recent discussions about the "replication crisis"—the growing realization that in some fields, a distressing number of results vanish in replication studies (e.g., Open Science Collaboration, 2015). Many reforms have been suggested in response to the replication crisis, and some, like preregistration and incentivizing replication attempts, are welcome. Another proposal is to ban either p values or hypothesis tests outright, replacing them with confidence intervals, Bayes factors (see Chapter 10), or nothing at all. Advocates of a ban will correctly point out the problems we have just discussed, and certainly the problems of NHST suggest the need for major changes to standard practice. But this section contains no criticisms of hypothesis tests *per se*; it is about misunderstandings of, misuses of, and overreliance on hypothesis tests. Often, arguments for an outright ban entail the cynical premise that researchers are unable—even in principle—to understand p values, to use them appropriately, or to place them in context with other information.

The problems of NHST are about ways in which hypothesis testing is (mis)used, not with hypothesis tests themselves. Most accepted procedures for significance testing work as advertised—meaning that they produce probability statements that have the properties described in this chapter, provided that their assumptions are met. A careful, knowledgeable user need not fear them. Problems reside in the ways in which people try to translate statistical results into claims about the world, not in the statistical procedures themselves.

Exercise Set 7-5

(1) In this exercise, you will explore the effects of one form of p hacking. Imagine that your research group is interested in the effects of a lifestyle intervention on cardiovascular health. You recruit participants and randomly assign them either to engage in a regular exercise regimen or to remain sedentary. After 8 weeks, you collect seven measurements of cardiovascular health: resting heart rate, heart rate variability, cholesterol levels (LDL and HDL), resting blood pressure (systolic and diastolic), and VO_2 max. Assume that all seven variables are scored so that larger values indicate better cardiovascular health. Your team decides to run seven hypothesis tests with type I error rate $\alpha = 0.05$, one for each measurement of cardiovascular health. The tests will all examine the null hypothesis that the exercise intervention has no effect on cardiovascular health. If any one of the tests is significant, then your group will report that the exercise intervention changes cardiovascular health.

To execute the code necessary for this exercise, you will need to install and load the `stfspack` package. See Exercise Set 2-2, Problem 3 for instructions.

(a) The `many.outcome.sim()` function simulates the experiment proposed above assuming that the null hypothesis is true. Call it with

```
many.outcome.sim(n, nsim, n.dv, correl)
```

where n is the size of each group in every simulation, `nsim` is the number of simulations to run, `n.dv` is the number of outcome variables measured on each person, and `correl` is the (shared) correlation among the outcomes. For example, you can run a simulation of the described setup with

```
ps <- many.outcome.sim(20, 10000, 7, .7)
```

In this version, there are 20 participants per group, 7 measurements of cardiovascular health mutually correlated at 0.7, and we run 10,000 trials. The output is a matrix of p values with 10,000 rows—one for each simulated trial—and 7 columns, one for each outcome variable. To determine significance, we compare every p value with the desired level (here, 0.05):

```
sigs <- ps < .05
```

To identify the proportion of tests significant for each measurement, we could run

```
colMeans(sigs)
```

And to find the proportion of trials in which the null hypothesis would be rejected for at least one measurement, we could run

```
mean(rowMeans(sigs) > 0)
```

Run these commands. For each measurement, how often is the null hypothesis rejected? How often would you reject the null hypothesis if you did as your lab planned and reported a positive finding if any *one* of the measurements produced $p < 0.05$?

(b) Repeat part (a) several more times, varying the number of measurements and the degree of correlation between the measurements. How do these factors affect the rate at which at least one of the tests produces $p < 0.05$?

(2) [Optional, but read the question and solution if you skip it] Another form of p hacking is repeated testing, also called optional stopping. Modifying the situation in Problem 1, assume that your research team assesses the effect of a lifestyle intervention using a single measurement, systolic blood pressure. However, they do not know how many participants they ought to recruit to the study. Your team will start with 20 participants in each of the two groups, and if the difference in mean blood pressure between the two groups is significant, then your team will publish the study and stop gathering data. If the difference is not significant, then your team will collect 10 more participants in each group and run the test again, including all participants. If the difference is significant, then the team will publish and stop collecting data, but if the difference is not significant, then they will collect 10 more subjects. If necessary, then they will repeat this procedure until they have collected up to 200 total participants (100 in each group). The `serial.testing.sim()` function[22] simulates this scenario. The default settings run 10,000 simulations of the scenario described above, where the p value is computed after 20, 30, 40, and 50 subjects are recruited to each of the two groups:

---

[22] Available in the book's R package, `stfspack` (see Exercise Set 2-2, Problem 3 For installation instructions).

```
serial.testing.sim(ns = c(20, 30, 40, 50), nsim = 10000)
```

Run the function with the default settings, and then again to simulate checking until the sample size is 200 total (100 per group). What proportion of the time will your team report an effect of the lifestyle intervention given that it in fact has no effect?

7.7 Frequentist inference III: power

In Section 7.4, we discussed one method of constructing a rejection-framework hypothesis test: calculate a p value and observe whether p is less than α, the level. If $p < \alpha$, then the null hypothesis is rejected. Provided that the model used to calculate p is appropriate, then this method ensures that if the null hypothesis is true, it will be rejected less than a proportion α of the time. The type I error rate is correctly calibrated.

But just to preserve the type I error rate is easy. We don't even need to collect data. Suppose that every time you want a hypothesis test, I draw an observation from a continuous $\text{Uniform}(0, 1)$ distribution. If the value of the random variable I draw is less than α, then I reject the null hypothesis; otherwise I fail to reject. This absurd test meets the minimal requirement of preserving the type I error rate—I will reject the null hypothesis α proportion of the time if the null hypothesis is true. The problem with the test is obvious, though—I will reject the null hypothesis with probability α regardless of whether it is true or false. To be useful, a hypothesis testing procedure needs to reject the null hypothesis when it is false with probability greater than α.

The probability that a hypothesis testing procedure leads to a rejection of the null hypothesis under the alternative hypothesis is called the **power** of the test. In the notation of Table 7-1b, the power is

$$P(R|H_a) = \frac{t_p}{t_p + f_n},\qquad(7.5)$$

where R is the event that the null hypothesis is rejected, H_a is a specific alternative to the null hypothesis, t_p is the true positive rate, and f_n is the false negative rate. This definition gives the right idea but is a little too simple. It gives the power of a test with a set type I error rate and fixed sample size against a specific alternative, H_a. That's a useful piece of information, but we also need to be able to describe how a test's power depends on variables including the sample size n, the type I error rate α, and the specific alternative to the null hypothesis that is being considered. In Exercise Set 7-4, Problem 1, you saw that a test's power increases as the sample size increases, as the type I error rate increases, and as the specific alternative hypothesis being considered deviates more extremely from the null hypothesis. We will arrive at a more precise description of these patterns soon, but first we need a way of measuring the degree to which a specific alternative hypothesis deviates from the null hypothesis.

One way to summarize the extent to which a specific alternative hypothesis deviates from the null hypothesis is to specify a standardized **effect size**.[23,24] The number used to

---

[23] The term "effect size" suggests that the independent is a cause of an effect, but in practice that need not be true. Further, even if there is a causal effect at play, it needn't be the same for all the units being observed—for example, a medication might be differently effective for two groups of people. For our purposes of power calculation, the key thing is that a standardized effect size measures a degree of departure from some null hypothesis.

[24] Not every hypothesis test can be easily paired with a notion of an effect or an effect size, though many can.

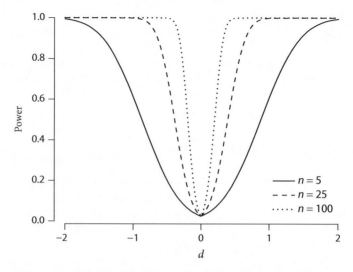

Figure 7-3 Power function (Equation 7.7) for the test of the null hypothesis that a sample is drawn from a normal distribution with a specified expectation. Power is on the vertical axis, and the horizontal axis shows the standardized effect size d defined in Equation 7.6. Three curves are shown, representing the power function at three different sample sizes. For all three, the type I error rate is set to $\alpha = 0.05$.

summarize the effect size depends on the study design and on the statistical test being used. For the test of the hypothesis that Arizonan women are shorter than average, we could have used

$$d = \frac{\mu_a - \mu_0}{\sigma}, \tag{7.6}$$

where μ_a is the assumed mean height of Arizonan women under a specific alternative hypothesis, μ_0 is the mean height of Arizonan women that is assumed by the null hypothesis, and σ is the standard deviation of heights among Arizonan women. The expression in Equation 7.6 is a version of "Cohen's d," a popular measurement of effect size—in words, it is equal to the number of standard deviations by which the mean height of Arizonan women differs from the mean height of women from other parts of the United States. The numerator quantifies the difference between the null hypothesis and the alternative hypothesis in the original units of the study; dividing by σ *standardizes* the effect size and allows comparison with effect sizes found in other contexts. Standardization is not always a good idea—when the study's measurement units are meaningful, standardization sometimes obscures the units of interest.

With a measurement of effect size in hand, we can define a test's **power function**

$$\pi(\alpha, n, \psi) = P\Big(R|H_a(\psi)\Big). \tag{7.7}$$

Here, π represents the power function, α is a specified significance level, n is a sample size, ψ is an effect size, R is the event of rejecting the null hypothesis, and $H_a(\psi)$ is an alternative hypothesis implying effect size ψ. That is a lot of notation, but what Equation 7.7 expresses in words is already familiar: the probability of rejecting the null hypothesis under a specific alternative—that is, the test's power—depends on the specified type I error rate, the number of samples drawn, and the degree to which the true state of the world differs from the null hypothesis, indexed by the effect size. Power functions can, in general, also depend on other factors, but these three are nearly always important. It is often possible to derive an exact expression for a test's power function, but in complicated situations, it is sometimes easier to estimate the power function by simulation.

Figure 7-3 shows the power function for the test we used to assess the hypothesis that Arizonan woman are shorter than other American women. The significance level is $\alpha = 0.05$, and the three curves shown represent three different possible sample sizes. The effect size in Equation 7.6 is on the horizontal axis, and power is on the vertical axis. The null hypothesis of no height difference is equivalent to setting $d = 0$—notice that when $d = 0$, the null hypothesis is rejected 5% of the time, regardless of the sample size. As the absolute value of the effect size increases, the power increases, regardless of the sample size. For any given (non-zero) effect size, larger samples have more power than smaller samples.

Good researchers consider the sample size in relation to the questions they'd like to answer, and one way to do that is to conduct a power analysis. In a power analysis, one studies a test's power function. Often this is done before a study is conducted, with the goal of deciding how many observations to gather. Disciplines vary in the degree of power they deem acceptable; conventional power standards range from 80% to 95%, usually in the context of a type I error rate of 5%. These numbers are hard to interpret on their own, though—as we have seen, power is not a single number; it is a function. One typically has higher power to detect large effects and lower power to detect smaller effects. One way to simplify the analysis is to pick a minimal effect size that would merit further study. For example, in a planned study of a chemical compound that might be useful as a medicine, one might claim that the candidate medicine only merits further study if it alleviates symptoms by at least as much as the standard treatment. One could then plan a study that has high power if the effect size is at least as large as that associated with standard treatment.

Power also matters when interpreting a study's results. If a study has low power against most reasonable alternatives because it is based on a small sample, then a failure to reject the null hypothesis is only weak evidence against the null hypothesis. For example, suppose a rival researcher thinks my hypothesis about Arizonan women's heights was false, so she samples a single Arizonan woman at random and fails to reject the null hypothesis that Arizonan women are as tall as other American women. I would rightly be unimpressed by the result—even if my hypothesis is true, such a small sample would be unlikely to lead to a rejection of the null hypothesis. But if my rival were to sample 10,000 women from Arizona and still fail to reject the null hypothesis, then I would have some explaining to do—such a large study would be nearly certain to reject the null hypothesis for all but the tiniest effect sizes. In other words, if I am right that Arizonan women are shorter on average, then either (a) there is a problem with my colleague's study or data analysis, (b) the sample drawn was extremely improbably tall on average, masking the true height difference, or (c) the degree by which Arizonan women are shorter must be very small indeed. You derived a mathematical statement expressing the situation more compactly in Exercise Set 7-4, Problem 1, part (e).

Low power affects the interpretation of studies that reject the null hypothesis, too. For example, one difficulty in interpreting significant results from tests with low power is the **winner's curse**—if the null hypothesis entails a true effect size of zero, estimates of effect size associated with significant results are likely to be overestimates. As you will see in an exercise, the winner's curse is a pronounced bias in underpowered studies—that is, studies with too-small samples—but becomes negligible as power approaches 1. The winner's curse can be a strong force in the presence of publication bias. An initial (underpowered) study might identify a significant effect, and if it does, the effect size estimate will likely be too large—if the effect size estimate were accurate, then the hypothesis wouldn't have been significant: that is roughly what "underpowered" means. The next researcher's power analysis relies on the inflated effect size estimate, leading to another study that is underpowered with respect to the true effect size. If the effect found in the next study is not significant, it will not be published, but if it is significant, then there will be another overestimated effect size in the literature, and so on. Publication bias and the winner's curse combine in a way that systematically inflates recorded effect size estimates.

Exercise Set 7-6

(1) In this problem, you will consider the power function of a two-sided test of the null hypothesis that a sample is drawn from a normal distribution with a specified expectation. The following function[25] gives a power estimate by simulating nsim random samples of n observations each, drawn from a normal distribution with an expectation that is d standard deviations away from the expectation under the null hypothesis. Each sample is tested against the null hypothesis, and the power is estimated by computing the proportion of the samples that generate *p* values less than lev.

```
power.sim.1sz <- function(n, nsim, d, lev = 0.05){
    simmat <- matrix(rnorm(n*nsim, d, 1), nrow = nsim)
    samp.means <- rowMeans(simmat)
    neg.devs <- -abs(samp.means)
    ps <- 2*pnorm(neg.devs, 0, 1/sqrt(n))
    mean(ps < lev)
}
```

Use the function power.sim.1sz() to plot the power function for values of $d \in [-2, 2]$ with a level of 0.05 and $n = 25$.

(2) [Optional] The function wincurse.sim.1sz()—available in stfspack (see Exercise Set 2-2, Problem 3)—extends power.sim.1sz() to simulate the "winner's curse"—the tendency of significant results from underpowered studies to be associated with overestimated effect sizes. The main inputs are the true effect size (d) defined as in Equation 7.6, the number of observations in each sample (n), and the significance level of the test (lev). The function simulates nsim samples from a normal distribution whose expectation deviates from the null expectation by an amount determined by d. For each sample, an effect size is estimated and a *p* value is computed. The function returns a vector with the true effect size, the mean estimated effect size from the samples associated with a rejection of the null hypothesis, and the proportion of samples associated with a rejection of the null hypothesis, or the power. It also produces a histogram of the estimated effect sizes from all studies, with the effect sizes from studies associated with a rejection of the null hypothesis colored in gray. (The "estimated" effect size is the mean of the estimated effect sizes plotted in gray on the histogram.) The arguments of the function are

```
wincurse.sim.1sz(n, nsim, d, lev = 0.05, abs.vals = FALSE, br = 50)
```

So, for example, to simulate the winner's curse with a true effect size of 0.1 standard deviation, a significance level of 0.05, and 50 observations per sample, you could use

---

[25] Available in the book's R package, stfspack (see Exercise Set 2-2, Problem 3 for installation instructions).

```
wincurse.sim.1sz(n = 50, nsim = 10000, d = 0.1, lev = 0.05)
```

(a) Use the function to explore some interesting values of d and n. What do you observe?

(b) Use the function to make plots relating (i) the effect size estimated from studies with significant results to the sample size used in the study and (ii) the size of the "winner's curse effect"—the difference in effect size estimated from studies with significant results vs. the true effect size—to the power of the study. You can use $d = 0.3$ as the effect size.

7.8 Putting it together: What happens when the sample size increases?

Statisticians use the tools of frequentist interval estimation and inference—including standard errors, confidence intervals, p values, the rejection framework, and power—to describe the degree of uncertainty inherent in their estimates. One major theme of this chapter is that it is easy to misinterpret the tools of frequentist statistics, for example by claiming that the p value is the probability that the null hypothesis is true given the data.

Another lesson is that the precision of most statistical procedures improves as the number of data used increases. Observing what happens as the sample size increases provides an opportunity to notice once more the strong connections between standard errors, confidence intervals, and hypothesis tests. Consider once more the example of testing the null hypothesis that the expected height of Arizonan women is the same as that of other women from the United States. In that example, we tested a null hypothesis about the expectation of a random variable, X, under the assumption that X has a known standard deviation, σ, and that the mean of a set of independent observations has a normal distribution. We tested the hypothesis by computing the sample mean $\hat{\theta} = \sum X_i/n$ as an estimator of the unknown expectation θ, and then comparing the value of $\hat{\theta}$ obtained with the distribution it would have if $\theta = \theta_0$. If we increase the sample size, three things will happen:

(1) The standard error of $\hat{\theta}$ will decrease—that is, repeated samples of a larger size would lead to less variable estimates of $\hat{\theta}$ than would repeated samples of a smaller size.

(2) The width of $1 - \alpha$ confidence intervals for θ will decrease for any fixed α.

(3) If the null hypothesis is false and $\theta = \theta_a$, then the power of a level-α test of the null hypothesis that $\theta = \theta_0$ will increase, for any fixed α, θ_0, and θ_a.

The patterns in (1)–(3) may seem like three separate ways in which the concepts of this chapter respond to increasing sample size, but they are actually all different consequences of the same response. This is because:

(4) The width of the $1 - \alpha$ confidence interval in (2) is proportional to the standard error of $\hat{\theta}$ from (1). So as the standard error decreases, the width of the confidence interval decreases.

(5) A level-α test of the null hypothesis that $\theta = \theta_0$ is significant whenever a $1 - \alpha$ confidence interval does not contain θ_0. If the null hypothesis is false and $\theta = \theta_a$, then narrow confidence intervals are more likely to exclude θ_0 and to be associated with tests that reject the null hypothesis. That is, the power of the test will increase.

The points in (1)–(5) are true for the example of a normally distributed sample mean, and versions of them hold in many settings. (We proved a statement related to (5) in greater generality in optional Section 7.5.) They are useful heuristics in many problems—larger samples are associated with less variable estimators, less-variable estimators allow for smaller confidence intervals, and with smaller confidence intervals come more powerful tests against false null hypotheses. In short, the precision of all our procedures increases.

Figure 7-4 displays simulations showing all these patterns together. In each of the nine panels, fifty simulated estimates of a parameter (points) and 90% confidence intervals (vertical lines) are shown. The true value of the parameter, θ, is indicated by a dashed horizontal line, and the value of the parameter under the null hypothesis, θ_0, is indicated by a solid horizontal line. Estimates and confidence intervals associated with rejections (at level $\alpha = 0.1$) of the null hypothesis are colored black; failures to reject are gray. The columns vary in the degree to which the null hypothesis is false—on the left, the null hypothesis is true, and the difference between θ and θ_0 increases left to right—and the rows vary in sample size, with sample size increasing from top to bottom.

The patterns listed in (1)–(3) above are visible as one looks down the columns of Figure 7-4. Going down the columns, estimates are more tightly clustered around the true value of the parameter, and confidence intervals are tighter. In the columns where the null hypothesis is false (middle and right), power increases with larger n—more of the intervals are colored black as n increases. It is also possible to spot the winner's curse in the figure. In the panels where power is relatively low (the central panel and the middle and right panels of the top row), the estimates associated with significant hypothesis tests are overestimates of θ, even though the estimates in aggregate (including gray and black) are on target.

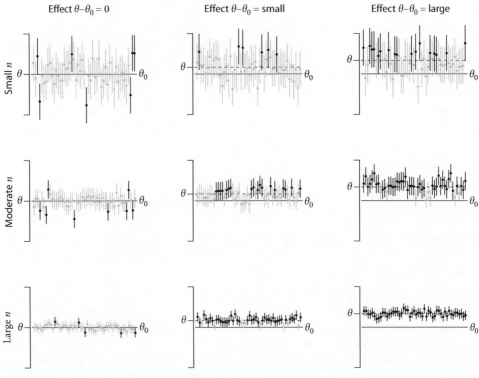

Figure 7-4 Simulations of estimates and 90% confidence intervals for various sample sizes n (varies by row) and degrees of discrepancy between the true value of a parameter and its value under the null hypothesis (varies by column). Estimates/confidence intervals associated with significant hypothesis tests at $\alpha = 0.1$ are colored black.

We have now finished our survey of the goals of statistical estimation and inference. Until now, we have said little about how to derive estimators, confidence statements, and hypothesis tests. In the following chapters, you will encounter three approaches to estimation and inference, each of which makes different assumptions and warrants different claims.

7.9 Chapter summary

Interval estimation is the attempt to define intervals that quantify the degree of uncertainty in an estimate. The standard deviation of an estimate is called a *standard error*. Confidence intervals are designed to cover the true value of an estimand with a specified probability.

Hypothesis testing is the attempt to assess the degree of evidence for or against a specific hypothesis. One tool for frequentist hypothesis testing is the p value, or the probability that, if the null hypothesis is in fact true, the data would depart as extremely or more extremely from expectations under the null hypothesis than they were observed to do. In the Neyman–Pearson rejection framework, the null hypothesis is rejected if p is less than a pre-specified value, often chosen to be $\alpha = 0.05$. A test's power function gives the probability that the null hypothesis is rejected given the significance level α, a sample size n, and a specified alternative hypothesis.

7.10 Further reading

There is an amazing amount of writing devoted to the interpretation of hypothesis tests. Here are a few key references that are useful, interesting, and entertaining.

Cohen, J. (1994). The earth is round ($p < .05$). *American Psychologist*, 49, 997–1003.
  An entertaining and clear tour of some major abuses of significance tests.

Ioannidis, J. P. (2005). Why most published research findings are false. *PLoS Medicine*, 2, e124.
  This provocative article argues that in many fields of research, the claims of most of the empirical articles published are false. The argument has generated controversy, but the important thing in this article is not the estimate of the proportion of studies that make false claims—instead, it is Ioannidis' synthesis of factors that affect the proportion of false studies in a given field. Studies in fields where typical effect sizes are small or zero, where researchers are strongly incentivized to publish and publishability hinges on statistical significance, and where studies are chronically underpowered are most likely to report false claims.

Meehl, P. E. (1978). Theoretical risks and tabular asterisks: Sir Karl, Sir Ronald, and the slow progress of soft psychology. *Journal of Consulting and Clinical Psychology*, 46, 806.
  This is one of my favorite academic articles on any subject. Meehl's thesis is that null hypothesis significance testing has held hindered progress in some fields of psychology by excusing researchers from conducting "risky" tests of their theories. A risky test is a test that a false theory is likely to fail. In contrast, Meehl argues that the null hypothesis significance test is an easy threshold to cross for any large psychological study.

Mayo, D. G. (2018). *Statistical Inference as Severe Testing: How to Get Beyond the Statistics Wars*. Cambridge University Press.
  A leading philosopher of statistics tours the logical underpinnings, history, and connections to scientific practice of many of the ideas encountered in this chapter.

Simmons, J. P., Nelson, L. D., & Simonsohn, U. (2011). False-positive psychology: Undisclosed flexibility in data collection and analysis allows presenting anything as significant. *Psychological Science*, 22, 1359–1366.
  This piece re-started a serious conversation within psychology about p hacking.

Wasserstein, R. L., & Lazar, N. A. (2016). The ASA's statement on p-values: context, process, and purpose. *The American Statistician*, 70, 129–133.

The American Statistical Association (ASA) released this brief statement on p values and their interpretation. The advice is good, and the 20 attached commentaries are thought provoking.

Switching gears to other topics represented in the chapter:

The R package pwr (Champely, 2009) can perform power calculations for most basic analyses. More advanced analyses may require bespoke code.

Kraemer, H. C., & Blasey, C. (2015). *How Many Subjects? Statistical Power Analysis in Research*. Sage Publications, Thousand Oaks, CA.

Guidance on pre-study power analysis. The book is aimed at medical researchers but covers methods and concepts useful in many fields.

CHAPTER 8

Semiparametric estimation and inference

> **Key terms:** Bootstrap, Empirical distribution function, Functional, Method of moments, Nonparametric, Parametric, Permutation test, Plug-in estimator, Resampling methods, Sample moment, Semiparametric

In the previous two chapters, we explored some fundamental concepts statisticians use to organize their thoughts about estimation and inference. Though we discussed some of the properties that good estimators and inference procedures have, we delayed the question of how one goes about developing methods for estimation and inference until now. In this and the next two chapters, we consider three general approaches to estimation and inference. The first, which requires the fewest assumptions of the three, is semiparametric estimation and inference. Before defining "semiparametric," it is best to start by clarifying the terms "parametric" and "nonparametric"—semiparametric models sit between these.

"**Parametric**" means "governed by parameters." For example, normal distributions are parametric—all the properties of a normally distributed random variable are determined by its parameters: θ and σ^2. This is a strong property. An infinity of values that describe the behavior of a random variable—the values of the probability density function and cumulative distribution function at every point, from which follow the expectation, variance, and many other properties—are determined by just two numbers. If we have a model for a process we care about that depends on a finite number of parameters, then we can learn about the process by trying to estimate or make inferences about the parameters that we postulate.

"**Nonparametric**" means the opposite of "parametric"—*not* governed by parameters. Suppose we wanted to learn about a random variable with a probability density function given by the curve shown in Figure 8-1. The curve is a valid probability density function—it is strictly non-negative, and the total area between it and the x axis is 1—but it is not a member of any well-studied distribution family.[1] There is no small set of parameters that concisely describes the curve and that we would realistically attempt to estimate. But we can still learn things about the random variable—we could estimate its expectation or variance, for example. Many of the methods covered in this chapter—plug-in estimation,

---

[1] The curve in Figure 8-1 can actually be characterized by a finite—albeit large and unwieldy—set of parameters, and so the distribution of a random variable with the curve as its density function is technically parametric. Nonetheless, it serves to illustrate the point. The "parameters" that characterize this function are not the kind that would interest anyone—they have been arbitrarily chosen to make a strange-looking function, not to describe a process concisely.

Statistical Thinking from Scratch: A Primer for Scientists. M. D. Edge, Oxford University Press (2019).
© M. D. Edge 2019. DOI: 10.1093/oso/9780198827627.001.0001

bootstrapping, and permutation tests—apply to nonparametric as well as semiparametric models. We usually apply nonparametric methods when we cannot justifiably assume that a random variable's distribution is a member of any particular distribution family.

The subject of this chapter is **semiparametric** estimation and inference, which applies when we want to estimate parameters of a model whose behavior is partially but not completely governed by parameters. For example, the model for linear regression that we developed at the end of Chapter 5 is semiparametric.[2] It relies on parameters α and β, the intercept and slope (see Equation 5.23), but the distributions of X, the independent variable, and ϵ, the disturbance, are left unspecified. Because the distributions of X and ϵ are not assumed to be members of a family that is described by a set of parameters, they are treated as nonparametric. The model as a whole is called "semiparametric" because it has a parametric part—the relationship between X, ϵ, and Y defined in Equation 5.23—and a nonparametric part, the distributions of X and ϵ.

In semiparametric and nonparametric statistics, the star of the show is the **empirical distribution function**. The empirical distribution function is a way of summarizing all the information that the data provide about a random variable's distribution, and its properties are the basis of both plug-in estimation and of bootstrapping, two of the major subjects of this chapter.

Recall that the cumulative distribution function of a random variable X gives the probability that X takes a value less than or equal to some constant: $F_X(z) = P(X \leq z)$. Similarly, the empirical distribution function $\widehat{F}_n(z)$ gives the proportion of *observations in a sample* that are less than or equal to z. More formally, if $X_1, X_2, \ldots, X_n$ represent n independent and identically distributed observations, then the empirical distribution function is

$$\widehat{F}_n(z) = \frac{1}{n}\sum_{i=1}^{n} I_{X_i \leq z}. \tag{8.1}$$

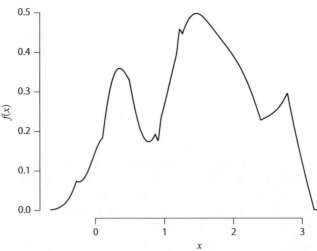

Figure 8-1 A probability density function of a random variable best treated with nonparametric (or semiparametric) methods.

[2] The simple regression model discussed in this chapter—though it is semiparametric by the definition given here—bears little resemblance to the class of models statisticians typically mean to evoke with the phrase "semiparametric regression." For a review of some recent research, see Ruppert, Wand, & Carroll (2009).

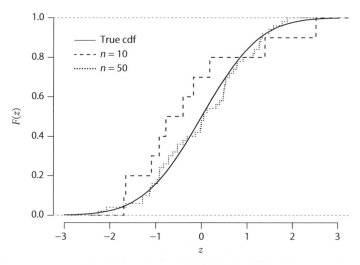

Figure 8-2 The true cumulative distribution function (cdf) for the Normal (0,1) distribution (solid curve), with two empirical distribution functions, one based on a sample of 10 observations and another based on a sample of 50 observations. Empirical distribution functions have a step-like shape. Every observation is represented by an upward tick in the empirical distribution function, with the length of each upward tick equal to the fraction of the observations that take the value corresponding to the number on the horizontal axis. Thus, as the number of observations included in the empirical distribution function increases, the number of steps increases and the steps become shorter.

$I_{X_i \leq z}$ is an *indicator variable*; it indicates whether observation X_i is less than or equal to z: $I_{X_i \leq z} = 1$ if $X_i \leq z$, and $I_{X_i \leq z} = 0$ if $X_i > z$. Figure 8-2 shows two empirical distribution functions drawn alongside a true cumulative distribution function.

The empirical distribution function is a consistent estimator of the cumulative distribution function. In the following exercises, you will study the consistency of the empirical distribution function through both simulations and mathematics. Because it is possible to estimate the cumulative distribution consistently, we can often derive consistent estimators for quantities that can be expressed in terms of the cumulative distribution function.

Exercise Set 8-1[3]

(1) The following R code makes a figure similar to Figure 8-2 using a standard normal distribution and a sample size of 20 for the empirical distribution function:

```
n <- 20 #size of sample for ecdf.
x.vals <- seq(-3, 3, length.out = 10000)
Fx <- pnorm(x.vals, 0, 1)
plot(x.vals, Fx, xlab = "z", ylab = "F(z)", type = "l")
x <- rnorm(n, 0, 1)
lines(ecdf(x), verticals = TRUE, do.points = FALSE, lty = 2)
```

(a) Run this code and examine the plot that results. Now run it several more times, changing the value of n from 20 to another number. Here are some values you can try: 2, 5, 10, 50, 100, 200, 500. What do you notice as the sample size gets larger?

---

[3] See the book's GitHub repository at github.com/mdedge/stfs/ for detailed solutions and R scripts.

(b) [Optional] Try to write a similar R script to perform the same task with a different distribution family, such as the Poisson.
(2) [Optional] In this problem, you will sketch a proof of the consistency of the empirical distribution function as an estimator of the true cumulative distribution function.
    (a) Consider $I_{X_i \leq z}$, the indicator variable in Equation 8.1. What is $P(I_{X_i \leq z} = 1)$ in terms of the cumulative distribution function, $F_X(z)$? What is $E(I_{X_i \leq z})$? What is $Var(I_{X_i \leq z})$?
    (b) Now consider the empirical distribution itself, $\widehat{F}_n(z) = \frac{1}{n}\sum_{i=1}^{n} I_{X_i \leq z}$ (Equation 8.1). What is $E[\widehat{F}_n(z)]$? What is $Var[\widehat{F}_n(z)]$? Remember that the $I_{X_i \leq z}$ are independent. What do these properties imply about the empirical distribution function? (Hint: see Equations 6.5 and 6.7.)

8.1 Semiparametric point estimation using the method of moments

The **method of moments** is a semiparametric approach to estimation. The method of moments can *also* be applied in fully parametric situations, but in fully parametric situations it is often overshadowed by the method of maximum likelihood, which is covered in the next chapter.

To carry out the method of moments, we follow these steps:

(1) Write equations that give the moments of several random variables—the *j*th **moment** of a random variable X is $E(X^j)$—in terms of the parameters to be estimated.
(2) Solve the equations from (1) for the desired parameters, giving expressions for the parameters in terms of the moments.
(3) Finally, estimate the moments and plug the estimates into the expressions for the parameters from (2).

Here is a simple, fully parametric example: Suppose we want to estimate μ and σ^2 using a sample of independent observations from a Normal(θ, σ^2) distribution. We already know that the first moment of X, if X has a Normal(θ, σ^2) distribution, is $E(X) = \mu$. We also know that the variance of X is $Var(X) = \sigma^2$. Recall that for all random variables, $Var(X) = E(X^2) - [E(X)]^2$, which means that the second moment of X is $E(X^2) = Var(X) + [E(X)]^2 = \sigma^2 + \mu^2$. We thus have equations for the first two moments in terms of the parameters (step 1), and the next step is to solve those equations so that the parameters are expressed in terms of the moments. In this case, the solution is

$$\mu = E(X),$$

$$\sigma^2 = E(X^2) - [E(X)]^2.$$

Thus, if we can estimate the first two moments, $E(X)$ and $E(X^2)$, then we can also estimate the two parameters μ and σ^2.

This is sensible, but to apply the method of moments to semiparametric situations, we need to be able to estimate the moments of arbitrary distributions. Whenever the moments of a distribution exist,[4] we can estimate them using plug-in estimators.

---

[4] Remember that some random variables have distributions in which some moments cannot be computed. In some cases in which moments do not exist, the parameters *can* be expressed in terms of other quantities, such as the median. In that case, the parameters can be estimated using a method similar to the method of moments.

8.1.1 *Plug-in estimators*

The main idea of **plug-in estimation** is intuitive. Suppose we want to estimate some property of a random variable X's distribution—its expectation, median, 25th percentile, or something else—and that we will collect a sample of observations $X_1, X_2, \ldots, X_n$ that were independently drawn from the same distribution as X. When we compute a plug-in estimator, we simply treat the sample as if it were exactly the same as the underlying distribution. For example, to estimate the expectation, $E(X)$—which is, you'll recall, a "mean" of the random variable's distribution—we would simply compute the mean of the sample. To estimate the median of X's distribution, we would compute the median of the sample. To estimate the 25th percentile of X's distribution, we would compute the 25th percentile of the sample.

An optional, more formal description of plug-in estimation

In plug-in estimation, we are taking some operation that we would like to be able to perform on a random variable's cumulative distribution function—such as applying a function that would give the expectation of the random variable—and we are performing that same operation on the empirical distribution function that arises from a sample of observations.

More formally, plug-in estimators are estimators of quantities that are *functions* of a cumulative distribution function. Functions of other functions are sometimes called *functionals*. For example, the expectation is a function of the cumulative distribution function, and thus a functional. To see why, recall that a continuous random variable X has expectation $E(X) = \int_{-\infty}^{\infty} x f_X(x) dx$ (Equation 5.2). Remember that the probability density function is the derivative of the cumulative distribution function—i.e., $f_X(x) = F_X'(x)$ (Equation 4.7)—so the expectation, though we write it as a function of the probability density function, can also be viewed as a function of the cumulative distribution function. All the moments of a random variable are functions of its cumulative distribution function, as are the median and all quantiles.

To compute the plug-in estimator of a functional, we estimate the cumulative distribution function with the empirical distribution function. Because we can estimate the cumulative distribution function consistently, we can also estimate many *functions* of the cumulative distribution function—i.e., **functionals**—consistently.

Given a functional of the cumulative distribution function, $T[F_X(z)]$, the plug-in estimator of $T[F_X(z)]$ is $T[\widehat{F}_n(z)]$. In words, we compute plug-in estimators by treating the sample as if it represented the entire population. A more concrete example may help.

> **Example: The sample mean as the plug-in estimator of** $E(X)$ Suppose a discrete random variable[5] X has cumulative distribution function F_X and probability mass function f_X. By Equation 5.1, the expected value of X is $E(X) = \sum_i x_i f_X(x_i)$, where the sum is over all the values X can take. To write the expectation in terms of the cumulative distribution function, denote the ith-smallest value that X can take x_i, so the smallest value X takes is x_1, the second-smallest is x_2, and so on. Define $F_X(x_0) = 0$, where x_0 is some value smaller than x_1. Then, by Equation 4.4, we can write the probability mass function in terms of the cumulative distribution function as $f_X(x_i) = F_X(x_i) - F_X(x_{i-1})$. We can then write the expected value of X as $E(X) = \sum_i x_i [F_X(x_i) - F_X(x_{i-1})]$.

---

[5] For continuous random variables, the approach is similar, but there are some technical issues surrounding the relationship between sums and integrals that we can skip.

To compute the plug-in estimator of $E(X)$, which we can write as $\tilde{\mu}$, we imagine drawing a sample of n observations, with each drawn independently from the same distribution as X. We substitute the empirical distribution function for the cumulative distribution function in the expression for the expectation, giving $\tilde{\mu} = \sum_i x_i[\widehat{F}_n(x_i) - \widehat{F}_n(x_{i-1})]$. This may not yet look familiar, but notice that $\widehat{F}_n(x_i) - \widehat{F}_n(x_{i-1}) = j_i/n$, where n is the number of observations in the sample and j_i is the number of times that the value x_i was observed. Thus, $\tilde{\mu} = \frac{1}{n}\sum_i x_i j_i$, where the sum has one term for each unique value observed. This is equal to the sample mean: we take each number we observed, multiply by the number of times we observed it, sum these products, and divide by the total number of observations. So the sample mean is the plug-in estimator of the expectation.

The more formal part is now over.

Using the same reasoning as in the example, we can write an expression for the plug-in estimator for the kth moment of a random variable. Suppose X is a random variable and that we sample n independent observations with the same distribution as X; label these observations $X_1, X_2, \ldots, X_n$. Then the plug-in estimator of the kth moment of X, or $E(X^k)$, is the kth **sample moment**,

$$\overline{X^k} = \frac{1}{n}\sum_{i=1}^{n} X_i^k. \tag{8.2}$$

In words, the kth sample moment is calculated by raising the value of each observation to the kth power, summing the numbers that result, and dividing by the number of observations.[6] The method also extends to *joint* moments of two or more random variables. For example, suppose that X and Y are two jointly distributed random variables and that we have n paired observations $(X_1, Y_1), (X_2, Y_2), \ldots, (X_n, Y_n)$, where each pair is an independent observation from the joint distribution of X and Y. The plug-in estimator of the joint moment $E(XY)$ is the average of the products of each instance of X multiplied by its corresponding instance of Y, or $\frac{1}{n}\sum_{i=1}^{n}X_i Y_i$. One can also compute plug-in estimators of functions of moments—such as the variance, which is $E(X^2) - [E(X)]^2$, a function of the first and second moment of X—by applying the same function to plug-in estimators of the moments.

How do plug-in estimators fare by the criteria we established in Chapter 6? One admirable quality of plug-in estimators is that they are usually consistent:[7] as the number of observations increases, plug-in estimators of functionals of the cumulative distribution approach the true values of the functionals they estimate. Nonetheless, plug-in estimators can be biased, and in some circumstances, they have larger risk than some alternatives. Their mean squared error is often larger than—and, for large samples, is no smaller than—the variance of maximum-likelihood estimators, which we will discuss in the next chapter.

Despite being overshadowed by maximum-likelihood estimators in many contexts, plug-in estimators can be used in some situations in which maximum-likelihood estimators are unavailable or difficult to compute. Notice, for example, that we do not need to specify the distribution family of a random variable in order to estimate its moments using sample moments. Plug-in estimators consistently estimate the moments of any random variable with any distribution, provided that those moments exist and that the sample actually reflects independent observations from the distribution of interest.

---

[6] Because we are talking about a plug-in *estimator*, we write this quantity as a function of random variables X_i. A plug-in *estimate* is an instance of this estimator calculated from instances of the random variables X_i, which instances we would write in lowercase as x_i.

[7] The plug-in estimators that you're most likely to see in practice are consistent, but consistency depends on some assumptions about the functional being estimated.

Exercise Set 8-2

(1) (a) Suppose $X_1, X_2, \ldots, X_n$ are independent and identically distributed observations. What is the plug-in estimator of the variance of X?

   (b) What is the plug-in estimator of the standard deviation of X?

   (c) Suppose $(X_1, Y_1), (X_2, Y_2), \ldots, (X_n, Y_n)$ are pairs of observations, with each pair drawn independently of the other pairs from an identical joint distribution. What is the plug-in estimator of the covariance of X and Y?

   (d) Under the assumptions in part (c), what is the plug-in estimator for the correlation of X and Y?

(2) Though the plug-in estimator of the variance (Problem 1a) is consistent, it is biased downward.

   (a) Demonstrate the downward bias of the plug-in estimator of the variance using simulation. In R, draw 100,000 independent samples of five observations each from a Normal$(0, 1)$ distribution. (Recall that the variance of a random variable with a Normal$(0, 1)$ distribution is 1.) Compute the plug-in estimator of the variance on each sample. What is the mean? Is it very close to 1?

   (b) Repeat the simulation you performed in part (a) for sample sizes of 2 through 10. Look at the pattern. Can you guess what the expectation of the plug-in estimator of the variance is, in terms of n, the sample size? Using this insight, suggest an unbiased estimator of the variance.

   (c) [Optional, hard] Prove that the expectation of the plug-in estimator of the sample variance is the expression you identified in part (b).

3) (Optional) Suppose $X \sim$ Normal$(0, 1)$. What, approximately, is $E(X^4)$? How about the 5th, 6th, 7th, and 8th moments? Use simulation. (Hint: Use R, remembering that mean() takes the sum of all the numbers in a vector and divides by the length of the vector.)

8.1.2 *The method of moments*

The method of moments takes advantage of the consistency of sample moments to produce consistent estimates of *parameters*. The principle of the method is simple—we express the parameters we want to estimate in terms of moments, and then we estimate the parameters by replacing the true moments with estimated moments.[8] Each equation expressing the value of a parameter in terms of moments is called a "moment condition." As with any system of equations, generally a unique solution exists only when the number of equations equals the number of unknown terms, so we use as many moment conditions as we have parameters to estimate.[9] To restate the steps from the beginning of Section 8.1 more formally, suppose we have independent, identically distributed observations $X_1, \ldots, X_n$ governed by k parameters $\theta_1, \ldots, \theta_k$. The steps are as follows:

(1) Identify k expressions, one for each of the first k moments (i.e., $E(X), E(X^2), \ldots, E(X^k)$), in terms of the parameters.

(2) Solve the k equations identified in (1) for the k parameters, giving expressions for the parameters in terms of the first k moments.

---

[8] Sometimes, the calculus necessary for writing a distribution's moments in terms of its parameters is tricky. But if we work with a well-studied distribution, the moments are likely to be available already.

[9] The *generalized* method of moments, which is beyond our scope, allows the use of more moment conditions than parameters.

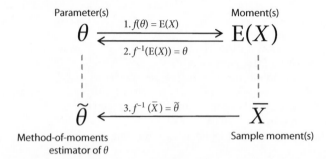

Parameter(s)   Moment(s)

θ $\rightleftarrows$ $\text{E}(X)$

1. $f(\theta) = \text{E}(X)$
2. $f^{-1}(\text{E}(X)) = \theta$

$\widetilde{\theta}$ $\xleftarrow{\quad 3.\, f^{-1}(\bar{X}) = \widetilde{\theta} \quad}$ $\overline{X}$

Method-of-moments   Sample moment(s)
estimator of θ

Figure 8-3 A schematic of the method of moments. The goal is to estimate a parameter (or, in practice, more than one)—represented by θ (upper left)—that governs the distribution of a random variable, X. The first step is to identify the moment(s) of X—that is, the values E(X), E(X²), E(X³),...—in terms of the parameter(s). Here, f(θ) is a function that takes the parameter(s) as input and returns the moments(s). Its inverse—that is, the function that take the moment(s) as its argument(s) and returns the parameter value(s)—is f⁻¹, and identifying it is the second step. If we knew the exact values of the moment(s), then we could compute the parameter(s) using the inverse function f⁻¹. We generally do not know the exact values of the moment(s), but we can estimate the moment(s) by computing the sample moment(s). In step 3, method-of-moments estimators are computed by feeding the sample moment(s) to f⁻¹. The dashed gray lines represent estimand–estimator relationships, with the unknown estimand above the estimator.

(3) Replace the moments in the expressions from step (2) with the sample moments (i.e., $\overline{X}, \overline{X^2}, \dots, \overline{X^k}$). These expressions are the method-of-moments estimators of the parameters.

Figure 8-3 illustrates the procedure.

We can apply the method of moments to the model for simple linear regression outlined at the end of Chapter 5. Recall that in that model, Y is a linear function of X plus a random disturbance ϵ, i.e. (Equation 5.23),

$$Y = \alpha + \beta X + \epsilon.$$

Recall that we made the linearity assumption (A1):

Assumption A1 $\text{E}(\epsilon \mid X = x) = 0$ *for all x.*

See Figure 8-4 for schematics of models that obey the linearity assumption but disobey some other common regression assumptions. The method-of-moments estimators we are about to derive would be consistent for any of the models sketched in Figure 8-4.[10]

After making Assumption A1, we derived the results in Equations 5.24 and 5.29:

$$\text{E}(Y) = \alpha + \beta \mu_X$$

and

$$\text{Cov}(X, Y) = \beta \sigma_X^2,$$

---

[10] The method-of-moments estimator is consistent for all these models, but that does not mean that it is the best possible choice. The Gauss–Markov theorem that gives an optimality result about homoscedastic models, like those on the left side. However, the Gauss–Markov theorem result only considers a limited class of estimators—for example, it does not consider the least-absolute-errors line.

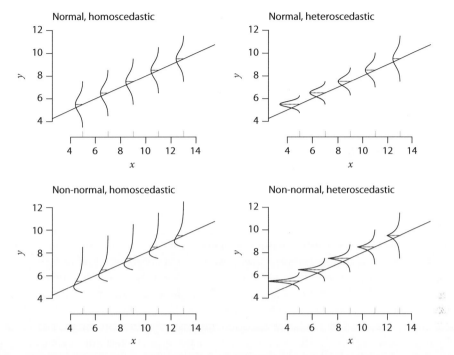

Figure 8-4 Schematics of four models that obey the linearity assumption. As in Figure 5-8, in each case, a "true" regression line is shown, along with density functions (rotated 90°) representing the distribution of Y conditional on X. In all four cases, the expectation of Y given that X = x is a linear function of x. The method-of-moments estimator we describe here is consistent in all these cases.

where $\mu_X = E(X)$ and $\sigma_X^2 = \text{Var}(X)$. Remembering that the covariance $\text{Cov}(X, Y) = E(XY) - E(X)E(Y)$ and that $\text{Var}(X) = E(X^2) - [E(X)]^2$, these two equations can be solved to give expressions for the parameters α and β in terms of the moments of X and Y, including the joint moment $E(XY)$. Specifically,

$$\alpha = E(Y) - \beta E(X)$$

and

$$\beta = \frac{\text{Cov}(X, Y)}{\sigma_X^2} = \frac{E(XY) - E(X)E(Y)}{E(X^2) - [E(X)]^2}.$$

The expression for β is entirely in terms of moments. To get the method-of-moments estimator of β, which we label $\widetilde{\beta}$, we replace the moments in the expression for β with sample moments. If we sample n independent pairs of observations (X_1, Y_1), $(X_2, Y_2), \ldots, (X_n, Y_n)$, then

$$\widetilde{\beta} = \frac{\frac{1}{n}\sum_{i=1}^{n} X_i Y_i - \frac{1}{n}\left(\sum_{i=1}^{n} X_i\right)\frac{1}{n}\left(\sum_{i=1}^{n} Y_i\right)}{\frac{1}{n}\sum_{i=1}^{n} X_i^2 - \left(\frac{1}{n}\sum_{i=1}^{n} X_i\right)^2}.$$

Multiplying the numerator and denominator by n simplifies the expression a little, giving

$$\widetilde{\beta} = \frac{\sum_{i=1}^{n} X_i Y_i - \frac{1}{n}\left(\sum_{i=1}^{n} X_i\right)\left(\sum_{i=1}^{n} Y_i\right)}{\sum_{i=1}^{n} X_i^2 - \frac{1}{n}\left(\sum_{i=1}^{n} X_i\right)^2}. \tag{8.3}$$

The expression for α above has β in it, but now that we have a method-of-moments estimator $\widetilde{\beta}$, we can substitute it for β to get an estimator for α. The method-of-moments estimator of α, which we call $\widetilde{\alpha}$, is

$$\widetilde{\alpha} = \frac{1}{n}\sum_{i=1}^{n} Y_i - \widetilde{\beta}\frac{1}{n}\sum_{i=1}^{n} X_i = \frac{\sum_{i=1}^{n} Y_i - \widetilde{\beta}\sum_{i=1}^{n} X_i}{n}, \tag{8.4}$$

where $\widetilde{\beta}$ is given in Equation 8.3. The estimators in Equations 8.3 and 8.4 are consistent. We can be assured of their consistency without making strong assumptions about the shape of the distribution of ϵ—we only need assume that its expectation is 0 regardless of the value X takes.

Now here is something interesting. Look back at the expressions for the coefficients of the least-squares line found in Equations 3.6 and 3.7, and compare them with Equations 8.3 and 8.4. You will notice that the slope of the least-squares line in Equation 3.7 is identical to the method-of-moments estimator of β in Equation 8.3, with the exception that the observations (lowercase x and y) are replaced by random variables (uppercase X and Y). Similarly, the intercept of the least-squares line is identical to the method-of-moments estimator of α. We started our investigation of probability with the goal of interpreting the least-squares line, and we have reached a major reward. Under the linearity assumption (A1), the least-squares line provides the method-of-moments estimates of α and β, which are consistent. (You performed simulations that agree with the consistency of the least-squares estimators in Exercise Set 6-9.)

To interpret the least-squares line as an estimator, we made assumptions about the way in which the data were generated. In particular, we assumed they were generated by a linear model. One way to think of estimation is as a way of imbuing data summaries, such the least-squares line, with meaning. A statistic, such as $\widetilde{b}$ (Equation 3.7), is merely a function of some observed measurements. But if we make assumptions about the source of the measurements, we can sometimes develop new and satisfying interpretations of statistics. Often, if we are willing to make stronger assumptions, then we can make richer interpretations. For now, we have made relatively mild assumptions, and the reward has been consistency—a nice but also rather minimal property. We have no reason to think that the least-squares estimator is a uniquely good choice.[11] In the next chapter, we will explore some conditions under which we can make stronger claims.

Exercise Set 8-3

(1) Assume that $X_1, X_2, \ldots, X_n$ are independent and identically distributed as Uniform$(0, b)$. What is the method-of-moments estimator of b?

(2) [Optional] Express the method-of-moments estimator of β (Equation 8.3) in terms of the plug-in estimators of the correlation and standard deviation of X and Y.

---

[11] For example, in Exercise Set 6-9, you showed that if the disturbance term has a Laplace distribution, then the least-absolute-errors line provides more efficient estimates than the least-squares line. Further, the least-absolute-errors line is more robust to data contamination than least squares.

8.2 Semiparametric interval estimation using the bootstrap

The method of moments provides consistent point estimators in semiparametric (and fully parametric) settings. We would also like to be able to assess their uncertainty and make interval estimates. The **bootstrap**[12]—named for the expression "to pull oneself up by one's bootstraps"—is a versatile tool for arriving at estimated standard errors and confidence intervals.

As with the method of moments, one way of understanding the bootstrap is to notice that with enough high-quality data, the empirical distribution function is a reasonable approximation to the true cumulative distribution function.

To illustrate the bootstrap, consider the case of estimating the standard error of the median of a sample of n observations drawn from an unknown distribution. Though it is fairly straightforward to estimate the standard error of the *mean* of such a sample, the standard error of the *median* is harder to pin down. (It's still possible to do it mathematically, but it's harder.) We can estimate it with the bootstrap.

If we had access to the true probability distribution function or the true cumulative distribution function, then we could estimate the standard error of the median of a sample of n observations easily. We would simulate many independent samples of n observations, take the median of each sample, and estimate the standard deviation of the set of sample medians. Because we can consistently estimate the standard deviation of any random variable,[13] such a method will converge to the correct answer, provided that we produce enough simulated samples. Most of the time, though, we cannot implement this lovely method—we do not know the true cumulative distribution function.

Though we do not know the cumulative distribution, we *can* estimate it consistently using the empirical distribution function. Thus, we can draw samples from the *estimated* cumulative distribution function. Such samples ought to be similar to the samples we would draw if we had access to the true cumulative distribution function. This is the key insight of the bootstrap: loosely, we treat the original data as if they reflect the underlying distribution exactly, and we draw samples from the original data—that is, we "resample"—to simulate the sampling distribution of a statistic of interest. Figure 8-5 shows a schematic of the procedure.

A little more formally, suppose that we want to approximate the distribution of a statistic $\widehat{\vartheta}(D)$. $\widehat{\vartheta}$ is a function—for example, the median, mean, or maximum—applied to a sample of independent and identically distributed observations D. The observations in D might be drawn from an unknown distribution. One way to write the bootstrap method is:

(1) Compute the empirical distribution function of the data D.
(2) Draw many simulated samples—say B of them—of n observations each from the empirical distribution function. Label these "bootstrap" samples $D_1^*, D_2^*, \ldots, D_B^*$.
(3) For each of the simulated samples, calculate the statistic of interest $\widehat{\vartheta}$. That is, calculate $\widehat{\vartheta}(D_1^*), \widehat{\vartheta}(D_2^*), \ldots, \widehat{\vartheta}(D_B^*)$. These values form the *bootstrap distribution* of $\widehat{\vartheta}(D)$.

Under some mild assumptions, the bootstrap distribution is a consistent estimator of the distribution of $\widehat{\vartheta}(D)$. The precision of the estimation depends on n, the number of observations from the original population, and B, the number of bootstrap samples. B is constrained only by computing time, and we can usually make B large cheaply: $B \geq 10,000$ is

---

[12] We will confine our discussion to what is sometimes called the "nonparametric" bootstrap, which is what most people mean when they say "bootstrap." There is also a "parametric" bootstrap.

[13] Provided that the standard deviation exists and that the data are not contaminated by outliers. We could estimate it using plug-in estimation, for example.

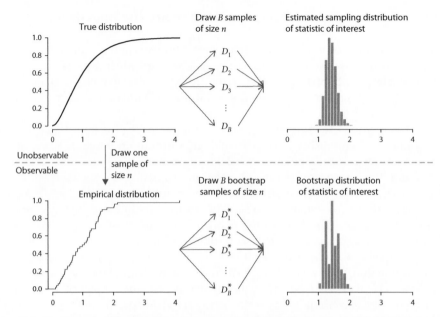

Figure 8-5 A schematic illustrating the bootstrap procedure. Suppose that there is a statistic of interest, $\hat{\vartheta}(D)$, whose sampling distribution we need to estimate for samples of size n. In particular, we are interested because we have access to a sample of size n, and we want to make interval estimates. The upper panel shows a nearly foolproof strategy that would be available if we had access to the true distribution of the data—we could draw B samples from the underlying distributions and compute the statistic on each one, estimating the sampling distribution directly. However, the true distribution from which the data are drawn is usually unobservable. The bootstrap strategy is illustrated in the lower panel. We do not have access to the true distribution, but we do have an empirical distribution, computed from the sample we have just drawn. We draw B bootstrap samples from the empirical distribution by sampling n observations with replacement from the original data B times. For each bootstrap sample, we compute the statistic of interest. In many cases, the resulting bootstrap distribution approximates the unknown sampling distribution, as it does here.

not uncommon. It is harder to make one's dataset larger than to take more bootstrap samples, so the accuracy of the bootstrap distribution is limited much more by n, which determines how closely the empirical distribution function approximates the cumulative distribution function, than by B.

One step in the above procedure may seem mysterious—what exactly does it mean to sample from the empirical distribution function? As you will see in an optional exercise below, sampling n observations from the empirical distribution function is equivalent to sampling n observations *with replacement* from the original observations. That is, we randomly select n observations one at a time, and each time we make a selection, every observation from the original set has an equal chance of being chosen, regardless of whether it has been chosen for this bootstrap sample already.[14] We can rewrite the bootstrap procedure as

---

[14] For all but the smallest values of n, a sample drawn with replacement from the original dataset will almost certainly include duplicates of some of the observations and exclude others. One can show that for large samples, approximately $e^{-1} \approx 37\%$ of the observations will be excluded from a given bootstrap sample. e^{-1} is $P(X = 0)$ if X has a Poisson(1) distribution.

(1) Take B "bootstrap" samples, with each sample being a set of n observations drawn with replacement from the original observations. Label these samples $D_1^*, D_2^*, \ldots, D_B^*$.

(2) For each of the simulated samples, calculate the statistic of interest $\hat{\vartheta}$. That is, calculate $\hat{\vartheta}(D_1^*), \hat{\vartheta}(D_2^*), \ldots, \hat{\vartheta}(D_B^*)$. These values form the *bootstrap distribution* of $\hat{\vartheta}(D)$.

This version of the procedure is equivalent to the three-step procedure above. It is a little harder to see how this version works, but it is easier to see exactly how one would carry it out.

It's time for an example. We can calculate the bootstrap distribution of $\tilde{\beta}$, the method-of-moments estimator in Equation 8.3. Specifically, we will calculate the bootstrap distribution of $\tilde{\beta}$ applied to the agricultural data we have been considering throughout this book, shown first in Figure 1-1. To apply the bootstrap to $\tilde{\beta}$, we need to assume that the 11 x, y pairs in the sample represent independent draws from a joint distribution:[15]

Assumption A3-S *For all i and all $j \neq i$, X_i, Y_i are independent of X_j, Y_j. Call this the "independence of units" assumption.*[16]

Assumption A4-S *For all i, X_i, Y_i are drawn from the same joint distribution $F_{X,Y}(x, y)$. Call this the "distribution" assumption.*

In addition to Assumptions A3-S and A4-S,[17] we will adopt the linearity assumption (A1) in all the discussion below.[18]

The bootstrap is an asymptotic procedure, meaning that its guarantees of accuracy depend on a sample size approaching infinity. In practice, the bootstrap can work well for moderately sized samples, but—as we will explore in an exercise—one ought to have a sample larger than 11 observations, which is what we have. Nonetheless, we'll apply the bootstrap, bearing in mind its limitations for small samples.

Table 8-1 shows the original data on the left, sorted by fertilizer consumption. On the right, on a gray background, one bootstrap sample is shown. Each observation in the bootstrap sample appears in the original data. Some observations from the original data appear more than once in the bootstrap sample, and some do not appear at all. Notice also that in this case, one "observation" consists of a *pair* of numbers—a fertilizer consumption measurement and a cereal yield measurement. Because the slope estimate depends on the pairings of the x and y observations, we preserve the pairings of the original observations in each bootstrap sample.

Calculating $\tilde{\beta}$ (Equation 8.3) for the original data yields $\tilde{\beta} = 0.50$, whereas calculating $\tilde{\beta}$ using this bootstrap sample yields $\tilde{\beta}^* = 0.64$. (The $*$ superscript indicates that this is a result from a bootstrap sample.) To get the bootstrap distribution of $\tilde{\beta}$, we will program the computer to repeat this procedure many times, drawing a bootstrap sample, calculating $\tilde{\beta}$ for the bootstrap sample, and recording it.

---

[15] These are the assumptions for what is called a "case-based" bootstrap, which is what we will do here. One can also perform a "residual-based" bootstrap, in which residuals are resampled with replacement. The residual-based bootstrap makes more restrictive assumptions than the case-based bootstrap.

[16] The "units" here are the things being measured to produce x, y pairs of observations. In our running agricultural example, the units are countries, which are measured to obtain average fertilizer use and crop yield observations. We will use it from here until chapter 10, but the independence assumption deserves to be treated with care. In our example, one might want to consider the spatial arrangement of countries, perhaps treating neighbors as related rather than independent. We will return to this briefly in the Postlude.

[17] The "S" is for "semiparametric—we will invoke different versions of these assumptions in the next chapter.

[18] Assumptions A3-S and A4-S do not entail Assumption A1, the linearity assumption. There is a subtle distinction to make here. It is perfectly possible to use the bootstrap to estimate the distribution of values that $\tilde{\beta}$ might take in different samples, even if the linearity assumption is false. The trouble is that if the linearity assumption is false, then $\tilde{\beta}$ is not necessarily a good estimator of β. Therefore any uncertainty-related statements—like confidence intervals—would not necessarily be confidence intervals for β. One possible interpretation of such a confidence interval would be that it summarizes uncertainty about the value to which $\tilde{\beta}$ would converge as the sample size becomes infinitely large.

Table 8-1 Original fertilizer use and cereal yield data, along with one bootstrap sample. In R, you can produce the original data with

```
> cbind(anscombe$x1, anscombe$y1)[order(anscombe$x1),]
```

To see the bootstrap data, use

```
> set.seed(8675309)
> cbind(anscombe$x1, anscombe$y1)[sample(1:11, replace = TRUE),]
```

To see different bootstrap samples that might be drawn, re-run the second command without resetting the seed.

Original data		One bootstrap sample	
Fertilizer consumption (kg/hectare)	Cereal yield (100 kg/hectare)	Fertilizer consumption (kg/hectare)	Cereal yield (100 kg/hectare)
4	4.26	8	6.95
5	5.68	14	9.96
6	7.24	12	10.84
7	4.82	12	10.84
8	6.95	13	7.58
9	8.81	4	4.26
10	8.04	5	5.68
11	8.33	7	4.82
12	10.84	7	4.82
13	7.58	11	8.33
14	9.96	7	4.82

We now need a way to draw bootstrap samples. Here is a function,[19] boot.samp(), that takes either a vector or a matrix and returns a bootstrap sample of either the entries in the vector or the rows in the matrix. We will be using boot.samp() to sample rows from a two-column matrix. The core of boot.samp() is the sample() function, which samples entries from a vector with or without replacement:

```
#Draw a bootstrap sample
boot.samp <- function(x) {
  #If x is a vector, convert it to a matrix with one column.
  if(is.null(dim(x))) {
    x <- matrix(x, ncol = 1)
  }
  n <- nrow(x)
  boot.inds <- sample(1:n, replace = TRUE)
  x[boot.inds,]
}
```

---

[19] This function and the next are available in the book's R package, stfspack (see Exercise Set 2-2, Problem 3 for installation instructions).

We also need a function to compute $\widetilde{\beta}$ (Equation 8.3) from two vectors. Here is a function that translates Equation 8.3 into R syntax:[20]

```
beta.mm <- function(x, y){
  n <- length(x)
  (sum(x*y) - (1/n)*sum(x)*sum(y)) /
    (sum(x^2) - (1/n)*sum(x)^2)
}
```

Now we can draw a bootstrap sample and compute the estimator in which we're interested. All we need is a way to repeat the procedure many times. The following code repeats the procedure 10,000 times using a `for()` loop:[21]

```
set.seed(8675309) #Optional.22
B <- 10000
boot.dist <- numeric(B)
dat <- cbind(anscombe$x1, anscombe$y1)
for(i in 1:B){
  samp <- boot.samp(dat)
  boot.dist[i] <- beta.mm(samp[,1], samp[,2])
}
```

The variable `boot.dist` now stores the bootstrap distribution of $\widetilde{\beta}$.

The bootstrap distribution approximates the sampling distribution of $\widetilde{\beta}$. There are several ways to extract useful information from the bootstrap distribution. First, we can plot it, as in Figure 8-6, which shows a histogram of 10,000 values of $\widetilde{\beta}^*$ calculated from bootstrap samples. The histogram reveals that the $\widetilde{\beta}^*$ values are symmetric and approximately normally distributed[23] around a mean of 0.5. Next, we can use the bootstrap distribution to estimate the standard error of $\widetilde{\beta}$. Because the bootstrap distribution is an estimate of the distribution of $\widetilde{\beta}$ that would arise from repeated sampling, the standard deviation of the bootstrap distribution is an estimate of the standard error of $\widetilde{\beta}$—recall that the standard error is the standard deviation of an estimator's sampling distribution. The command

```
> sd(boot.dist)
[1] 0.1209691
```

gives the relevant estimate.

We can use the bootstrap distribution to develop approximate confidence intervals for $\widetilde{\beta}$. There are many methods for deriving approximate confidence intervals from bootstrap distributions. We will consider three. First, if the sampling distribution can be assumed to be normal, then we can use the bootstrap estimate of the standard error to identify a confidence interval. Denoting the standard deviation of the bootstrap distribution $\widehat{\omega}$, we adapt Equation 7.3 slightly to get the approximate $1 - \alpha$ confidence interval for a quantity of interest θ as

[20] One could also use `lm()`, but `beta.mm()` is slightly faster, which comes in handy in some of the simulation exercises below.

[21] As noted in several places in this book, most R users eschew `for()` loops wherever possible in favor of vectorized approaches. It would not be hard to vectorize this bootstrapping procedure, but the `for()` loop makes the procedure transparent and will work just fine in this case.

[22] Setting the seed allows replication of the exact results reported here.

[23] The normality is not exact—the tails are too heavy. Nonetheless, the violations of normality are relatively mild for our purposes.

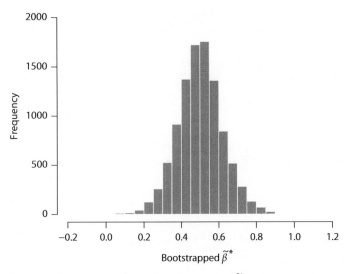

Figure 8-6 Histogram showing the distribution of values of $\widetilde{\beta}^*$ calculated from 10,000 bootstrap samples.

$$(\widehat{\theta} - \widehat{\omega} z_{\alpha/2}, \widehat{\theta} + \widehat{\omega} z_{\alpha/2}), \qquad (8.5)$$

where $\widehat{\theta}$ is the estimate of θ from the original data, $z_{\alpha/2} = \Phi^{-1}(1 - \alpha/2)$ and Φ^{-1} is the inverse of the cumulative distribution function of the Normal$(0, 1)$ distribution. For a 95% confidence interval, $z_{\alpha/2} = 1.96$, and in our example, the approximate 95% confidence interval is $(0.263, 0.737)$. Some people call this a *normal* bootstrap confidence interval.

A second approach is to use the percentiles of the bootstrap distribution itself to bound a confidence interval. To obtain a $1 - \alpha$ *percentile* confidence interval, one merely identifies the values at the $\alpha/2$ and $1 - \alpha/2$ percentiles of the bootstrap distribution. In our case, to find a 95% confidence interval, we can use

```
> quantile(boot.dist, c(0.025, 0.975))
    2.5%      97.5%
0.2617351 0.7474635
```

The result is very similar to what we obtained with the normal interval. The percentile approach works well when the bootstrap distribution is symmetric and median-unbiased, meaning that the expectation of its median is equal to the true value being estimated. Figure 8-6 suggests that the bootstrap distribution is roughly symmetric. A third approach that works in more general situations is given in (optional) Box 8-1.

Box 8-1 BOOTSTRAP PIVOTAL INTERVALS

A third approach to building confidence intervals from the bootstrap distribution is the so-called *basic* or *pivotal* bootstrap confidence interval. The main idea of the pivotal interval is to approximate the true (but unknown) sampling distribution of the estimator *around the estimand* using the distribution of the bootstrapped statistics *around the original estimate*. More formally, denote the qth percentile of the (unknown) distribution of the estimator— from which the estimate we are considering, $\widehat{\theta}$, is drawn—as $Q(q)$, and denote the qth percentile of the bootstrap distribution as $Q^*(q)$. The approximation we invoke is

(Continued)

Box 8-1 (Continued)

$$\theta - Q(q) \approx \widehat{\theta} - Q^*(q).$$

That is, the difference between the true value and the qth percentile of the estimator's sampling distribution is assumed to be approximately the same as the difference between the estimated value and the qth percentile of the bootstrap distribution. Loosely, we are assuming that the bootstrap distribution's spread around the estimate reflects the spread of the sampling distribution of the estimator around the true estimand.

Consider an artificial situation in which we know the value of the estimand and its distribution. If we knew θ and $Q(q)$ (for any value of q), then we could construct a confidence interval with any desired level of confidence. (These confidence intervals would not be of much use if we already knew θ, but stay with me.) For example, we could construct a 95% confidence interval by designing a procedure that captures θ whenever $\widehat{\theta}$ falls in the "middle" 95% of its distribution—meaning that $Q(0.025) \leq \widehat{\theta} \leq Q(0.975)$. By definition, the probability that $\widehat{\theta}$ falls in the "middle" 95% of its distribution is 0.95.

We want an interval around the estimate $\widehat{\theta}$ that will capture the true parameter value whenever $\widehat{\theta}$ is in the middle 95% of its distribution. Let's break this into two problems. First, if the estimate is too small but within the middle 95% of its distribution, then $Q(0.025) \leq \widehat{\theta} < \theta$. Because $Q(0.025) \leq \widehat{\theta}$, we know that $\theta - Q(0.025) \geq \theta - \widehat{\theta}$, and adding $\widehat{\theta}$ to both sides gives $\widehat{\theta} + (\theta - Q(0.025)) \geq \theta$. Thus, an interval with an upper bound of $\widehat{\theta} + (\theta - Q(0.025))$ is guaranteed to capture θ if $\widehat{\theta}$ is too small but in the middle 95% of its distribution. Similarly, if the estimate is too large but within the middle 95% of its distribution, then $\theta \leq \widehat{\theta} < Q(0.975)$, meaning that $\widehat{\theta} - (Q(0.975) - \theta) \leq \theta$, and an interval with a lower bound of $\widehat{\theta} - (Q(0.975) - \theta)$ will capture θ. For the general case of a $1 - \alpha$ confidence interval, we simply replace 0.025 with $\alpha/2$ and 0.975 with $1 - \alpha/2$, giving the confidence interval

$$[\widehat{\theta} - (Q(1 - \alpha/2) - \theta), \widehat{\theta} + (\theta - Q(\alpha/2))].$$

We generally cannot identify a confidence interval in this way, because θ and the function Q are unknown, but if the approximation $\theta - Q(q) \approx \widehat{\theta} - Q^*(q)$ holds, then one can obtain an approximate interval by replacing $Q(1 - \alpha/2) - \theta$ with $Q^*(1 - \alpha/2) - \widehat{\theta}$ and replacing $\theta - Q(\alpha/2)$ with $\widehat{\theta} - Q^*(\alpha/2)$. Doing so gives

$$\left[\widehat{\theta} - \left(Q^*(1 - \alpha/2) - \widehat{\theta}\right), \widehat{\theta} + \left(\widehat{\theta} - Q^*(\alpha/2)\right)\right],$$

which simplifies to

$$[2\widehat{\theta} - Q^*(1 - \alpha/2), 2\widehat{\theta} - Q^*(\alpha/2)].$$

In the case of the bootstrap distribution shown in Figure 8-6, the pivotal interval is

```
> b.est <- beta.mm(anscombe$x1, anscombe$y1)
> 2*b.est - quantile(boot.dist, c(0.975, 0.025))
    97.5%       2.5%
0.2527183 0.7384468
```

In this case, the results are again similar to what we obtained using normal or percentile-based intervals. Pivotal bootstrap confidence intervals strike a good balance between simplicity and generality of application, and they are a good default choice when constructing a confidence interval using the bootstrap.

There are many other things one can do with bootstrap distributions, including deriving confidence intervals that correct for issues we have not considered.[24] The best thing about these and other bootstrap methods is the way they cut through difficult problems—they can be applied to statistics whose distributions are difficult or impossible to derive mathematically.

Typically, bootstrap methods make less stringent assumptions than many other approaches to interval estimation—for example, the distribution from which the data are drawn is not assumed to be a member of any known distribution family. Nonetheless, the data must meet the model's assumptions in order for conclusions drawn from the bootstrap to be valid. In our case, this means that each pair of points must be a random, independent sample from a population or process as described by Assumptions A3-S and A4-S. Further, the guarantees associated with bootstrap methods are about consistency, which is fundamentally a large-sample property. The nonparametric bootstrap is not guaranteed to work well with small samples, which produce imprecise estimates of the empirical distribution function. It also does not work well with statistics whose value can change a lot if the data change slightly. (See Problem 2 in the following exercise set for an example.)

Because of the connection between confidence intervals and p values discussed in Chapter 7, it is sometimes possible to conduct hypothesis tests using bootstrap confidence intervals. In the next section, we will cover another **resampling method** that can be used to test a range of hypotheses.

Exercise Set 8-4

(1) In this problem, you will explore the ways in which the precision of the bootstrap distribution depends on n, the sample size, and B, the number of bootstrap samples. So that we can easily assess how well the bootstrap performs, we will apply it to an example we already understand, that of estimating the standard error of the mean of a sample of independent observations from a normal distribution. The following function[25] takes a vector x of data, draws B bootstrap samples from it, and returns the mean (by default) of each sample:

```
#Return the value of a statistic (by default, the mean),
#computed from B bootstrap samples from a vector x.
boot.dist.1d <- function(x, B, FUN = mean,…) {
  boot.samps <- sample(x, length(x)*B, replace = TRUE)
  boot.mat <- matrix(boot.samps, ncol = B)
  apply(boot.mat, 2, FUN,…)
}
```

Use boot.dist.1d() to explore the way in which the accuracy and precision of the bootstrap samples depend on n, the sample size, and B, the number of bootstrap samples. Try n values of 5, 20, 50, and 100; try B values of 10, 100, 1,000, and 5,000. Remember that the mean of a sample of n independent normal observations is normally distributed with the same expectation as the distribution from which the data are drawn and with standard error $\sigma/\sqrt{n}$, where σ is the standard deviation of the distribution from which the data are drawn. Draw simulated samples using rnorm().

---

[24] See the so-called "bias-corrected and accelerated" bootstrap (Efron, 1987).
[25] Available in the book's R package, stfspack (see Exercise Set 2-2, Problem 3 for installation instructions).

You may wish to use the helper function[26] `wrap.bm()`, which simulates a normal sample, calls `boot.dist.1d()`, and summarizes the results by plotting a histogram of the bootstrap sample and returning a list containing its mean and standard deviation.

```
#Simulate a sample from a normal distribution, take B bootstrap
#samples. Plot a histogram and return the mean and standard
#deviation of the bootstrap distribution.
wrap.bm <- function(n, B, mu = 0, sigma = 1, FUN = mean, …) {
  sim <- rnorm(n, mu, sigma)
  boots <- boot.dist.1d(sim, B, FUN = FUN, …)
  hist(boots, main = "", xlab = "Bootstrap Means")
  list("boot m"= mean(boots), "boot se"= sd(boots))
}
```

(2) [Optional] The midrange is a measure of central tendency equal to the mean of the minimum and maximum values in a sample of data. In R, you can compute the midrange for a numeric vector with the function

```
midrange <- function(x) {
  (min(x) + max(x)) / 2
}
```

For independent samples of size n from a Normal$(0, 1)$ distribution, the standard error of the midrange is approximately $\pi/\sqrt{24\ln(n)}$, where ln is the natural logarithm.[27] Repeat Problem 1 for the midrange by defining the `midrange()` function and adding `FUN = midrange` to your calls of `wrap.bm()`. How well does the bootstrap standard error approximate the expected standard error?

(3) [Optional] Explain why sampling from the empirical distribution function and sampling from the original set of observations with replacement are equivalent processes. (Hint: See Exercise Set 4-6, Problem 4(b) to understand what it means to sample from a distribution function. In particular, make the figure suggested in the solution to that exercise.)

8.3 Semiparametric hypothesis testing using permutation tests

To conduct a hypothesis test, one compares the value of a test statistic with the distribution that the test statistic would have if the null hypothesis were true. If the test statistic obtained from the original data is extreme compared with the distribution it has when the null hypothesis is true, then we have some evidence against the null hypothesis. For some test statistics, it is possible to derive the null distribution mathematically. But what if we do not know the distribution family from which the data are drawn? Is there any way to obtain—or at least to approximate—the distribution that the test statistic would have if the null hypothesis were true?

In many situations, we can obtain the distribution we require by randomly shuffling—or permuting—the original data. In particular, we try to permute the observations in such a way that if the null hypothesis is true, then the permuted dataset is just as probable to

---

[26] Available in the book's R package, `stfspack` (see Exercise Set 2-2, Problem 3 for installation instructions).

[27] This is an asymptotic (i.e., large-sample-size) result (Stuart & Ord, 1987, Section 14.28), but it is approximately correct for small samples.

observe as the original dataset. If the original dataset is unusual compared with the permuted versions, then that is evidence against the null hypothesis.

Consider the agricultural data first shown in Figure 1-1. We have chosen to study these data using the model developed at the end of Chapter 5 and represented by Equation 5.23: $Y = \alpha + \beta X + \epsilon$, where Y represents cereal yield, X represents fertilizer consumption, ϵ is a disturbance, and α and β are constants. We will make the assumptions that we've already used during this chapter.

Assumption A1, linearity $E(\epsilon|X = x) = 0$ *for all x.*
Assumption A3-S, independence of units *For all i and all j≠i, X_i, Y_i are independent of X_j, Y_j.*
Assumption A4-S, distribution *For all i, X_i, Y_i are drawn from the same joint distribution $F_{X,Y}(x,y)$.*

We will also need an additional assumption:

Assumption A5 *For all i, X_i and ϵ_i are independent. Call this the "independence of disturbances and X" assumption.*

Assumption A5 also implies the homoscedasticity assumption from Chapter 5 (Assumption A2), but we will not make direct use of homoscedasticity.

Suppose we want to test the hypothesis that $\beta = 0$, or, in words, that there is no association between fertilizer consumption and cereal yield. In this model, setting $\beta = 0$ implies that

$$Y = \alpha + \epsilon. \tag{8.6}$$

In words, the cereal yield Y is the sum of α, a constant, and ϵ, a random variable with expectation 0 that is independent of X. Further, the value of ϵ contributing to each individual observation of Y is an independent draw from the same unknown distribution. Under the null hypothesis, the only random component of Y is ϵ, and because ϵ is independent of X, it follows that Y is also independent of X. The independence of X and Y under the null hypothesis will allow us to test the null hypothesis using a **permutation test**.

The surest way to see this is with some notation. Call the set of observed fertilizer consumption values $x_1, x_2, \ldots, x_{11}$. Similarly, call the set of observed cereal yield values $y_1, y_2, \ldots, y_{11}$. The observations are paired, and we can denote the pairs as $\{x_1, y_1\}, \{x_2, y_2\}, \ldots, \{x_{11}, y_{11}\}$. The probability of observing these 11 pairs of values could be represented using a joint density function

$$f_D(\{x_1, y_1\}, \{x_2, y_2\}, \ldots, \{x_{11}, y_{11}\}),$$

where the subscript D stands for "data" and is a shorthand for the joint density function governing all the observations. Because we do not know the distribution of either X or Y, we cannot be very precise about the form of f_D. But we can say enough to make progress.

First, remember that we assumed that every X, Y pair is independent of every other such pair (Assumption A3-S). Recall that in Exercise Set 5-2, Problem 3, you showed that the joint probability mass function of a set of independent random variables is the product of those random variables' individual probability mass functions. The same applies to joint probability density functions. Thus, we have

$$f_D(\{x_1, y_1\}, \{x_2, y_2\}, \ldots, \{x_{11}, y_{11}\}) = f_{X,Y}(x_1, y_1) * f_{X,Y}(x_2, y_2) * \cdots * f_{X,Y}(x_{11}, y_{11}).$$

In words, the joint density function of *all* the data is the product of the values of the joint density function for each pair of observations. Because X and Y are also independent under the null hypothesis, we can also split up the joint density of X and Y into a product: $f_{X,Y}(x, y) = f_X(x) * f_Y(y)$. So if the null hypothesis is true, then

$$f_D(\{x_1, y_1\}, \{x_2, y_2\}, \ldots, \{x_{11}, y_{11}\}) = f_X(x_1) * f_Y(y_1) * f_X(x_2) * f_Y(y_2) * \cdots * f_X(x_{11}) * f_Y(y_{11}).$$

Now notice if the null hypothesis is true, *any* pairing of $x_1, x_2, \ldots, x_{11}$ and $y_1, y_2, \ldots, y_{11}$ with one unique x and one unique y in each pair would have the same joint density function— the product of the marginal densities of each observation. As an example,

$$f_D(\{x_1, y_6\}, \{x_2, y_1\}, \ldots, \{x_{11}, y_4\}) = f_X(x_1) * f_Y(y_1) * f_X(x_2) * f_Y(y_2) * \cdots * f_X(x_{11}) * f_Y(y_{11}).$$

If the null hypothesis is true, then any way of grouping the 11 fertilizer observations and 11 cereal yield observations into pairs—with a fertilizer observation and a cereal yield observation in each pair—is as probable as any other.

This observation implies that we can generate many hypothetical samples that, if the null hypothesis is true, are just as likely to be observed as the original data. If the original data produces an extreme test statistic compared with these hypothetical samples, then we reject the null hypothesis. The procedure is as follows:

(1) Choose a test statistic and calculate it using the original data. Call the resulting value s_d.
(2) Randomly permute the data in a way that, if the null hypothesis is true, produces a hypothetical dataset that is just as probable as the original data.
(3) Calculate the test statistic using the permuted data and save the value.
(4) Repeat steps (2) and (3) many times.[28] The resulting test statistics are called a "permutation distribution."
(5) Compare the original statistic s_d with the permutation distribution.

Another way to explain permutation testing is in terms of the empirical distribution function. The paired observations (X_i, Y_i) define a joint empirical distribution, which encodes information about X, about Y, and about the association of X and Y. The information about the association of X and Y is encoded by the *pairings* of the observations. By randomly re-pairing the X and Y observations, one forms a new joint empirical distribution in which information about X and Y is preserved exactly—that is, the X and Y values have not changed—but any association between X and Y that was present in the original dataset is broken.[29] The permuted sample is thus "like" a sample that might be observed if X and Y were independent. If the degree of association observed in the original dataset is similar to that observed in permuted datasets, then there is little evidence against the hypothesis that X and Y are independent.

Table 8-2 shows the original data and a randomly permuted version of the data that, if the null hypothesis is true, is just as probable as the original data. Notice that every cereal yield observation from the original data appears exactly once in the permuted dataset, but in most cases, it is paired with a different fertilizer consumption value. Calculating $\widetilde{\beta}$ (Equation 8.3) for the original data yields $\widetilde{\beta} = 0.50$, whereas calculating $\widetilde{\beta}$ using the permuted sample yields $\widetilde{\beta}_p = -0.19$, a quite different result. (The subscript p indicates that this is a result from a permuted sample.) To identify the permutation distribution of $\widetilde{\beta}$ under the

---

[28] When working with small samples, it is sometimes possible to include every possible permutation in the permutation distribution. Usually though, we just use a large number of randomly chosen permutations.

[29] Because permutation distributions consider only reshufflings of the observed data, they are a type of *conditional* inference—the inference is conditional on the marginal empirical distributions of X and Y. Conditional inference raises some interesting puzzles in statistical theory. For a technical treatment, see Lehmann & Romano (2005) Chapter 10, especially Example 10.2.5.

Table 8-2 Original fertilizer use and cereal yield data, along with one randomly permuted sample. In R, to see the permuted data, use

```
> set.seed(8675309)
> cbind(anscombe$x1, anscombe$y1[sample(1:11, replace = FALSE)])[order
(anscombe$x1),]
```

To see different possible permutations, re-run the second command without resetting the seed.

Original data		One randomly permuted sample	
Fertilizer consumption (kg/hectare)	Cereal yield (100 kg/hectare)	Fertilizer consumption (kg/hectare)	Cereal yield (100 kg/hectare)
4	4.26	4	8.81
5	5.68	5	4.26
6	7.24	6	9.96
7	4.82	7	8.04
8	6.95	8	8.33
9	8.81	9	10.84
10	8.04	10	6.95
11	8.33	11	5.68
12	10.84	12	7.58
13	7.58	13	7.24
14	9.96	14	4.82

null hypothesis that $\beta = 0$, we will program the computer to repeat this procedure many times, each time permuting the sample, calculating $\tilde{\beta}$ for the permuted sample, and recording it.

We need a function that will permute the sample in the required way, shuffling the pairings of observations. Here is a function[30] that takes a matrix and permutes its columns independently.[31] As with boot.samp(), the core of the function is the sample() function, which we now apply without replacement:

```
#Permute the columns of a matrix independently.
perm.samp <- function(x) {
  apply(x, 2, sample)
}
```

We will use beta.mm() to compute $\tilde{\beta}$ (Equation 8.3) for each permuted sample. The below code repeats the procedure 10,000 times using a for() loop.

```
set.seed(8675309) #Optional.32
nperms <- 10000
perm.dist <- numeric(nperms)
dat <- cbind(anscombe$x1, anscombe$y1)
for(i in 1:nperms) {
  samp <- perm.samp(dat)
  perm.dist[i] <- beta.mm(samp[,1], samp[,2])
}
```

---

[30] Available in the book's R package, stfspack (see Exercise Set 2-2, Problem 3 for installation instructions).
[31] We only need to permute one of the columns in this case, but there's no harm in doing both.
[32] Setting the seed allows replication of the exact results reported here.

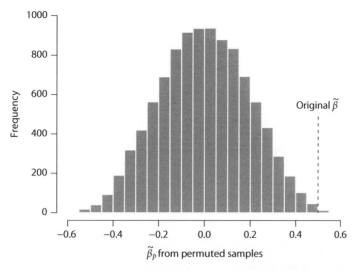

Figure 8-7 The permutation distribution of $\tilde{\beta}$. The value of $\tilde{\beta}$ computed from the original data is extreme compared with the permutation distribution, which is evidence against the null hypothesis.

The variable `perm.dist` now stores the permutation distribution of $\tilde{\beta}$, which is displayed as a histogram in Figure 8-7.

To compute a two-sided p value testing the null hypothesis that $\beta = 0$, calculate the proportion of the permuted samples for which the absolute value of $\tilde{\beta}_p$ is greater than the absolute value of $\tilde{\beta}$ calculated from the original data:

```
> b.orig <- beta.mm(anscombe$x1, anscombe$y1)
> b.orig
[1] 0.5000909
> mean(abs(perm.dist) >= b.orig) #p value
[1] 0.0033
```

Thus, $p \approx 0.003$, which most people would take as some evidence against the null hypothesis.

We have tested the hypothesis that $\beta = 0$, and we have done it without assuming that the distribution of the disturbance term ϵ belongs to any particular family. It is nice to be able to avoid the assumption that ϵ is, say, normally distributed.

Overenthusiastic advocates of permutation tests sometimes describe them as "assumption-free." They are not. For our permutation test procedure, we assumed linearity, independence of units, a common distribution from which all observations are drawn, and independence of the independent variable X_i from the disturbance term ϵ_i. Assuming all these claims, we compute p under the null hypothesis that $\beta = 0$. To say that we are testing the null hypothesis that $\beta = 0$ is a somewhat misleading shorthand. It would be fairer to say that we are testing the null hypothesis that $\beta = 0$ *and* that Assumptions A1, A3-S, A4-S, and A5 hold. A low p could, in principle, mean either that $\beta \neq 0$, that any one of our assumptions is false, or simply that a rare event has occurred. This issue is not unique to permutation tests, and we discussed it in Chapter 7—I mention it here to emphasize that even permutation tests and other nonparametric and semiparametric tests make important assumptions.

Permutation tests are valuable and merit careful study. Their flexibility is especially useful—permutation test can be applied to many hypotheses that involve independent sets of observations. Permutation tests' power also tends to be comparable to that of alternatives that make more stringent distributional assumptions. In the following exercises and in the next chapter, you will have the chance to develop your ability to apply permutation tests and to compare them with alternatives.

Exercise Set 8-5

(1) A client brings you yield data from 20 wheat fields. Your client is interested in the effects of a fertilizer additive called substance Z. She randomly assigned 10 of the fields to receive fertilizer supplemented with substance Z; the other 10 fields received the same fertilizer without substance Z added. The yields from distinct fields can be assumed to be independent of each other. Design a permutation procedure to assess the claim that adding substance Z to the fertilizer changes the wheat yield. What are the null and alternative hypotheses being tested?

(2) The sim.perm.B function (in stfspack, see Exercise Set 2-2, Problem 3) uses the sim.lm() function (Exercise Set 5-5, Problem 2) as well as perm.samp() and beta.mm() (main text) to simulate data according to the simple linear model from the end of Chapter 5, computing two-sided permutation p values that test the null hypothesis that X and Y are independent. Its parameters and defaults are

```
sim.perm.B(n, nsim, a, b, nperm = 500, sigma.disturb = 1, mu.x = 8, sigma.x =
2, rdisturb = rnorm, rx = rnorm, het.coef = 0)
```

n is the number of pairs of observations to draw per simulation, nsim is the number of simulations to run, and nperm is the number of random permutations to compute to form the permutation distribution. The other parameters are the same as in sim.lm(); the key arguments are a, the intercept in the simulated linear model, and b, the slope β. The output is a vector of nsim permutation p values, one for each simulation. For example, to generate 20 p values from simulations with intercept 3, slope 0.5, and 10 pairs of observations per trial, use

```
sim.perm.B(n = 10, nsim = 20, a = 3, b = 0.5)
```

Use sim.perm.B() to examine the type I error rate and power of a two-sided permutation test when the significance level is 0.05. Test the null hypothesis that $\beta = 0$ using simulated data with $\beta = 0$, $\beta = 0.1$, and $\beta = 0.2$, and with the variance of the disturbances set to 1 (the default). Use $n = 10$, $n = 50$, and $n = 100$ pairs of observations in each simulated dataset, and permute each simulated dataset 500 times (the default). Generate 500 datasets for each combination of n and β values. You can choose any value of a you want and leave all other parameters at their default values.

8.4 Conclusion

It has long been possible to develop consistent point estimates using plug-in estimation and the method of moments. The ideas behind nonparametric and semiparametric interval estimation and inference are old, too. Permutation tests trace back at least to the 1930s.[33] The bootstrap, though not as old as permutation testing, was conceived by Bradley Efron in

---

[33] In fact, permutation tests are in some sense the original hypothesis testing procedure. In 1936, R. A. Fisher suggested that the goal of the methods he developed was to recover the conclusions that could have been obtained by permutation testing, if only it were more practical. He wrote, "the statistician does not carry out this very simple and very tedious process, but his conclusions have no justification beyond the fact that they agree with those which could have been arrived at by this elementary method" (quoted in Ernst, 2004).

the 1970s, and a similar method called the jackknife is even older. But resampling methods were impractical until most researchers had easy access to computers. Before that, researchers who wanted to avoid distributional assumptions had to rely on procedures that were lighter on computation.[34] Though these methods are clever, they are not as versatile as bootstrapping and permutation testing.

Though they are more important now than ever, resampling methods are still underused in many fields. Resampling methods entail fewer assumptions than some other methods and are often conceptually simple. An appropriately designed bootstrapping or permutation procedure can prevent hand-wringing about distributional assumptions or spare you difficult math. The computational cost is often easy to bear.

Nonetheless, it's possible to be too cavalier about resampling methods, and their genuine virtues are surrounded by mythical ones. It is sometimes claimed that bootstrap methods are especially good when very few observations are available—this is the opposite of the truth. Permutation tests, for their part, test null hypotheses that are more general than the user sometimes appreciates.

In the next chapter, we will cover parametric methods, which make more involved assumptions about the process generating the data. In exchange for more stringent assumptions, parametric methods offer computational ease, guarantees of efficiency, and more power.

8.5 Chapter summary

Nonparametric and semiparametric statistical methods assume models whose properties cannot be described by a finite number of parameters. For example, a linear regression model that assumes that the disturbances are independent draws from an unknown distribution is semiparametric—it includes the intercept and slope as regression parameters but has a nonparametric part, the unknown distribution of the disturbances.

Nonparametric and semiparametric methods focus on the empirical distribution function, which, assuming that the data are really independent observations from the same distribution, is a consistent estimator of the true cumulative distribution function. With plug-in estimation and the method of moments, we estimate functionals or parameters by treating the empirical distribution function as if it were the true cumulative distribution function. Such estimators are consistent for a broad class of estimands. To study the sampling distributions of point estimates, we use bootstrapping to resample from the empirical distribution function. For hypothesis testing, one can either use a bootstrap-based confidence interval or conduct a permutation test, which can be designed to test null hypotheses of independence.

Resampling methods—including bootstrapping and permutation testing—are flexible and easy to implement with a little programming expertise. Though they are not a panacea, they can be extremely useful.

8.6 Further reading

Efron, B., & Tibshirani, R. (1986). Bootstrap methods for standard errors, confidence intervals, and other measures of statistical accuracy. *Statistical Science*, 1, 54–75.

---

[34] There are many nonparametric older nonparametric inference procedures that do not require intensive resampling, such as such as the Kruskal-Wallis test and the Mann–Whitney *U* test.

An introduction to bootstrap methods and to some of the things that one can do with a bootstrap distribution. The math is more demanding than in this book, but you will be able to read the most important parts. The first author, Efron, is the inventor of the bootstrap (Efron, 1979).

Ernst, M. D. (2004). Permutation methods: a basis for exact inference. *Statistical Science*, 19, 676–685.

A clear introduction to permutation testing. The math is at a slightly higher level than in this book.

Hothorn, T., Hornik, K., Van De Wiel, M. A., & Zeileis, A. (2008). Implementing a class of permutation tests: the `coin` package. *Journal of Statistical Software*, 28, 1–23.

The `coin` package is popular and provides a set of methods for running permutation tests in R.

There is also the package `boot` (Davison & Hinkley, 1997; Canty & Ripley, 2017) for bootstrapping.

CHAPTER 9

Parametric estimation and inference

Key terms: Fisher information, Functional invariance property of maximum-likelihood estimators, Likelihood function, Likelihood-ratio test, Logarithm (or log), Log-likelihood, Maximum-likelihood estimation, Wald test, Wilks' theorem

---

We continue to take on assumptions in exchange for richer meanings. In this chapter, we consider fully parametric models—those in which the behavior of the entire model, including the distributions of all random variables, can be described by a finite number of parameters. Parametric models are more completely structured than the semiparametric and nonparametric models of the previous chapter, but they are less completely structured than the Bayesian approaches of the next chapter.

Most introductions to statistics focus on parametric methods like those in this chapter. The reasons for privileging these methods are not as strong as they were before computers were widely available, making resampling methods and Bayesian methods much more practical than they once were. Nonetheless, there are still good reasons for learning these methods—provided that their assumptions are met, they are consistent, efficient, and powerful. Further, even if you tend to prefer nonparametric or Bayesian methods, understanding parametric methods will help you communicate with your fellow data analysts.

Whereas in Chapter 8, the most important mathematical object was the empirical distribution function, the star of this chapter is the **likelihood function**, or just "likelihood." The likelihood allows us to compare values of a parameter θ in terms of their ability to explain the observed data. Roughly, the likelihood answers the question, "How likely are the observed data d if the true model parameter is θ?"

We interrogate the likelihood by providing candidate parameter values as arguments. Suppose the observed data d are taken to be an instance of a random variable[1] D, with probability mass function or probability density function $f_D(d|\theta)$. The likelihood $L(\theta)$ is equal to the probability mass function or probability density function evaluated at the observed data,

$$L(\theta) = f_D(d|\theta). \tag{9.1}$$

If the likelihood function is equal to the probability mass or density function of the observed data, then why define it as a separate idea? Why not just refer to the density—a

---

[1] This random variable can be a vector, which means it can contain as many observations as we want.

Statistical Thinking from Scratch: A Primer for Scientists. M. D. Edge, Oxford University Press (2019).
© M. D. Edge 2019. DOI: 10.1093/oso/9780198827627.001.0001

construct with which we are already familiar? The reason is that it is convenient to think of the likelihood as a function of the parameter (or parameters) rather than as a function of the data. We view the likelihood this way to ask questions about the plausibility of the data—which are now assumed to be fixed—assuming different values of the parameter(s).

Much of the time, we do not work directly with likelihoods—instead, we work with their logarithms (see Box 9-1). The reason is that likelihoods are often large products, which are mathematically inconvenient. Taking the logarithm of the likelihood lets us work with large sums instead or products—much easier. For example, suppose the data are n independent observations $x_1, x_2, ..., x_n$, each with density function $f_X(x)$, which depends on a parameter θ. Because the observations are independent and identically distributed, the joint density function is

$$f_{X_1, X_2, ..., X_n}(x_1, x_2, ..., x_n) = f_X(x_1) * f_X(x_2) * \cdots * f_X(x_n) = \prod_{i=1}^{n} f_X(x_i).$$

(The symbol $\prod$ denotes multiplication in the same way that $\sum$ denotes addition.) Such a large product is typically difficult to manipulate mathematically. The log is easier. The log of a product is the sum of the logs of the terms being multiplied (see Equation 9.3 in Box 9-1), so

$$\ln[f_X(x_1) * f_X(x_2) * \cdots * f_X(x_n)] = \sum_{i=1}^{n} \ln[f_X(x_i)].$$

Motivated by this relationship, we define the log of the likelihood function, or **log-likelihood** function, which is written with a lowercase *l*:

$$l(\theta) = \ln[L(\theta)]. \tag{9.2}$$

The log-likelihood is easier to work with than the likelihood function, and if we ever need to consult the likelihood function itself, we can recover it by inverting the relationship in Equation 9.2. See Figure 9-1 for a depiction of a likelihood and a log-likelihood.

Box 9-1 LOGARITHMS

To work with likelihood functions, we will need **logarithms**, which haven't appeared yet in this book. The logarithm of a number z is the power to which some number, called a "base," needs to be raised in order to equal z. Here, we are only concerned with "natural" logs—logarithms that use Euler's number $e \approx 2.718...$ as their base. The natural log function is written ln, and we can define ln with the equation

$$z = e^{\ln(z)}.$$

For our purposes, logarithms have two important properties.

First, the logarithm of a product is the sum of the logarithms of the terms being multiplied. In symbols,

$$\ln(yz) = \ln(y) + \ln(z). \tag{9.3}$$

To prove this, notice that by the definition of the natural logarithm, $yz = e^{\ln(y)}e^{\ln(z)}$, and recall from algebra that if a, b, and c are real numbers, then $a^b a^c = a^{b+c}$. These facts imply that $e^{\ln(y)}e^{\ln(z)} = e^{\ln(y)+\ln(z)}$ and therefore that $\ln(yz) = \ln(y) + \ln(z)$.

(Continued)

Box 9-1 (Continued)

Second, recall the "up and under" rule for integration of polynomials of the form ax^n (Exercise Set A-3 in Appendix A, Problem 1). The "up and under" rule does not apply when $n = -1$, because division by zero is not allowed. It turns out that the natural log function is the solution that completes our ability to integrate polynomials. The integral of ax^{-1} is a natural log function:

$$\int ax^{-1}dx = a\ln(x) + C. \tag{9.4a}$$

Because differentiation and integration are inverse processes (Equation A.5), Equation 9.4a implies that the derivative of $a\ln(x)$ with respect to x is

$$\frac{d}{dx}a\ln(x) = ax^{-1} = a/x. \tag{9.4b}$$

We will need this fact often, since we need to take the derivative of log-likelihood functions in order to maximize them.

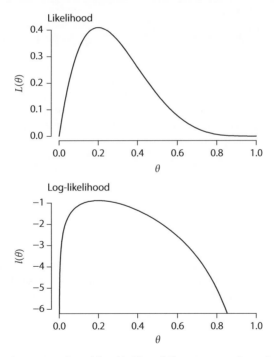

Figure 9-1 A likelihood (top panel) and log-likelihood (bottom panel) on the vertical axis, against candidate values of a parameter on the horizontal axis. The data used are five independent flips of a coin with unknown probability θ of showing heads on each flip. In the data shown, one of the five flips was heads, and the other four were tails. The likelihood and log-likelihood resemble each other—they both start low, rise to a single peak at $\theta = 0.2$, and fall again as θ approaches 1. One difference is that the likelihood function cannot be negative, whereas the log-likelihood is unbounded from below.

The likelihood provides a framework for estimation and inference. We will consider parametric point estimation via the method of maximum likelihood, in which estimates are chosen to maximize the probability of (or the probability density associated with) observing the data in hand. We will also sketch the calculation of standard errors of maximum likelihood estimates. Finally, we will consider likelihood-ratio tests, which test competing hypotheses against each other by comparing the likelihoods that result from them.

Exercise Set 9-1[2]

(1) Identify whether the italicized statement is true or false. If it is false, modify it in a way that makes it true. *The value of θ that maximizes L(θ) is the most probable value of θ given the observed data.*

(2) Write all of the following in the simplest form you can:
   (a) The probability density function of X, a Normal(μ, σ^2) random variable.
   (b) The log-likelihood of μ given an observation x, which is assumed to be an instance of X.
   (c) The joint density of X_1 and X_2, two independent Normal(μ, σ^2) random variables.
   (d) The log-likelihood of μ given observations x_1 and x_2, which are assumed to be instances of X_1 and X_2.
   (e) The log-likelihood of μ given observations $x_1, x_2, ..., x_n$, which are assumed to be instances of n independent random variables, each with a Normal(μ, σ^2) distribution.

9.1 Parametric estimation using maximum likelihood

The method of maximum likelihood is probably the most popular method of estimation. In words, the **maximum-likelihood estimate** of a parameter is the value of the parameter that maximizes the probability of observing the data. More formally, suppose we wish to estimate a parameter θ using data d. As in the previous section, the data are viewed as an instance of a random variable D, with probability mass function or probability density function $f_D(d|\theta)$. Following the definition in Equation 9.1, the likelihood is $L(\theta) = f_D(d|\theta)$. The maximum-likelihood estimate of θ, which we can call $\widehat{\theta}$, is

$$\widehat{\theta} = \underset{\theta}{\text{argmax}}\, L(\theta). \tag{9.5}$$

"argmax" is short for "argument of the maximum." Here, it means that $\widehat{\theta}$ is the value of θ that maximizes $L(\theta)$, assuming that $L(\theta)$ has a unique maximum. $\widehat{\theta}$ is *not* the maximum value that $L(\theta)$ takes, which would be written $\max L(\theta)$. Conveniently, the value of θ that maximizes the likelihood function is the same value that would maximize the log of the likelihood function.[3] Thus, we can equivalently write

$$\widehat{\theta} = \underset{\theta}{\text{argmax}}\, l(\theta). \tag{9.6}$$

In practice, we usually use Equation 9.6 rather than Equation 9.5 to find maximum-likelihood estimators.

---

[2] See the book's GitHub repository at github.com/mdedge/stfs/ for detailed solutions and R scripts.
[3] Remember that $z = e^{\ln(z)}$, so increasing the value of z always increases the value of $\ln(z)$. Also see Figure 9-1 for an example.

To identify a maximum-likelihood estimator, use three steps:

(1) Write down the likelihood function $L(\theta)$.
(2) Take the log of the likelihood function to get $l(\theta)$, and simplify it as much as possible.
(3) Maximize the log-likelihood function $l(\theta)$ in terms of θ. The value of θ that maximizes $l(\theta)$ also maximizes $L(\theta)$; it is the maximum-likelihood estimator $\hat{\theta}$.

Let's turn to an example to clarify these steps. Suppose that we sample n independent, identically distributed observations from an Exponential(λ) distribution (see Table 4-2). Exponential distributions are often used to model waiting times, such as the time it takes for components of a machine to fail. We can imagine that we have observed the failure times of n lightbulbs, and we want to estimate the failure rate, λ. The n observations are independent random variables $X_1, X_2, \ldots, X_n$, each with probability density function $f_X(x) = \lambda e^{-\lambda x}$ for $x > 0$ and $f_X(x) = 0$ otherwise—this is the density function for the exponential family. We wish to estimate λ using the method of maximum likelihood.

The first step is to write down the joint probability density function of the observations. Because the observations are independent, their joint density is the product of their individual probability density functions. Thus, the joint probability density function describing $X_1, X_2, \ldots, X_n$ is

$$f_X(x_1, x_2, \ldots, x_n | \lambda) = \prod_{i=1}^{n} \lambda e^{-\lambda x_i}.$$

To remind ourselves that we seek to maximize this function with respect to a *parameter*, we relabel it as a likelihood function:

$$L(\lambda) = \prod_{i=1}^{n} \lambda e^{-\lambda x_i}.$$

This is the function we will maximize.

As written, $L(\lambda)$ is difficult to maximize. Thus, the second step is to take the log-likelihood function l. Let's break this into steps. Remember that, by Equation 9.3, the natural log of a product is the sum of the natural logs of the terms being multiplied. We therefore convert the product from i to n into a sum from i to n:

$$l(\lambda) = \ln[L(\lambda)] = \ln(\prod_{i=1}^{n} \lambda e^{-\lambda x_i}) = \sum_{i=1}^{n} \ln(\lambda e^{-\lambda x_i}).$$

Take this one step further by noticing that $\lambda e^{-\lambda x_i}$ is itself a product, so $\ln(\lambda e^{-\lambda x_i}) = \ln(\lambda) + \ln(e^{-\lambda x_i})$. And, by the definition of the natural logarithm, $\ln(e^{-\lambda x_i}) = -\lambda x_i$. Putting this together gives a simplified form of the log-likelihood function,

$$l(\lambda) = \sum_{i=1}^{n} [\ln(\lambda) - \lambda x_i].$$

We can simplify still further. The first term in the sum, $\ln(\lambda)$, does not depend on i, so we can pull it out of the sum and replace it with $n \ln(\lambda)$. The second term features a λ that does not depend on i and can be removed from the sum. Thus, we have

$$l(\lambda) = n \ln(\lambda) - \lambda \sum_{i=1}^{n} x_i.$$

Now remove the sum entirely by remembering that the sample mean is $\bar{x} = \frac{1}{n} \sum_{i=1}^{n} x_i$, which lets us replace $\sum_{i=1}^{n} x_i$ with $n\bar{x}$,

$$l(\lambda) = n \ln(\lambda) - \lambda n \bar{x}.$$

The log-likelihood is now in a form that we can maximize.

The third step is to find the value of λ that maximizes the log-likelihood function. Recall that one way to optimize a function is to find the values of the function's arguments that set the derivative of the function equal to zero. That is what we did in Chapter 3 to minimize the squared line errors, and it is what we will do in this case. The log-likelihood function is written as the sum of two functions of λ, and, by Equation A.4, the derivative of the sum is the sum of the derivatives. By Equation 9.4b, the derivative with respect to λ of $n \ln(\lambda)$ is $n\lambda^{-1} = n/\lambda$. The derivative of $\lambda n \bar{x}$ is $n \bar{x}$ (by Equation A.2). Thus,

$$\frac{d}{d\lambda} l(\lambda) = \frac{n}{\lambda} - n\bar{x}.$$

To set the derivative of the log-likelihood function to zero, we solve the equation

$$0 = \frac{n}{\lambda} - n\bar{x}$$

for λ. Dividing both sides by n gives

$$0 = \frac{1}{\lambda} - \bar{x}.$$

Adding $\bar{x}$ to both sides and taking the reciprocal of both sides gives the solution,

$$\lambda = \frac{1}{\bar{x}}.$$

This is the only optimum of the log-likelihood function. To confirm that it is a maximum, we check the second derivative of the log-likelihood function (Exercise Set A-1, Problem 2), which is

$$\frac{d^2}{d\lambda^2} l(\lambda) = -\frac{n}{\lambda^2}.$$

Because n and λ^2 are always positive, the second derivative is always negative, which implies that $\lambda = 1/\bar{x}$ maximizes the log-likelihood function. Thus, for independent, identically distributed observations from an exponential distribution, the maximum-likelihood estimate of λ is the reciprocal of the sample mean,

$$\hat{\lambda} = \frac{1}{\bar{x}}.$$

Though it is not always possible to maximize the log-likelihood using calculus, it is often convenient to maximize it numerically using a computer. Maximum-likelihood estimators have many desirable properties under relatively weak conditions. We will ignore some of the technical assumptions, but remember the most important one: we assume that the data are actually being generated by the model specified by the likelihood function. If the data are produced in another way, then we will have good estimates of the wrong quantities, which may be less helpful than suboptimal estimates of the right quantities.

(1) Maximum-likelihood estimators are consistent—they converge to the true value of the estimand as the sample size increases. (Consistency implies asymptotic unbiasedness, but maximum-likelihood estimators can be biased in finite samples.)
(2) Maximum-likelihood estimators are asymptotically normally distributed. This means that as the number of samples drawn approaches infinity, the distribution of the maximum-likelihood estimator approaches a normal distribution. In practice, many

maximum-likelihood estimators are approximately normal for moderate-to-large samples.

(3) In the large-sample limit, there are no consistent estimators with lower mean squared error than maximum-likelihood estimators. That is, maximum-likelihood estimators are asymptotically efficient.

(4) Maximum-likelihood estimators are **functionally invariant**. That is, suppose that you want to estimate a function of a parameter, call it $\tau = g(\theta)$, where θ is a parameter. Then the maximum-likelihood estimator of τ is $\widehat{\tau} = g(\widehat{\theta})$, where $\widehat{\theta}$ is the maximum-likelihood estimator of θ.

Three caveats are in order. First, and once again, the maximum-likelihood estimate is *not* the value of the parameter that is most probable given the data. It is the parameter value that makes *the data* most probable. Remember Bayes' theorem (Equation 4.2). Loosely, if D represents the data, then the maximum-likelihood estimator of θ maximizes $P(D = d|\theta)$. It does *not* maximize $P(\theta|D = d)$. (More on this in the next chapter.)

Second, much of the appeal of maximum-likelihood estimation comes from the guarantee of efficiency—at least with large samples, maximum-likelihood estimators obtain, in some sense, the maximum level of precision possible for an accurate estimator. But precision is being measured with respect to mean squared error. This is a reasonable default, but if squared errors do not describe the costs of mistaken estimates—for example, if it is much more desirable to avoid underestimates than overestimates—then estimators that minimize mean squared error may not be appropriate.

Third, it bears repeating that maximum-likelihood estimation assumes that the postulated model actually generated the data. If the model is a poor fit for the situation under study, then maximum-likelihood estimates from that model might be meaningless. Before considering maximum-likelihood estimation for simple linear regression, we pause for some exercises.

Exercise Set 9-2

(1) Suppose that $X_1, X_2, ..., X_n$ are independent and identically distributed as Bernoulli(p), meaning that independently for all i, $P(X_i = 1) = p$ and $P(X_i = 0) = 1 - p$. The probability mass function for each observation is therefore $P(X_i = x_i) = p^{x_i}(1 - p)^{1-x_i}$ for $x_i \in \{0, 1\}$.
  (a) What is the likelihood function $L(p)$?
  (b) What is the log-likelihood function $l(p)$?
  (c) [Optional] Use R to draw ten independent observations from a Bernoulli(0.6) distribution. (Remember that a Bernoulli(p) random variable has the same distribution as a Binomial(1, p) random variable, and use rbinom().) Make two plots with p on the x axis for $p \in (0, 1)$. On one of the plots, draw the likelihood function on the y axis, and on the other, draw the log-likelihood function on the y axis. Hint: Use prod() to calculate the product of all entries in a vector.
  (d) [Optional] Compare the plot you made in part (c) with the results of summary() applied to the observations you drew in each part. What do you think is the maximum-likelihood estimator of p?

(2) Suppose that $X_1, X_2, ..., X_n$ are independent and identically distributed as Normal(θ, σ^2).
  (a) What is the maximum-likelihood estimator of θ? (Refer to Exercise Set 9-1, Problem 2 for the log-likelihood.)
  (b) [Optional] If $Y = e^X$ and X is normally distributed, then Y is said to have a "log-normal" distribution. If $X \sim$ Normal(θ, σ^2), then $E(Y) = e^\theta$. Suppose that $Y_1, Y_2, ..., Y_n$ are independent and have an identical log-normal distribution with expectation e^θ. What is the maximum-likelihood estimator of e^θ? (Hint: use your answer to part (a) and the properties of the maximum-likelihood estimator listed in the main text.)

9.1.1 *Maximum-likelihood estimation for simple linear regression*

We now apply maximum likelihood to a version of the model for simple linear regression. Specifically, we assume that we have n pairs of observations of an independent variable x and a dependent variable Y produced by the model

$$Y_i = \alpha + \beta x_i + \epsilon_i. \tag{9.7}$$

As before, α and β are fixed constants. In addition, the observations x_i of the independent variable are no longer considered to be random—they are now fixed constants. The only random variable remaining on the right side of the equation is the disturbance term, ϵ_i. We assume that the n disturbance terms—one for each pair of observations (x_i, Y_i)—are independent and identically distributed as Normal$(0, \sigma^2)$. Because the only random variable in the model, ϵ_i, is assumed to have a parametric distribution—that is, a distribution completely described by parameters—this model is fully parametric.

There are two major differences between this model and the one to which we applied the method of moments in the previous chapter. First, the independent variable x_i is no longer viewed as random.[4] If you like, you can view this model as conditional on the independent variable.

Second, the assumptions have changed. In Chapter 8, we made the following assumptions, invoking A5 only to perform a permutation test:

Assumption A1: linearity $E(\epsilon|X = x) = 0$ *for all x.*

Assumption A3-S: independence of units *For all i and all j≠i, X_i, Y_i are independent of X_j, Y_j.*

Assumption A4-S: distribution *For all i, X_i, Y_i are drawn from the same joint distribution $F_{X,Y}(x, y)$.*

Assumption A5 *For all i, X_i and ϵ_i are independent. Call this the "independence of disturbances and X" assumption.*

And now we assume the following:

Assumption A1: linearity $E(\epsilon|X = x) = 0$ *for all x.*

Assumption A2: homoscedasticity $Var(\epsilon|X = x) = \sigma^2$.

Assumption A3-P: independence of units *For all i and all j≠i, ϵ_i is independent of ϵ_j.*

Assumption A4-P: distribution *For all i, ϵ_i is drawn from a normal distribution.*

Whereas Assumption A1 is the same as before, we have added a new assumption (A2), and we have made a more stringent distributional assumption (A4-P). These two new assumptions—normality of disturbances and constant variance of the disturbances—are substantive and warrant inspection whenever applied to data.[5] Figure 9-2 is a reproduction of Figure 8-4, showing sketches of linear models that follow different assumptions.

We seek to estimate α and β by maximum likelihood. To obtain the likelihood, notice that the only random component of Y_i is ϵ_i, which is normally distributed with $E(\epsilon_i) = 0$ and

---

[4] Both models assume that the independent variable is measured without error. "Errors-in-variables" models consider the possibility that the independent variable is measured imperfectly, which is true in most real-life applications.

[5] Note that these assumptions cannot be investigated directly, however, because we do not *observe* the disturbances. We observe only the (x_i, Y_i) pairs. More on this in the Postlude.

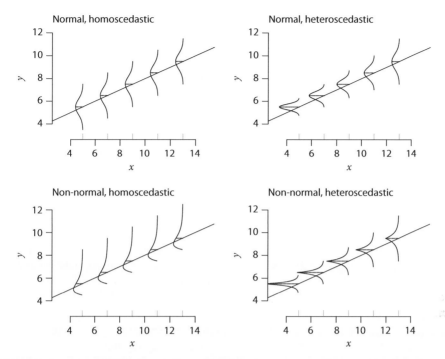

Figure 9-2 Schematics of four possible linear models. As in Figures 5-8 and 8-4, in each case, a "true" regression line is shown, along with density functions (rotated 90°) representing the distribution of Y conditional on X. The likelihood-based methods we develop in this chapter apply only to the normal, constant variance (i.e. homoscedastic) model in the upper-left panel. In principle, we could develop maximum-likelihood methods for the models in the other panels, and we might do so if we had good reason.

$\mathrm{Var}(\epsilon_i) = \sigma^2$. Because $\mathrm{E}(\epsilon|X = x) = 0$ (Assumption A1), the expectation of Y_i given x_i, α, and β must be $\mathrm{E}(Y_i) = \alpha + \beta x_i + 0 = \alpha + \beta x_i$. Thus, given x_i, α, and β, the dependent variable is distributed as $Y_i \sim \mathrm{Normal}(\alpha + \beta x_i, \sigma^2)$. The likelihood function is equal to the joint probability density function of the observations of Y, which, because the disturbances are independent, is the product of the probability density functions of the observations, or

$$L(\alpha, \beta) = \prod_{i=1}^{n} \frac{1}{\sigma\sqrt{2\pi}} e^{-\frac{(y_i - \alpha - \beta x_i)^2}{2\sigma^2}}.$$

The log-likelihood is then

$$l(\alpha, \beta) = \ln[L(\alpha, \beta)] = n \ln\left(\frac{1}{\sigma\sqrt{2\pi}}\right) - \frac{1}{2\sigma^2} \sum_{i=1}^{n} (y_i - \alpha - \beta x_i)^2. \tag{9.8}$$

Notice that the first term of the log-likelihood in Equation 9.8 does not depend on α or β. Nor does the $1/(2\sigma^2)$ multiplying the second term. We only care about maximizing $l(\alpha, \beta)$ with respect to α and β, so we can get rid of terms that are constant in α and β, noticing that

$$\underset{\alpha, \beta}{\mathrm{argmax}}\, l(\alpha, \beta) = \underset{\alpha, \beta}{\mathrm{argmax}} \sum_{i=1}^{n} \left[-(y_i - \alpha - \beta x_i)^2 \right]. \tag{9.9}$$

We could proceed by taking derivatives of the sum on the right of Equation 9.9 to find the values of α and β that maximize it, but we would be repeating ourselves. The quantity on the right is the negative sum of squared line errors, with α and β substituted for a and b (Equation 3.4). Thus, maximizing the likelihood is equivalent to minimizing the sum of the squared line errors, which we already know how to do. The maximum-likelihood estimators of α and β are equal to the expressions that specify the least-squares line in Equations 3.6 and 3.7. In particular,

$$\hat{\alpha} = \frac{\sum_{i=1}^{n} Y_i - \hat{\beta}\sum_{i=1}^{n} x_i}{n} \tag{9.10}$$

and

$$\hat{\beta} = \frac{\sum_{i=1}^{n} x_i Y_i - \frac{1}{n}\left(\sum_{i=1}^{n} x_i\right)\left(\sum_{i=1}^{n} Y_i\right)}{\sum_{i=1}^{n} x_i^2 - \frac{1}{n}\left(\sum_{i=1}^{n} x_i\right)^2}. \tag{9.11}$$

The only difference between these expressions and those in Equations 3.6 and 3.7 is that the Y terms are now capitalized to convey that they are viewed as random variables.[6] Thus, we have yet another interpretation of the least-squares line: it is a maximum-likelihood estimator of the "true" line under the assumption of normally distributed, independent, homoscedastic disturbances.

In this case, the method of maximum likelihood produced the same estimators as the method of moments in Chapter 8. The method-of-moments and maximum-likelihood estimators are given by the same expressions, but they are distinct because of the assumptions that justify them and because of the guarantees they provide. The maximum-likelihood estimators were justified by assumptions that we did not use when we applied the method of moments—namely, normality and constant variance of the disturbances. Given that these assumptions are satisfied, the maximum-likelihood estimators come with guarantees that do not accompany the method-of-moments estimators: the maximum-likelihood estimators, for example, are asymptotically efficient and functionally invariant.

Exercise Set 9-3

(1) Assuming the model used in this section, show that the maximum-likelihood estimators of α and β in Equations 9.10 and 9.11 are unbiased. Hints: Start with Equation 9.11. Remember that $Y_i = \alpha + \beta x_i + \epsilon_i$ and make the appropriate substitution early.

(2) [Optional] What is the variance of $\hat{\beta}$ (Equation 9.11)? Hints: Make the substitution $Y_i = \alpha + \beta x_i + \epsilon_i$, and use Equations 5.8 and 5.9. You may also use the identity

$$\sum_{i=1}^{n} (x_i - \bar{x})^2 = \sum_{i=1}^{n} x_i^2 - \frac{1}{n}\left(\sum_{i=1}^{n} x_i\right)^2,$$

where $\bar{x} = \frac{1}{n}\sum_{i=1}^{n} x_i$. This identity is a sample-moment analogue of Equation 5.7.

(3) [Optional] Estimating σ^2. We have estimated two parameters—α and β—using maximum likelihood. You may have noticed that the model has another parameter, σ^2, the variance of the disturbances. What is the maximum-likelihood estimator of σ^2? First hint: In your derivation, it may be helpful to replace σ^2 with a term that doesn't have an exponent, such as v. Second hint: Remember Equation 9.3: the log of a product is the sum of the logs of the terms being multiplied. Third hint: You will need the derivative

---

[6] Once again, the estimator—which is a function of random variables—is random, but estimates—which are functions of observed data—are not random.

$$\frac{d}{dv}\ln(\sqrt{v}).$$

The chain rule for derivatives says that if $u = g(v)$, then

$$\frac{d}{dv}f(u) = f'(u)g'(v).$$

The chain rule was not covered in the section on derivatives, and you do not need it for anything in this book besides this problem. Remembering that $\sqrt{v} = v^{1/2}$, applying the chain rule with Equations A.2 and 9.4 gives

$$\frac{d}{dv}\ln(\sqrt{v}) = \frac{1}{\sqrt{v}}\frac{1}{2\sqrt{v}} = \frac{1}{2v}.$$

(4) [Optional] Here are two facts about normally distributed random variables:

(i) If $X \sim \text{Normal}(\mu, \sigma^2)$ and $Y = a + bX$, then $Y \sim \text{Normal}(a + \mu, b^2\sigma^2)$.
(ii) If $X \sim \text{Normal}(\mu, \sigma^2)$, $Y \sim \text{Normal}(\tau, \omega^2)$, X and Y are independent, and $Z = X + Y$, then $Z \sim \text{Normal}(\mu + \tau, \sigma^2 + \omega^2)$.

Use (i) and (ii) to prove that $\hat{\beta}$ (Equation 9.11) is normally distributed.

9.2 Parametric interval estimation: the direct approach and Fisher information

One attractive feature of maximum-likelihood estimators is that their asymptotic variance is smaller than—or at least no larger than—the consistent alternatives. But what *is* their variance? There are typically three families of options for estimating the variance of a maximum-likelihood estimator. The first approach is to use a resampling method, such as bootstrapping. (The bootstrap is not *just* for nonparametric and semiparametric models.) The second approach is to analyze mathematically the specific model under study, considering the estimator as a random variable and computing its variance using the same methods we would use to compute the variance of any random variable (see Chapter 5). If you completed Problem 2 in the previous exercise set, then you have already carried out the direct approach for the maximum-likelihood estimator of the slope in simple linear regression. The third approach is to rely on a general property of maximum-likelihood estimators—their asymptotic variance is given by a function called the Fisher information. We do not have quite enough math to show off the Fisher information approach for simple linear regression—a little matrix algebra is required—but we will take it up in a simpler case.

9.2.1 *The direct approach*

In parametric models, we often make enough assumptions to allow direct computation of an estimator's distribution, or at least its variance. For example, we consider $\hat{\beta}$ (Equation 9.11), the maximum-likelihood estimator of β, the slope parameter in the simple linear regression model in Equation 9.7. If you performed the optional Problem 2 in Exercise Set 9-3, you found that

$$\mathrm{Var}(\widehat{\beta}) = \frac{\sigma^2}{\sum\limits_{i=1}^{n} (x_i - \bar{x})^2},\qquad(9.12)$$

where $\sigma^2 = \mathrm{Var}(\epsilon_i)$, the variance of the disturbances. The derivation of Equation 9.12 relies on the independence and constant variance of the disturbances, but it does not require that the disturbances be normally distributed.

Because $\widehat{\beta}$ is a maximum-likelihood estimator,[7] it is asymptotically normal and asymptotically unbiased. Thus, for large sample sizes, $\widehat{\beta}$ approximately obeys

$$\widehat{\beta} \sim \mathrm{Normal}\left(\beta, \frac{\sigma^2}{\sum\limits_{i=1}^{n} (x_i - \bar{x})^2}\right).\qquad(9.13)$$

In fact, we do not need to rely on asymptotic properties to get the distribution in 9.13. By Problem 1 in Exercise Set 9-3, $\mathrm{E}(\widehat{\beta}) = \beta$, and, by Problem 4, $\widehat{\beta}$ is normally distributed, even in small samples, under the assumptions of linearity, homoscedasticity, independence of units, and normality of disturbances.

We usually do not know σ^2, the variance of the disturbances, but we can estimate it. Here are two estimators that differ from each other only slightly (Problem 3, Exercise Set 9-3):

$$\widehat{\sigma^2} = \frac{1}{n} \sum_{i=1}^{n} (Y_i - \widehat{\alpha} - \widehat{\beta} x_i)^2,\qquad(9.14a)$$

$$\widetilde{\sigma^2} = \frac{n}{n-2} \widehat{\sigma^2} = \frac{1}{n-2} \sum_{i=1}^{n} (Y_i - \widehat{\alpha} - \widehat{\beta} x_i)^2.\qquad(9.14b)$$

Here, $\widehat{\alpha}$ and $\widehat{\beta}$ are the maximum-likelihood estimators of α and β (Equations 9.10 and 9.11). Thus, both estimators are based on the minimum possible sum of squared line errors. The function in Equation 9.14a is the maximum-likelihood estimator of σ^2. The maximum-likelihood estimator is biased downward; the estimator in 9.14b corrects the bias at the cost of lower efficiency. In large samples, the two estimators are nearly equal. To estimate the variance of $\widehat{\beta}$, substitute either $\widehat{\sigma^2}$ or $\widetilde{\sigma^2}$ for σ^2 in Equation 9.12.

Thus, if the data are generated by the model outlined in Equation 9.7, then $\widehat{\beta}$ has a normal distribution with a variance we can estimate. We can therefore modify the expression for a $1 - \alpha$ confidence interval of a normally distributed estimator $\widehat{\theta}$ with standard error ω (Equation 7.3):

$$(\widehat{\theta} - \omega z_{\alpha/2}, \widehat{\theta} + \omega z_{\alpha/2}).$$

As before, $z_{\alpha/2} = \Phi^{-1}(1 - a/2)$ and Φ^{-1} is the inverse of the cumulative distribution function of the Normal$(0, 1)$ distribution: $z_{\alpha/2}$ is chosen so that the probability that an observation drawn from a normal distribution falls within $z_{\alpha/2}$ standard deviations of the mean is $1 - \alpha$. To apply the expression to the slope in simple linear regression, substitute $\widehat{\beta}$ for $\widehat{\theta}$, and estimate the standard error ω by substituting either $\widehat{\sigma^2}$ (Equation 9.14a) or $\widetilde{\sigma^2}$

---

[7] And because the model meets some technical conditions that we'll ignore—such conditions are safe to ignore in most standard models.

(Equation 9.14b) for σ^2 in Equation 9.12 and taking the square root. Using $\widetilde{\sigma^2}$ (Equation 9.14b) gives

$$\left(\widehat{\beta} - z_{\alpha/2} \sqrt{\frac{\widetilde{\sigma^2}}{\sum\limits_{i=1}^{n}(x_i - \bar{x})^2}}, \widehat{\beta} + z_{\alpha/2} \sqrt{\frac{\widetilde{\sigma^2}}{\sum\limits_{i=1}^{n}(x_i - \bar{x})^2}} \right). \tag{9.15}$$

With a large dataset, $\widetilde{\sigma^2}$ will be a precise estimator of σ^2, and the expression in Equation 9.15 will give a $1 - \alpha$ confidence interval for β.[8]

9.2.2 [Optional subsection] The Fisher information approach

The direct approach works well in the simple linear regression case, but it is not always easy to identify the variance or distribution of an estimator directly. In this section, we describe an indirect approach for to identifying the variance of the sampling distribution of a maximum-likelihood estimator.

For many models, the asymptotic variance of a maximum-likelihood estimator is a function of its Fisher information. Loosely, the **Fisher information** is a measurement of the amount of information about a parameter contained in the likelihood function. Look at Figure 9-3, which shows likelihoods and log-likelihoods. The functions all have the same maximum, but those on the right are much more sharply peaked around their maxima. The sharply peaked likelihood functions inspire more confidence that the estimates arising from them are nearly correct. In fact, the plots on the right reflect five times as much data, so the estimates resulting from the functions on the right are less prone to error.

One way to assess the degree to which a log-likelihood function is peaked is to take its second derivative. Recall that the first derivative of a function gives the slope of the graph of a function. The second derivative encodes how quickly the slope of the graph of a function changes. If a function is sharply peaked at its maximum, then it rapidly changes from having a large positive slope to having a large negative slope. Thus, the second derivative near the maximum of a sharply peaked function is negative with large absolute value.

One expression[9] for the Fisher information $\mathcal{J}(\theta)$ applicable to many statistical models is

$$\mathcal{J}(\theta) = -\mathrm{E}_X \left(\frac{\partial^2}{\partial \theta^2} \ln[f_X(X|\theta)] \right). \tag{9.16}$$

Here, X is a random variable, f_X is the probability density or probability mass function of X, and f_X depends on a parameter, θ. The subscript X in E_X indicates that an expectation is being taken over the possible values of X. The $\frac{\partial^2}{\partial \theta^2}$ indicates that a second derivative is being taken with respect to a parameter. Finally, notice that $\ln[f_X(X|\theta)]$ is almost the log-likelihood, with the exception that X is random here, whereas in the log-likelihood, x represents data that have already been observed.

Recall that the more strongly negative the second derivative of the log-likelihood, the more sharply peaked is the function. In words, Equation 9.16 expresses that the Fisher

---

[8] If few data are available, then the variability of $\widetilde{\sigma^2}$ will cause the given confidence interval to be slightly too small on average. The fix is to use a t distribution (with degrees of freedom parameter $n - 2$) rather than a standard normal distribution.

[9] A more general definition that's often used is $\mathcal{J}(\theta) = E_X \left[\left(\frac{d}{d\theta} \ln[f_X(X|\theta)] \right)^2 \right]$, but this definition is equivalent to the expression in Equation 9.16 in many cases.

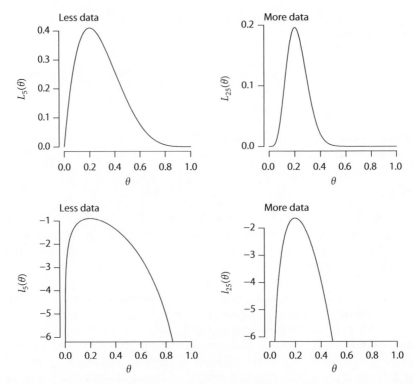

Figure 9-3 With more data, the likelihood and log-likelihood functions become more steeply peaked near their maxima. On the left, the likelihood (top) and log-likelihood (bottom) from Figure 9-1 are shown—θ is the unknown probability that a coin shows heads, and the data are 5 flips of the coin, 1 of which showed heads. The right side is analogous, but with more data. The right-side likelihood and log-likelihood are those that result when the coin is flipped 25 times and shows heads 5 times—again, a 1/5 ratio. In all four plots, the maximum occurs at $\theta = 0.2$, but the functions on the right fall off more sharply from their maxima. (The values of the functions at their maxima are also different—likelihoods become smaller when more data are used—but this difference is less important.)

information is larger when the log-likelihood is more sharply peaked at θ. For a maximum-likelihood estimator $\widehat{\theta}$ of a parameter θ, the asymptotic variance is approximately

$$\mathrm{Var}(\widehat{\theta}) \approx 1/\mathcal{J}(\theta). \tag{9.17}$$

Because we don't know θ—after all, it is what we are trying to estimate—we usually replace it with its estimate, yielding

$$\mathrm{Var}(\widehat{\theta}) \approx 1/\mathcal{J}(\widehat{\theta}). \tag{9.18}$$

$\mathcal{J}(\widehat{\theta})$ is also called the *observed information*.

As an example, consider the exponential distribution, which we saw when discussing maximum-likelihood estimation. Again suppose that we consider sampling n independent, identically distributed observations from an Exponential(λ) distribution (see Table 4-2). We saw that the maximum-likelihood estimate of λ is $\widehat{\lambda} = 1/\bar{x}$, where $\bar{x}$ is the mean of the n samples. What is the standard error of $\widehat{\lambda}$?

We saw before that the simplified log-likelihood is

$$l(\lambda) = \sum_{i=1}^{n} \ln(\lambda) - \lambda x_i.$$

Recall that the term $\ln f_X(X|\theta)$, which appears in the expression for Fisher information (Equation 9.16), is the log-likelihood with the observed data replaced with random variables. Thus,

$$\ln[f_X(X_1, X_2, ..., X_n|\lambda)] = \sum_{i=1}^{n} \ln (\lambda) - \lambda X_i.$$

We need the second derivative of this expression. Using Equations A.2, A.4, and 9.4, the first derivative is

$$\frac{d}{d\lambda} \ln[f_X(X_1, X_2, ..., X_n|\lambda)] = \sum_{i=1}^{n} \frac{1}{\lambda} - X_i = \frac{n}{\lambda} - n\overline{X}.$$

By Equation A.2, the second derivative is then

$$\frac{d^2}{d\lambda^2} \ln[f_X(X_1, X_2, ..., X_n|\lambda)] = -\frac{n}{\lambda^2}.$$

The Fisher information is the negative expectation of this quantity. Because the second derivative is nonrandom—i.e., none of the X_i variables appear in it—the Fisher information is

$$\mathcal{J}(\lambda) = -\mathrm{E}_X\left(\frac{d^2}{d\theta^2} \ln[f_X(X_1, X_2, ..., X_n|\theta)]\right) = \frac{n}{\lambda^2}.$$

By Equation 9.17, the asymptotic variance of $\widehat{\lambda}$ is

$$\mathrm{Var}(\widehat{\lambda}) \approx \frac{\lambda^2}{n},$$

which we can estimate by plugging in $\widehat{\lambda} = 1/\overline{x}$ for λ. Because $\widehat{\lambda}$ is a maximum-likelihood estimator, it is asymptotically unbiased and asymptotically normally distributed. For large samples, we can use the estimated variance to build confidence intervals.

The Fisher information approach can also be applied when several parameters are being estimated simultaneously, as in the case of linear regression. But to use Fisher information with several variables, it helps to know a little about matrices, and in particular, how to invert them. Since matrix algebra is not assumed or covered in this book, we will not cover the multidimensional case, but it is conceptually similar. Instead of identifying a single Fisher information, we calculate a Fisher information matrix and invert it to get variance estimates.

In statistical theory, the Fisher information is an important piece in the argument in favor of maximum-likelihood estimation. According to a result called the Cramér–Rao bound, the variance of an unbiased estimator is at least as large as the expression in Equation 9.17, $1/\mathcal{J}(\theta)$. Because maximum-likelihood estimators are asymptotically unbiased with variance $1/\mathcal{J}(\theta)$—the smallest possible variance for an unbiased estimator—maximum-likelihood estimators are said to be asymptotically efficient. No asymptotically unbiased estimator has a smaller asymptotic variance.

The efficiency of maximum-likelihood estimators is a powerful argument for their use. But the caveats we listed for maximum-likelihood estimators also apply to their variances. Further, the variance guarantee from Fisher information is an asymptotic property—for small samples, the distribution of the estimator might not behave as promised.

Exercise Set 9-4

(1) [Optional] The Poisson distribution an option for describing count data. If $X \sim \text{Poisson}(\theta)$, then the probability mass function of X is

$$P(X = x) = \frac{e^{-\theta}\theta^x}{x!},$$

for $x \in 0, 1, 2, 3, \ldots$. Further, $E(X) = \theta$, and $\text{Var}(X) = \theta$. Imagine that we have n observations $x_1, x_2, \ldots, x_n$, which are assumed to be independent and identically distributed as $\text{Poisson}(\theta)$.
(a) What is $\hat{\theta}$, the maximum-likelihood estimator of θ?
(b) What is $\text{Var}(\hat{\theta})$? Do not use Fisher information.
(c) Use Fisher information to identify the asymptotic variance of $\hat{\theta}$.

9.3 Parametric hypothesis testing using the Wald test

We have just learned that there are at least three options for quantifying uncertainty about a maximum-likelihood estimate: we can resample, we can do probability calculations using the model under study, or we can rely on general results for maximum-likelihood estimators. As you might expect given the close connection between interval estimation and hypothesis testing, we have the same three options when performing inference with a maximum-likelihood estimate: we can use resampling (such as a permutation test), we can design a test specific to the model under study, or we can use a test that's applicable to a variety of maximum-likelihood problems. In this section and the next (optional) one, we will consider two approaches that apply to a wide range of maximum-likelihood models, the **Wald test** and the **likelihood-ratio test**.

Before we proceed to the tests, let me remind you of the χ^2 (pronounced "chi-squared") distribution (Table 4-2). A random variable with a $\chi^2(k)$ distribution has the same distribution as the sum of the squares of k independent, $\text{Normal}(0, 1)$ random variables. The parameter k is called the "degrees of freedom," and it can take any positive integer value.

9.3.1 *The Wald test*

Suppose we want to test the hypothesis that a parameter θ is equal to a hypothesized value θ_0. For many models, the asymptotic distribution of the maximum-likelihood estimator $\hat{\theta}$ is normal, with expectation equal to the true parameter value θ. We can define the statistic W as

$$W = \frac{\hat{\theta} - \theta_0}{\sqrt{\text{Var}(\hat{\theta})}}. \tag{9.19}$$

If the null hypothesis is true—meaning in this case that $\theta = \theta_0$—then for large sample sizes, the quantity in the numerator of the right side of Equation 9.19 is normally distributed with expectation 0 and variance 1. We often do not know $\text{Var}(\hat{\theta})$, which is necessary for computing W, but we can estimate $\text{Var}(\hat{\theta})$ using either direct methods or using the observed Fisher information (Equation 9.18). Replacing $\text{Var}(\hat{\theta})$ with an estimate, $\widehat{\text{Var}}(\hat{\theta})$, gives a test statistic W^*:

$$W \approx W^* = \frac{\hat{\theta} - \theta_0}{\sqrt{\widehat{\text{Var}}(\hat{\theta})}}. \tag{9.20}$$

For large samples, the estimated variance of $\hat{\theta}$ will be very accurate, and W^* will have an approximate Normal$(0, 1)$ distribution if the null hypothesis is true.[10]

Because W^* has an asymptotic Normal$(0, 1)$ distribution if the null hypothesis is true, we can test the null hypothesis by comparing W^* with the Normal$(0, 1)$ distribution. A value of W^* that is extreme when compared with the Normal$(0, 1)$ distribution is evidence against the null hypothesis. The approximate[11] two-sided p arising from an observed value of W^* is

$$p_W = 2\Phi(-|W^*|), \tag{9.21}$$

where Φ is the cumulative distribution function of the Normal$(0, 1)$ distribution. The "2" is there because the test is two-sided—one wants to account for the possibilities that are both much larger and much smaller than expected under the null hypothesis (see Figure 7-2).

The Wald test is a simple way to test hypotheses about whether a single parameter is equal to a specified value. The next procedure, the likelihood-ratio test, is even more versatile because it can be used to test hypotheses that consider the values of several parameters jointly.

Exercise Set 9-5

(1) Compute the test statistic and p value for the Wald test of the hypothesis that $\beta = 0$ using the agricultural data. Use Equation 9.14b to estimate σ^2, and then plug your estimate of σ^2 into Equation 9.12 to estimate $\text{Var}(\hat{\beta})$. Remember that you can access the fertilizer consumption data with `anscombe$x1` and the cereal yield data with `anscombe$y1`. (The p values obtained from the Wald test are smaller than they should be when the sample size is as small as it is in the example. For small sample sizes, a t test is more appropriate than the Wald test.)

(2) Some people prefer to report the Wald test statistic as W^2, where W is as in Equation 9.19. Under the null hypothesis, what is the asymptotic distribution of W^2?

(3) Use the `sim.Wald.B()` function in the book's R package, `stfspack` (Exercise Set 2-2, Problem 3) to examine the type I error rate and power of the Wald test when the significance level is $\alpha = 0.05$. Use `help(sim.Wald.B)` to see the syntax, which is similar to `sim.perm.B()` (Exercise Set 8-5, Problem 2). Test the null hypothesis that $\beta = 0$ using simulated data with $\beta = 0$, $\beta = 0.1$, and $\beta = 0.2$, and with the variance of the disturbances set to 1. Use $n = 10, n = 50$, and $n = 100$ pairs of observations in each simulated dataset. Generate 1,000 datasets for each combination of n and β values. Compare your results with those you obtained in Problem 2 of Exercise Set 8-5, in which you examined the permutation test.

9.4 [Optional section] Parametric hypothesis testing using the likelihood-ratio test

Unlike the Wald test, the likelihood-ratio test can be used to make joint inferences about several parameters at once. Specifically, likelihood-ratio tests can be used to test null hypotheses that involve *nested* models. Consider the model in Equation 9.7, repeated here:

---

[10] For small samples, the uncertainty of the estimated variance makes large values of W^* more probable than they would be if the true variance were known. One can correct for the excess of large values that arise from our uncertainty about $\text{Var}(\hat{\theta})$ by comparing W^* with a t distribution rather than with the Normal$(0, 1)$ distribution.

[11] Even if the variance of $\hat{\theta}$ is known, the p is approximate because $\hat{\theta}$ is not necessarily exactly normal—rather, it converges to a normal distribution as the sample size gets larger. The approximation improves as more data are collected.

$$Y_i = \alpha + \beta x_i + \epsilon_i.$$

In an estimation context, α and β are treated as *free* parameters. Their values are assumed to be unknown, but, by estimation, we can identify candidate values that fit the data. Suppose that your colleague proposes the null hypothesis that $\beta = \beta_0 = 1/3$. Under this hypothesis, β is no longer free; it is constrained to be $1/3$. The model then becomes

$$Y_i = \alpha + x_i/3 + \epsilon_i. \tag{9.22}$$

The model in Equation 9.22 is nested within the model in Equation 9.7 because all the free parameters in the model of Equation 9.7 appear in the model of Equation 9.22, but some of the free parameters in Equation 9.7 are constrained to take specific values in Equation 9.22. Now suppose that your colleague posits the null hypothesis that $\beta = \beta_0 = 1/3$ *and* that $\alpha = \alpha_0 = 2$. The model then becomes

$$Y_i = 2 + x_i/3 + \epsilon_i. \tag{9.23}$$

The model implied by Equation 9.23 is nested within *both* the model implied by Equation 9.22 *and* the model implied by Equation 9.7. One of the parameters that is free in Equation 9.22 is constrained in Equation 9.23, and two of the free parameters in Equation 9.7 are constrained in Equation 9.23. When comparing nested models, we keep track of the difference in the number of free parameters between the models, which we can call k.

Many null hypotheses can be expressed as comparisons of nested models. For example, one can constrain parameters by hypothesizing that they are equal to 0, effectively erasing them from the model.

The likelihood-ratio test statistic compares the maximum possible likelihoods attainable by two nested models. Specifically, we compute the maximum value of the likelihood possible under the model with more free parameters, or $L(\hat{\theta})$, as well as the maximum value of the likelihood possible when some of the parameters are constrained, or $L(\widehat{\theta_0})$. The intuition is that a large ratio of $L(\hat{\theta})$ to $L(\widehat{\theta_0})$—i.e., a much lower likelihood under the null hypothesis—is evidence against the null hypothesis under which some of the parameters are constrained (Figure 9-4).

One statistic suggested by the name "likelihood-ratio test" is

$$LR = \frac{L(\hat{\theta})}{L(\widehat{\theta_0})},$$

the ratio of the likelihood at its maximum to the likelihood at the hypothesized value of θ. Though this test statistic is reasonable, it is usually transformed before use. The typical statistic is[12]

$$\Lambda = 2\ln\left(\frac{L(\hat{\theta})}{L(\widehat{\theta_0})}\right) = 2\left(l(\hat{\theta}) - l(\widehat{\theta_0})\right). \tag{9.24}$$

In words, Λ is twice the natural log of the likelihood ratio. Recall that the log of a quotient is a difference in logs, leading to the simplification on the right. One reason for the popularity of the statistic Λ is **Wilks' theorem.**

---

[12] Some sources switch the numerator and the denominator in the likelihood ratio function, in which case Λ is *minus* twice the log of the likelihood ratio.

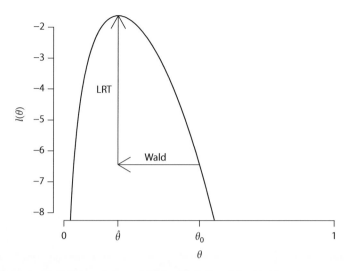

Figure 9-4 A geometric representation of the Wald test and the likelihood-ratio test (LRT). The log-likelihood in the figure is the one from the bottom-right panel of Figure 9-3—it arises when a coin is flipped 25 times independently and shows heads 5 times. The maximum likelihood estimate is $\hat{\theta} = 1/5$. Suppose we want to test the null hypothesis that $\theta = 0.5$. The Wald test statistic is based on the horizontal distance between $\theta_0 = 0.5$ and $\hat{\theta} = 0.2$, scaled by the standard error of $\hat{\theta}$, which in turn depends on the curvature of the log-likelihood function. In contrast, the likelihood-ratio test statistic is based on the difference between the value of the log-likelihood function at θ_0 and at $\hat{\theta}$, seen here as a vertical distance. In many situations, the two tests are asymptotically equivalent but can give different results in small samples.

> **Wilks' theorem (loosely stated)** Suppose that data X are generated according to a model[13] with probability mass function or probability density function $f_X(X|\theta_1, \theta_2, ..., \theta_p)$, where $\theta_1, \theta_2, ..., \theta_p$ are parameters. Suppose further that we wish to test the null hypothesis that k of the parameters are equal to specific values, e.g., $H_0: \theta_1 = \theta_{01}; \theta_2 = \theta_{02}; ...; \theta_k = \theta_{0k}$. Then, if the null hypothesis is true, the statistic Λ (Equation 9.24) is asymptotically distributed as $\chi^2(k)$.

Wilks' theorem provides a general approach for testing null hypotheses that involve nested models. To test the null hypothesis involving a nested model, one compares Λ with the $\chi^2(k)$ distribution, where k is the number of parameters that are constrained in the nested model but free in the comparison model. The p value is the (approximate) probability of obtaining a value of Λ as large or larger than the one observed given that the null hypothesis is true, or

$$p_{LRT} = 1 - F_{\chi^2(k)}(\Lambda), \tag{9.25}$$

where $F_{\chi^2(k)}$ is the cumulative distribution function of the $\chi^2(k)$ distribution.[14]

In some scenarios, it is not possible to devise a more powerful test than an appropriately constructed likelihood-ratio test.[15] In an exercise below, you will compute the Wald test

---

[13] There are some constraints on the type of model that can be used—for example, maximum-likelihood estimators for the model have to exist. Most commonly used models meet the requirements.

[14] In R, the cumulative distribution function can be interrogated with pchisq().

[15] The formal result that backs up this informal claim is called the Neyman–Pearson lemma.

and the likelihood-ratio test for the agricultural data, and you will verify that for simple linear regression, the Wald test and the likelihood-ratio test are equivalent for tests of hypotheses involving a single parameter. Though the Wald test and likelihood-ratio test are not equivalent for all models, they are *asymptotically* equivalent for a wide class of models. (See Figure 9-4 for a depiction of the relationship between the Wald and likelihood-ratio tests.) As the sample size goes up, their conclusions agree more closely. The results of the Wald test and the likelihood-ratio test may disagree in small samples,[16] which can create ambiguities that are hard to resolve.

Exercise Set 9-6

(1) In this problem, you will explore the likelihood-ratio test for simple linear regression. As with the Wald test, the p values obtained from the likelihood-ratio test are too small when the sample size is as small as it is in the example.
   (a) Prove that under the null hypothesis that $\beta = 0$, the likelihood is maximized by setting $\alpha = \bar{y}$. (The log-likelihood for simple linear regression is given in Equation 9.8.)
   (b) Compute the test statistic and asymptotic p value for the likelihood-ratio test of hypothesis that $\beta = 0$ using the agricultural data. Use Equation 9.14b to estimate σ^2 and plug your estimate into the likelihood function. Use pchisq() to interrogate the cumulative distribution function of the χ^2 distribution.
   (c) What is the relationship between the results of part (b) and of Problem 1 from Exercise Set 9-5? (This relationship between the Wald and likelihood-ratio tests does not hold for all models.)

(2) [Optional, hard] In this exercise, you will prove that for simple linear regression with independent, normally distributed disturbances of constant variance, Λ (Equation 9.24) has a $\chi^2(1)$ distribution under the null hypothesis that $\beta = 0$.
   (a) Prove that, for simple linear regression,

$$\sum_{i=1}^{n} (Y_i - \bar{Y})^2 = \sum_{i=1}^{n} (Y_i - \widehat{Y}_i)^2 + \sum_{i=1}^{n} (\widehat{Y}_i - \bar{Y})^2,$$

where $\bar{Y} = (1/n)\sum_{i=1}^{n} Y_i$ and $\widehat{Y}_i = \widehat{\alpha} + \widehat{\beta}x_i$. This claim is called the "ANOVA identity." You will want to use the expressions for $\widehat{\alpha}$ and $\widehat{\beta}$ suggested by Equations 3.8 and 3.9, namely,

$$\widehat{\alpha} = \bar{Y} - \widehat{\beta}\bar{x}$$

and

$$\widehat{\beta} = \frac{\sum_{i=1}^{n}(x_i - \bar{x})(Y_i - \bar{Y})}{\sum_{i=1}^{n}[(x_i - \bar{x})^2]}.$$

Hint: Start by writing $\sum_{i=1}^{n}(Y_i - \bar{Y})^2 = \sum_{i=1}^{n}([Y_i - \widehat{Y}_i] + [\widehat{Y}_i - \bar{Y}])^2$, which is true because adding $\widehat{Y}_i$ to and subtracting $\widehat{Y}_i$ from the quantity being squared does not change its value. (In a somewhat more general form, the statement you will prove is important in the theory of analysis of variance, which is a method closely related to simple linear regression.)

---

[16] The Wald test and likelihood-ratio test are designed for large samples. In the linear regression case, the t test is a small-sample analogue of the Wald test, and the F test is an analogue of the likelihood-ratio test. Linear regression procedures in most software packages, including R's lm() function, report results of the t and F tests rather than the Wald and likelihood-ratio tests.

(b) Use the statement you proved in part (a) to show that for simple linear regression with independent, normally distributed disturbances of constant variance, Λ (Equation 9.24) has a $\chi^2(1)$ distribution under the null hypothesis that $\beta = 0$.

9.5 Chapter summary

If it is reasonable to assume that the data are generated by a fully parametric model, then maximum-likelihood approaches to estimation and inference have many appealing properties. Maximum-likelihood estimators are obtained by identifying parameters that maximize the likelihood function, which can be done using calculus or using numerical approaches. Such estimators are consistent, and they are also efficient compared with their consistent competitors.

The sampling variance of a maximum-likelihood estimate can be estimated in several ways. As always, one possibility is the bootstrap. In many models, the variance of the maximum-likelihood estimator can be derived directly once its form is known. A third approach is to rely on general properties of maximum-likelihood estimators and use the Fisher information.

Similarly, there are many ways to test hypotheses about parameters estimated by maximum likelihood. In this chapter, we discussed the Wald test, which relies on the fact that the sampling distribution of maximum-likelihood estimators is normal in large samples, and the likelihood-ratio test, which is a general approach for testing hypotheses relating nested pairs of models.

9.6 Further reading

Eliason, S. R. (1993). *Maximum Likelihood Estimation: Logic and Practice*. Sage Publications, Newbury Park, CA.

  A short monograph providing a general exposition of maximum-likelihood estimation.

Muggeo, V. M., & Lovison, G. (2014). The "three plus one" likelihood-based test statistics: unified geometrical and graphical interpretations. *The American Statistician*, 68, 302–306.

  Provides a unified framework for understanding the two hypothesis tests discussed in this chapter— the Wald test and the likelihood-ratio test—along with another traditional test applicable in maximum-likelihood situations, the score test.

Stigler, S. M. (2007). The epic story of maximum likelihood. *Statistical Science*, 22, 598–620.

  An interesting history of thought about maximum-likelihood methods, extending back long before R. A. Fisher put the method at center stage. The article also discusses the conditions necessary for Fisher's proofs about consistency and efficiency to hold.

CHAPTER 10

Bayesian estimation and inference

Key terms: Bayes estimator, Bayes factor, Conjugate prior, Credible interval/region, Markov chain Monte Carlo (MCMC), Posterior distribution, Prior distribution, Rejection sampling

We have now developed several views of the least-squares line. For the exploratory data analyst, the least-squares line simply passes through a cloud of points in a specified way. From a semiparametric perspective, the least-squares line is a method-of-moments estimator of theoretical parameters, and we can make more substantial claims about the resulting estimates using bootstrapping or permutation approaches. From a parametric point of view that assumes normally distributed disturbances of constant variance, the least-squares line is a maximum-likelihood estimator, and we can make claims about it by examining the likelihood function. As we have added assumptions, we have seen the least-squares line imbued with extra properties—first consistency, then efficiency. The Bayesian perspective will provide another step in this process, relating the least-squares line to estimators with appealing decision-theoretic properties.

In Bayesian statistics, *parameters* can have probability distributions, whereas in frequentist statistics, parameters are fixed, not random—only data have probability distributions. In Chapters 8 and 9, we focused on the empirical distribution function and the likelihood function, respectively. The star of the Bayesian show is the **posterior distribution**, the distribution of the parameters being estimated given the data observed.

Suppose we observe data D that follow a probability density or probability mass function $f_D(d|\theta)$. The density or mass function depends on the parameter θ. For example, the data might be independent observations from a Normal$(\theta, 1)$ distribution. Notice that $f_D(d|\theta)$ is the likelihood of θ, $L(\theta)$ (Equation 9.1). We could maximize it to estimate θ, but the Bayesian wants more. Specifically, the Bayesian wants to know $f_\theta(\theta|D = d)$, the density function of the parameter given the observed data. This is the *posterior* density. To get it, we need a density representing the **prior distribution**, denoted $f_\theta(\theta)$—a density function taken to "describe" θ before any data are observed.[1]

---

[1] What does "describe" mean? The classic "subjective" interpretation is that the prior describes someone's beliefs about plausible values of θ. There are other frameworks that do not view the prior as describing a belief to which a data analyst is committed, but it can become difficult to say what the prior and its downstream results mean in these cases. For more context, see Mayo (2018, Excursion 6), and for a compelling alternative view, see Gelman & Shalizi (2013). We will mostly sidestep these issues, though a bit more is coming in the next section.

Statistical Thinking from Scratch: A Primer for Scientists. M. D. Edge, Oxford University Press (2019).
© M. D. Edge 2019. DOI: 10.1093/oso/9780198827627.001.0001

If we have a prior, then we can use Bayes' theorem (Equation 4.2) to "update" the prior on the basis of the observed data. More formally, we calculate the posterior density as

$$f_\theta(\theta | D = d) = \frac{f_D(d|\theta)f_\theta(\theta)}{f_D(d)} = \frac{f_D(d|\theta)f_\theta(\theta)}{\int_{-\infty}^{\infty} f_D(d|\theta)f_\theta(\theta)d\theta}. \qquad (10.1)$$

Two expressions for the posterior density function are given in Equation 10.1. The first is a direct adaptation of Bayes' theorem. The second expression relies on the fact that

$$f_D(d) = \int_{-\infty}^{\infty} f_D(d|\theta)f_\theta(\theta)d\theta.$$

In words, the probability of observing the data we have is the probability of observing the data given the parameter, averaged across all possible parameter values weighted by the probability (or density) of each parameter value.[2]

Two related questions immediately raise their heads. First, how do we choose a prior distribution? Second, given that the prior distribution is known, how can we compute the posterior distribution? We will consider these questions before considering how to use the posterior distribution for estimation and inference.

10.1 How to choose a prior distribution?

The prior distribution $f_\theta(\theta)$ can be used to summarize uncertainty about θ before any data are collected.[3] Two different researchers might set $f_\theta(\theta)$ differently, and if they do, then their conclusions will also typically differ, at least slightly. Thus, $f_\theta(\theta)$ is a route for subjectivity to enter statistical analysis,[4] and the choice is potentially important.

In principle, one way to choose a prior distribution is to use a posterior distribution calculated from a previous study addressing the same problem. If past work is available, then using the posterior distribution from a previous study as a prior allows the integration of new and old work. This approach is associated with a saying, "Today's posterior is tomorrow's prior," which, by the standards of statistical idiom, is catchy. Though this approach is attractive, it obviously cannot be applied if there are no previous studies on the question of interest, and it is unusual in practice. How can we set the prior distribution to be used for the first analysis on a question?

Subjective approaches suggest that prior distributions should aim to reflect the beliefs of experts. Expert knowledge can be a useful guide, and setting the prior on the basis of experts' beliefs allows their knowledge to be incorporated into the analysis. Critics of the subjective approach note that it is unclear how best to use expert knowledge. Experts' opinions may not be precise enough to be translated into a probability distribution, or they may not be accurate enough to motivate us to try.

Another approach is to use *uninformative* or *weakly informative* priors, which are associated with so-called *objective* or *default* Bayesianism. Such a prior places approximately the same

---

[2] This is a generalization of the strategy we used in Chapter 3 to solve the Bayes' theorem example, and it is an instance of the law of total probability, which is available in all of the probability textbooks recommended at the end of Chapter 4.

[3] The notation $f_\theta(\theta)$ suggests that f_θ is a proper probability density function, with total area under its curve equal to 1. In fact, the prior distribution need not integrate to 1 or even to a finite value, so long as the posterior can be scaled to equal 1 later. This allows, for example, "flat" priors that are constant for all possible values of θ.

[4] Certainly not the *only* route.

prior density on all possible values of a parameter, meaning that before any data come in, no hypothesis is strongly favored over others. Though this approach seems simple, it is associated with some mathematical complexities that are beyond our scope. Further, whereas posteriors in the subjective approach reflect updated beliefs, it is less clear how one ought to interpret posterior distributions that arise from default priors. At this writing, most Bayesian analyses are in the "objective" Bayesian tradition.

Two more considerations influence the use of both subjective and uninformative priors. First, many Bayesian statisticians are interested in the robustness of their inferences against changes to the prior distribution. If many different sets of priors lead to similar conclusions after the data are considered, then the conclusion reached is robust against differences in prior beliefs. Statisticians taking the robustness view need not subscribe to any particular set of prior beliefs. They simply use the machinery of Bayesian statistics to study which posterior beliefs follow when the data are combined with any particular prior. Second, mathematical convenience—though less important now than it has been in the past—can be a consideration in the choice of a prior. The choice of prior distribution can determine whether the posterior distribution (Equation 10.1) is easy or impossible to compute analytically.

One (imperfect) justification for some applied statisticians' relatively cavalier attitude toward setting priors is that for many models, as the number of data increase, Bayesian estimates tend to converge to the same conclusions, regardless of the initial choice of prior distribution.[5] This is not to say that prior choice doesn't matter—it can markedly influence results from analyses of small and medium-sized samples, as well as even very large samples, if the model is complex enough.

10.2 The unscaled posterior, conjugacy, and sampling from the posterior

Lack of consensus on the question of how to set priors is one reason that Bayesian methods were less popular than frequentist methods during the twentieth century. Another reason is that the integral in the denominator of Equation 10.1, $f_D(d) = \int_{-\infty}^{\infty} f_D(d|\theta)f_\theta(\theta)\, d\theta$, is often hard to calculate. Most of the time, $f_D(d)$ cannot be expressed in a convenient mathematical form and needs to be approximated by simulation. Bayesian analyses were therefore inconvenient until sufficient computational resources became widely available. Now virtually every data analyst has access to powerful computers and flexible software for Bayesian analyses.

For most of this chapter, we will let an R package integrate the posterior for us. Here, though, we will consider two methods for studying the posterior. The first method, use of conjugate priors, gives a mathematical solution but is applicable rarely. The second method, rejection sampling, requires a computer but is much more generally applicable.

The trick to most methods for studying the posterior distribution is to avoid computing the integral $\int_{-\infty}^{\infty} f_D(d|\theta)f_\theta(\theta)d\theta$ directly. Recall Equation 10.1:

$$f_\theta(\theta|D=d) = \frac{f_D(d|\theta)f_\theta(\theta)}{f_D(d)} = \frac{f_D(d|\theta)f_\theta(\theta)}{\displaystyle\int_{-\infty}^{\infty} f_D(d|\theta)f_\theta(\theta)d\theta}.$$

[5] The formal statement behind this loose claim is called the Bernstein–von Mises theorem.

We want to calculate the posterior density, $f_\theta(\theta|D = d)$, a function of θ. As mentioned above, the hard part is usually calculating the denominator, $f_D(d) = \int_{-\infty}^{\infty} f_D(d|\theta)f_\theta(\theta)\,d\theta$. Notice, though, that the denominator $f_D(d)$ is not a function of θ—given the observed data and the prior distribution, it is a constant, like 2 or e. Its purpose is merely to ensure that the total area under the posterior distribution is equal to 1—in other words, $f_D(d)$ is a scaling factor. Thus, though it may be difficult to compute the posterior density analytically, it is often easy to compute a function that is *proportional to* the posterior density—we just ignore the denominator. This insight is expressed by the relation

$$f_\theta(\theta|D = d) \propto f_D(d|\theta)f_\theta(\theta), \tag{10.2}$$

where the $\propto$ symbol means "is proportional to." We'll call the quantity on the right of Equation 10.2 the *unscaled posterior*. It is usually not too hard to compute.[6] The unscaled posterior has the same general shape as the actual posterior, and it shares some properties with it—for example, it is maximized by the same value(s) of θ.

For example, assume that we are interested in a posterior distribution for θ after drawing n independent observations $x_1, x_2, \ldots, x_n$ from a Normal(θ, σ^2) distribution. Assume that the variance σ^2 is known. Imagine that the prior for θ is another normal distribution—specifically a Normal(γ, τ^2) distribution. Remembering that the probability density function of a Normal(θ, σ^2) random variable X is

$$f_X(x) = \frac{1}{\sigma\sqrt{2\pi}}e^{-\frac{(x-\theta)^2}{2\sigma^2}},$$

$f_D(d|\theta)$, which is the joint density of n independent observations, is the product of n normal densities:

$$f_D(d|\theta) = \left(\frac{1}{\sigma\sqrt{2\pi}}\right)^n \prod_{i=1}^n e^{-\frac{(x_i-\theta)^2}{2\sigma^2}} = \left(\frac{1}{\sigma\sqrt{2\pi}}\right)^n e^{-\sum_{i=1}^n \frac{(x_i-\theta)^2}{2\sigma^2}}.$$

Similarly, the density function of a Normal(γ, τ^2) random variable representing the prior for θ is

$$f_\theta(\theta) = \frac{1}{\tau\sqrt{2\pi}}e^{-\frac{(\theta-\gamma)^2}{2\tau^2}}.$$

The unscaled posterior density of θ given the data is the product of these two expressions:

$$f_D(d|\theta)f_\theta(\theta) = \left(\frac{1}{\sigma\sqrt{2\pi}}\right)^n e^{-\sum_{i=1}^n \frac{(x_i-\theta)^2}{2\sigma^2}} \frac{1}{\tau\sqrt{2\pi}}e^{-\frac{(\theta-\gamma)^2}{2\tau^2}}. \tag{10.3}$$

Figure 10-1 shows an unscaled posterior along with the prior distribution and likelihood function from which it was calculated. You'll notice that the center of the unscaled posterior falls between the prior distribution and the likelihood function. In this case, you also might guess that the unscaled posterior is proportional to a normal density. This guess would turn out to be correct. In fact, when the unscaled posterior in Equation 10.3 is multiplied by the appropriate scaling constant, it is equal to the density function of a normal random variable with expectation

---

[6] In some cases, it *is* hard to compute the unscaled posterior density because the likelihood, $f_D(d|\theta)$, is intractable. Approximate Bayesian computation (ABC) is sometimes possible in such situations. For a review of ABC, see Beaumont (2010).

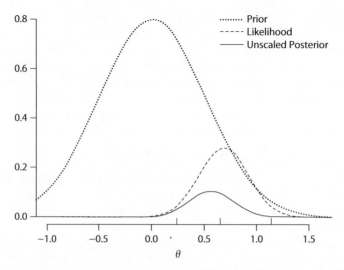

Figure 10-1 An example prior density ($f_\theta(\theta)$), likelihood function ($f_D(d|\theta) = L(\theta)$, Equation 9.1), and unscaled posterior density ($f_D(d|\theta)f_\theta(\theta)$, Equation 10.2). The prior density is that of a normal random variable with expectation 0 and standard deviation $1/2$. The likelihood is calculated from three independent observations (shown as upward tick marks at the base of the plot) from a normal distribution with expectation $3/5$ and known standard deviation $2/5$. The sample mean of the three observations—and therefore the maximum-likelihood estimate of θ—is 0.68.

$$\mathrm{E}(\theta|D = d) = \left(\gamma\frac{1}{\tau^2} + \bar{x}\frac{n}{\sigma^2}\right)\bigg/\left(\frac{1}{\tau^2} + \frac{n}{\sigma^2}\right) \tag{10.4}$$

and variance

$$\mathrm{Var}(\theta|D = d) = 1\bigg/\left(\frac{1}{\tau^2} + \frac{n}{\sigma^2}\right). \tag{10.5}$$

The posterior expectation (Equation 10.4) is a weighted average of the expectation of the prior distribution, γ, and the maximum-likelihood estimate, $\bar{x}$. The weights are functions of the prior variance and the number of observations—the smaller the prior variance, the greater the influence of the prior expectation, and the larger the sample size, the greater the influence of the maximum-likelihood estimate. This pattern—in which the posterior expectation converges to the maximum-likelihood estimate as the sample size increases—recurs in many Bayesian estimation settings.

In the example, both the prior distribution and the posterior distribution were members of the normal distribution family. If the prior distribution $f_\theta(\theta)$ is a member of the same distribution family as its associated posterior, $f_\theta(\theta|D = d)$, then $f_\theta(\theta)$ is called a **conjugate prior**. If a conjugate prior for a parameter can be identified, then the posterior density can be calculated explicitly. But the conjugate prior strategy is limited—it limits us to priors that belong to a particular family, and we might want to use a prior from a different distribution family.

There is another option. We can use a computer to draw simulated data from the posterior. As we draw more simulated data from the posterior, we can estimate the posterior density—or a function of it—with as much precision as desired. One general method for drawing samples from a posterior is **rejection sampling**. The idea of rejection sampling is to simulate data from a distribution from which it is easy to draw samples—perhaps the prior distribution—and then to prune the simulated data so that the data that remain after pruning reflect the posterior distribution. For example, to conduct rejection sampling for the example shown in Figure 10-1, we could sample from the prior distribution but reject samples that do not reflect the posterior distribution well. Figure 10-2 is a sketch of the situation.

Here is one algorithm for sampling from a posterior density that will work if (i) it is possible to sample from the prior distribution and (ii) the unscaled posterior is less than the prior density for all values of θ, as it is in the example in Figure 10-1. This is not the rejection sampling algorithm you would use in real applications—a more efficient and more general version appears in an exercise—but if (i) and (ii) are true, it will give you equivalent answers to the real thing with enough computational effort:

(1) Sample an observation θ^* from the prior distribution.
(2) Compute m, the quotient of the unscaled posterior divided by the prior at θ^*. (Notice that in this case, $m = f_D(d|\theta^*)f_\theta(\theta^*)/f_\theta(\theta^*) = f_D(d|\theta^*) = L(\theta^*)$, the likelihood at θ^*.)
(3) Independently, sample a random variable X from a continuous Uniform$(0, 1)$ distribution.
(4) If $X \leq m$, then "accept" θ^* as an observation from the posterior distribution and save its value. Otherwise "reject" θ^* and discard it.
(5) Repeat steps 1–4 until the desired number of simulated observations from the posterior distribution are gathered.

The key step in the algorithm is step 4, in which we reject a candidate observation with a probability proportional to the ratio of the unscaled posterior and the prior—this ratio is also the likelihood. Samples from the prior distribution that are also associated with high likelihood are accepted with high probability, whereas samples from the prior distribution associated with low likelihood are usually rejected.

This algorithm for rejection sampling can be modified to deal with cases in which the unscaled posterior is sometimes greater than the prior. It can also be made more efficient (see the lower panel of Figure 10-2). For example, if this algorithm were applied to the prior density and unscaled posterior in Figure 10-2, then most of the samples drawn would be rejected—the area of the gray region is small compared with the area of the black region. You will see how to achieve both these goals in an exercise. It's also possible to modify the algorithm to accommodate cases in which the prior distribution cannot be directly sampled.

Rejection sampling is a straightforward way to sample from a posterior distribution, but it is not the most general or most popular. At this writing, the most widely used algorithms for sampling from the posterior are versions of **Markov chain Monte Carlo (MCMC)**. The ways in which MCMC algorithms work are beyond our scope. Their main advantage over rejection sampling is that they work better in "high-dimensional" problems—problems in which there are many parameters.[7,8] MCMC methods are implemented in many software

---

[7] MCMC methods also have some disadvantages relative to rejection sampling. Sequential samples from an MCMC are generally not independent. MCMC methods also have to draw many samples before the drawn samples begin to reflect the desired posterior distribution. These initial samples, which must later be discarded, are called "burn-in" samples. There is also some art to determining whether enough burn-in samples have been drawn. With enough computation, these disadvantages can usually be overcome.

[8] MCMC works well for a broader range of models than rejection sampling, but when models get complex enough, MCMC methods also start to perform poorly. More modern approaches, such as hybrid Monte Carlo (also called Hamiltonian Monte Carlo) are becoming more popular in part because of their ability to handle very

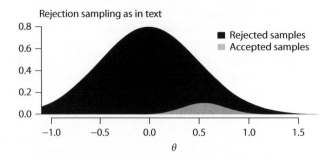

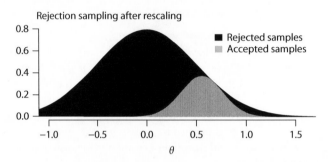

Figure 10-2 A sketch of a rejection sampling framework for the prior distribution and unscaled posterior distribution shown in Figure 10-1. Samples are drawn from the prior distribution, the density of which is represented by the union of the black and gray regions. To simulate samples from the posterior distribution, we keep samples from the prior distribution with a probability proportional to the ratio of the height of the gray region to the height of the black region. For example, a draw from the prior distribution of -0.5 is almost certain to be rejected as unrepresentative of the posterior. In contrast, a sample of 0.5 from the prior distribution stands a reasonable chance of being accepted. The upper panel shows the algorithm described in the text. In practice, the gray acceptance region would be scaled to be as large as possible, as in the lower panel. The algorithm with rescaling will be discussed in the next set of exercises.

packages. Three leading packages that can be accessed using R are BUGS (Spiegelhalter et al., 1996), JAGS (Plummer, 2003), and Stan (Gelman et al., 2015). For the rest of the chapter, we will use MCMC routines implemented in the R package MCMCpack (Martin et al., 2011). MCMCpack is less flexible than BUGS, JAGS, or Stan, but it is flexible enough for our purposes, it is easy to use, and it doesn't require the installation of any extra software. In the exercises, we will confirm that in one simple example, conjugate priors, rejection sampling, and MCMC all give the same posterior distribution.

Once we have a parameter θ's posterior distribution, we have all the information about θ that is implied by the prior distribution, the likelihood function, and the data. Nonetheless, the posterior distribution itself can be somewhat unwieldy—it is often best communicated as a picture that represents a large or infinite set of numbers. By highlighting particular features of the posterior distribution, we can supply point and interval estimates that can be communicated efficiently. In the next two sections, we consider summaries of posterior distributions that respond to questions about estimation.

complex models. For great visual descriptions of both MCMC and Hamiltonian Monte Carlo, see McElreath (2017).

Exercise Set 10-1[9]

(1) In this problem, we'll work through a simple Bayesian example using conjugate priors, rejection sampling, and MCMC, seeing that they give the same answers. We will work with the situation of n independent observations $x_1, x_2, ..., x_n$ drawn from a Normal(θ, σ^2) distribution, assuming that the variance σ^2 is known. We will also assume that the prior distribution for θ is Normal(γ, τ^2). You can choose your own values for $\theta, \sigma^2, \gamma, \tau^2$, and n, but if the prior distribution is centered far away from the likelihood, rejection sampling will take a long time because most candidate samples will be rejected. In the solution, I work with $\theta = 2$, $\gamma = 0$, $\sigma^2 = \tau^2 = 1$, and $n = 20$. Start by drawing a random sample of data from your chosen normal distribution using rnorm().

   (a) The conjugate prior for the expectation parameter of a normal distribution with known variance is itself normal. This means that the posterior will also be normal, with parameters given by Equations 10.4 and 10.5. Compute the expectation and standard deviation of the posterior distribution using Equations 10.4 and 10.5. (You can write an R function to compute these if you'd like.)

   (b) Draw 10,000 samples from the posterior distribution using MCMC, and compute the mean and standard deviation of the samples from the posterior. One way to do this is using the MCMCpack function MCnormalnormal().

   (c) Draw 10,000 samples from the posterior distribution using rejection sampling, and compute the mean and standard deviation of the samples from the posterior. If you don't plan to do the (optional) next problem, use the reject.samp.norm() function in the book's R package, stfspack. See Exercise Set 2-2, Problem 3 for installation instructions, and once the package is loaded, use help(reject.samp.norm) to see the function's documentation. Otherwise, put this part on hold until you complete Problem 2, after which time you can use your own function.

(2) [Optional] In this exercise, you will revise the algorithm for rejection sampling given in the text to make it faster, as well as to make it apply to cases in which the unscaled posterior (Equation 10.2) might be greater than the likelihood. Both of these goals can be achieved by re-scaling the unscaled posterior so that it has exactly the same value of the prior for at least one value of θ and that it never takes any value larger than the prior, as in the lower panel of Figure 10-2.

   (a) We want to be able to rescale the unscaled posterior so that its ratio to the prior equals 1 for at least one value of θ but is never larger than 1. What is the ratio of the unscaled posterior to the prior?

   (b) By what value c would you need to multiply the unscaled posterior so that the product of c and the unscaled posterior is equal to the prior distribution for some value of θ but never larger than the prior?

   (c) Use your insights from parts (a) and (b) to modify the algorithm given in the text for rejection sampling.

   (d) Write an R function to do rejection sampling. Specifically, perform rejection sampling for data drawn independently from a Normal(θ, σ^2) distribution. Consider σ^2 known, but sample from the posterior distribution for θ, which has a normal prior distribution with specified mean and standard deviation.

10.3 Bayesian point estimation using Bayes estimators

One goal of many statistical analyses, both Bayesian and frequentist, is to supply a point estimate for a parameter. We have discussed some frequentist methods for point estimation

[9] See the book's GitHub repository at github.com/mdedge/stfs/ for detailed solutions and R scripts.

in previous chapters, and it remains to say how one might extract a point estimate from a posterior distribution. A natural guess—one that turns out to be useful—is to report a standard indicator of the posterior distribution's central tendency, such as the posterior mean, median, or mode. These quantities are defined analogously to other means, medians, and modes. Namely, for a parameter θ with posterior distribution $f_{\theta|D}(\theta|D = d)$, the posterior mean θ_{mean} is the expectation of θ conditional on the data,

$$\theta_{mean} = E(\theta|D = d) = \int_{-\infty}^{\infty} \theta f_{\theta|D}(\theta|D = d) \, d\theta, \tag{10.6}$$

a posterior median is a[10] value θ_{med} that falls at the center of the posterior distribution, satisfying

$$P(\theta \le \theta_{med}|D = d) = 1/2, \tag{10.7}$$

and a[11] posterior mode θ_{mode} is a value of θ that maximizes the posterior,

$$\theta_{mode} = \underset{\theta}{\text{argmax}} \, f_\theta(\theta|D = d), \tag{10.8}$$

where "argmax" is "argument of the maximum." The posterior mode is also called the maximum a posteriori (MAP) estimator. Figure 10-3 shows a posterior distribution and its mean, median, and mode.

The posterior mean, median, and mode are all sensible point estimators. Beyond their apparent reasonableness, each of these estimators can be justified as **Bayes estimators**, or estimators that minimize the expected value of a loss function.

If you read the decision theory section of Chapter 6, you will recall that a loss function—also sometimes called a cost function—conveys how bad it is for an estimator $\widehat{\theta}$ to be wrong in a specified way. For example, suppose that a firm produces copies of a book for $20 each and sells them for $30. The number of books that consumers will buy is $\theta = 1000$, but the

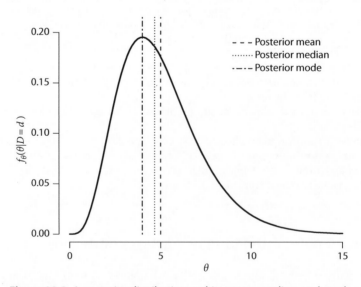

Figure 10-3 A posterior distribution and its mean, median, and mode.

---

[10] In some problems, there are multiple values that satisfy Equation 10.7, and thus multiple posterior medians.
[11] There is sometimes more than one posterior mode.

firm does not know that.[12] The firm estimates the number of books that will sell as $\widehat{\theta}$ and produces $\widehat{\theta}$ books. If $\widehat{\theta} = \theta = 1000$, then the firm spends \$20,000 producing books and earns \$30,000 in sales, for a profit of \$10,000—this is the best possible outcome. If the firm misestimates the demand, then its profit will decrease. If it produces fewer books than would sell—that is, if $\widehat{\theta} < \theta$—then it loses \$10 in profit for each book that would have sold. If it produces more books than will sell—that is, if $\widehat{\theta} > \theta$—then it loses \$20 in production costs for each book that doesn't sell. The loss function in this example—that is, the amount of money the firm will lose relative to the best case—is therefore

$$\lambda(\theta, \widehat{\theta}) = \begin{cases} 10(\theta - \widehat{\theta}) \text{ if } \widehat{\theta} < \theta, \\ 20(\widehat{\theta} - \theta) \text{ if } \widehat{\theta} > \theta. \end{cases}$$

Loss functions can take many forms. Table 6-2 lists some commonly used loss functions, including squared error loss, $\lambda(\theta, \widehat{\theta}) = (\widehat{\theta} - \theta)^2$, absolute error loss, $\lambda(\theta, \widehat{\theta}) = |\widehat{\theta} - \theta|$, and 0-1 ("zero-one") loss,

$$\lambda(\theta, \widehat{\theta}) = \begin{cases} 0 \text{ if } \widehat{\theta} = \theta, \\ 1 \text{ if } \widehat{\theta} \neq \theta. \end{cases}$$

When we considered loss functions in Chapter 6, we considered the estimand θ to be a fixed but unknown constant. Now that we are working in a Bayesian framework, θ is random and has a probability distribution—one that reflects the prior distribution for θ and the data. Given a loss function $\lambda(\theta, \widehat{\theta})$, we can choose a value of $\widehat{\theta}$—that is, a point estimate—that minimizes the expectation of the loss function with respect to the posterior distribution of θ.[13] More formally, a Bayes estimator $\widehat{\theta_B}$ satisfies

$$\widehat{\theta_B} = \underset{\widehat{\theta}}{\text{argmin}} \int_{-\infty}^{\infty} \lambda(\theta, \widehat{\theta}) f_{\theta|D}(\theta|D = d) \, d\theta. \tag{10.9}$$

Given the chosen prior distribution, the data, the likelihood, and the loss function, the Bayes estimator is the estimator that minimizes the expected loss. This involves a lot of "ifs"—*if* we choose a sensible prior, *if* we choose the right model for the data, *if* the loss function reflects the cost of mistakes, etc.—but it is an attractive property if we have models that describe how processes of interest work.

One reason for the popularity of the posterior mean, median, and mode is that they are Bayes estimators under intuitively appealing loss functions. In particular, the posterior mean is the Bayes estimator under squared-error loss $\left(\lambda(\theta, \widehat{\theta}) = (\widehat{\theta} - \theta)^2 \right)$, the posterior median is the Bayes estimator under absolute-error loss $\left(\lambda(\theta, \widehat{\theta}) = |\widehat{\theta} - \theta| \right)$, and the posterior mode is related to the Bayes estimator under zero-one loss, in which there is no loss for guessing exactly the correct value of the estimand, but there is a constant loss for any incorrect guess.[14] Thus, the posterior mean, median, and mode are not just convenient summaries of the posterior distributions, they are "best" summaries—under a

---

[12] Assume, for the sake of this example, that the number of books sold does not depend on the price.

[13] In Chapter 6, we discussed the "risk" as an expectation of a loss function (Equation 6.10). In that context, the estimand was treated as fixed, and the expectation was with respect to the distribution of the estimator. In the Bayesian context, the estimate $\widehat{\theta}$ is treated as fixed, and the expectation is taken with respect to the posterior distribution of the estimand. The quantity being minimized is in Equation 10.9 is sometimes called a "Bayes risk."

[14] The posterior mode is the Bayes estimator only if the parameter is forced to take one of a discrete set of possible values. If the parameter can take any value in a continuous space, then, under some weak conditions, the posterior mode is the limit of the Bayes estimator for a loss function that imposes 0 loss if the estimate falls within an interval of width w around the true value of the parameter and a constant loss otherwise, where the limit is taken as w goes to 0 (for a discussion of the necessary conditions, see Bassett & Deride, 2018).

particular definition of "best" and specific assumptions about the cost of incorrect estimates. In an optional exercise below, you will have the opportunity to prove these claims.

Exercise Set 10-2

(1) In this problem, we will use the `MCMCregress()` function in the `MCMCpack` package to
obtain posterior distributions for simple linear regression. `MCMCregress()` uses Gibbs
sampling—a form of MCMC—to sample from the posterior distribution of the parameters
from a linear regression model. `MCMCregress()` implements the most standard version of
linear regression, with normally distributed, homoscedastic, independent disturbances—
exactly the assumptions used in Chapter 9. For the user, the most salient difference from `lm()`
is that `MCMCregress()` requires as input a prior distribution for the model's coefficients and for
the variance of the disturbances. `MCMCregress()` requires that the prior distribution of the
model's coefficients be a member of the normal family and that the prior distribution of
the variance of the disturbances be a member of the inverse gamma family. To specify the
priors, the user enters parameters for these families.

  (a) To start, fit the model using maximum-likelihood with the `lm()` function. Use `anscombe$x1` as the independent variable and `anscombe$y1` as the dependent variable. Call
`summary()` on the output and note the estimated values for the coefficients.
  (b) To understand how choice of priors changes the results of the regression analysis, re-run
the regression using every combination of the following prior expectations and precisions
for the coefficients: expectations (0,0), (3,0), and (10, −5); precisions 0.0001, 1, 100. Keep
the defaults for all other settings. Precision is equal to $1/v$, where v is the variance of a
random variable. Using lower precision specifies a prior that is more like a flat, uniform
distribution. Record the estimated coefficients in each case by calling `summary()` on the
regression outputs. The estimates will be posterior means (posterior medians are reported
in the table of quantiles), but in this case, the posterior mean, mode, and median will be
effectively the same. To run the first case (after installing and loading `MCMCpack`), use

```
y <- anscombe$y1
x <- anscombe$x1
reg <- MCMCregress(y ~ x, b0 = c(0,0), B0 = 0.0001)
summary(reg)
plot(reg)
```

(2) [Optional, hard] Prove the following claims:
  (a) Show that the posterior mean is the Bayes estimator under squared error loss. (That is,
show that setting $\theta_0 = E(\theta|D)$ minimizes $\int_{-\infty}^{\infty}(\theta - \theta_0)^2 f_{\theta|D}(\theta) \, d\theta$ where $f_{\theta|D}(\theta)$ is the
posterior.)
  (b) Suppose that θ can take values in the set $\{\theta_1, \theta_2, ..., \theta_k\}$, with k finite. Show that
the posterior mode is the Bayes estimator under zero-one loss.
  (c) [Extra hard; requires calculus beyond what is covered in this book] Show that the
posterior median is the Bayes estimator under absolute-error loss. (Hint: Break
the integral into two pieces, and then use the Leibniz integral rule to reverse the order
of integration and differentiation. You may assume that $\lim_{\theta \to -\infty}(\theta_0 - \theta)f_\theta(\theta)$ and
$\lim_{\theta \to \infty}(\theta - \theta_0)f_\theta(\theta)$ do not diverge.)

10.4 Bayesian interval estimation using credible intervals

Just as in frequentist statistics, it is often useful to provide an interval that describes the
degree of uncertainty associated with the estimate. The main mechanism that frequentists

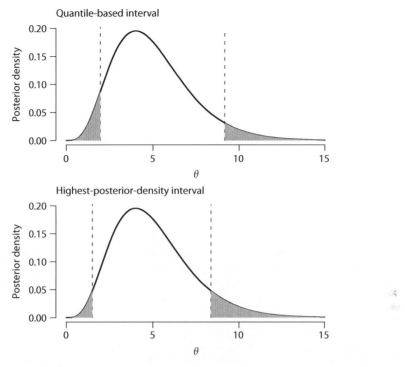

Figure 10-4 Two credible intervals computed on the basis of a posterior density (shown in black). In the upper panel, we have a 90% quantile-based credible interval, bordered by dashed lines. The areas of the two gray shaded regions are equal to each other. Notice, though, that there are parts of the unshaded credible interval with lower posterior density than part of the shaded gray region, which is excluded from the interval. In the lower panel, we have the same posterior density and a 90% highest-posterior-density credible interval, again bounded by dashed lines. The values of θ within the credible interval have posterior densities higher than all values of θ outside the interval. Though the posterior distribution shown is the same as in the upper panel, the bounds of the 90% credible interval are somewhat different.

use for this task is the confidence interval. The Bayesian analog of the confidence interval is the **credible interval**.[15]

A $1 - \alpha$ credible interval for an estimand θ satisfies

$$P(\theta \in v | D = d) = 1 - \alpha, \tag{10.10}$$

where v represents the interval. This is similar to the definition of a confidence interval, with the crucial difference that here θ is a random variable and the interval is treated as nonrandom. If the credible interval is bounded by v_1 and v_2, another way to write the definition is

$$\int_{v_1}^{v_2} f_\theta(\theta | D = d) \, d\theta = 1 - \alpha,$$

where $f_\theta(\theta | D = d)$ is the posterior distribution of θ given the data d. In words, the area under the posterior density between v_1 and v_2—which, by definition, is equal to the posterior probability that θ falls between v_1 and v_2—is $1 - \alpha$. Credible intervals have the

---

[15] Credible intervals are sometimes called posterior intervals, or, confusingly, Bayesian confidence intervals.

interpretation that many people give incorrectly to confidence intervals: the probability that the estimand falls in the interval is $1 - \alpha$. But one can only take this interpretation as seriously as one takes the prior distribution (plus the chosen model, data, etc.).

Given a posterior distribution, there are many different intervals that satisfy Equation 10.10 and thus could serve as credible intervals. One way forward is to select a credible interval that is symmetric, in that the probability θ falls to the right of the interval is equal to the probability that it falls to the left. More formally, if the interval is bounded by v_1 and v_2, we could choose v_1 and v_2 to satisfy $P(\theta < v_1 | D = d) = P(\theta > v_2 | D = d) = \alpha/2$. Such an interval is called a quantile-based credible interval, and one is shown in the upper panel of Figure 10-4.

There is an infelicity that can occur when quantile-based credible intervals are calculated using asymmetrical posterior distributions or posterior distributions with multiple peaks. Examining the upper panel of Figure 10-4, you can see that values of θ immediately to the left of the credible interval have higher posterior density than some values of θ that are included in the quantile-based interval. Values with higher posterior density are considered more plausible given the data and the prior, and it may not be desirable to exclude values with high posterior density in favor of values with lower posterior density. The solution is to use a "highest-posterior-density" credible set, which includes all values of θ with posterior density larger than a constant c, where c is chosen so that the regions of the posterior density larger than c integrate to $1 - \alpha$.[16] A highest-posterior-density credible interval is shown in the lower panel of Figure 10-4.

Exercise Set 10-3

(1) Using the outputs of the MCMCregress() function from Exercise Set 10-2, Problem 1, compute 95% credible intervals for the slope parameter under all the prior distributions you used in that problem. Use both the quantile approach and the highest-posterior-density approach. The information necessary for the quantile approach is provided by summary(), and you can compute highest-posterior-density intervals using the HPDinterval() function in the coda package (Plummer et al., 2006). Recall that the bootstrap confidence intervals for the slope parameter ranged from approximately 0.25 to 0.75.

10.5 [Optional section] Bayesian "hypothesis testing" using Bayes factors

Bayesian hypothesis testing is quite different from frequentist hypothesis testing. For frequentists, a "hypothesis" most often takes the form of a constraint imposed on a parameter in a model—for example, a null hypothesis might state that a particular parameter is equal to a specified constant. For Bayesians, a hypothesis often takes the form of model and a distribution of possible parameter values associated with the model.

We will sketch one approach to Bayesian hypothesis testing, the **Bayes factor**. Suppose we have data D and two hypotheses to explain the data, denoted H_0 and H_1. Each of these hypotheses entails a likelihood function *and* a prior distribution for the parameters that govern the likelihood—we'll label these with subscripts 0 and 1. The Bayes factor compares the likelihood of observing the data under each of the two models, averaged over the prior distributions of the model parameters. The Bayes factor is

$$B_{10} = \frac{P(D|H_1)}{P(D|H_0)}. \tag{10.11}$$

---

[16] If the posterior density has more than one peak, then highest-posterior-density credible sets may be discontinuous—imagine, for example, two peaks separated by a trough of very low density.

One interpretation of the Bayes factor comes from noticing that, by Bayes' theorem, if H_0 and H_1 are the only possible explanations of the data, the posterior probability of H_1 is

$$P(H_1|D) = \frac{P(D|H_1)P(H_1)}{P(D|H_0)P(H_0) + P(D|H_1)P(H_1)},$$

and similarly, the posterior probability of H_0 is

$$P(H_0|D) = \frac{P(D|H_0)P(H_0)}{P(D|H_0)P(H_0) + P(D|H_1)P(H_1)}.$$

Because these two expressions have the same denominator, their ratio is

$$\frac{P(H_1|D)}{P(H_0|D)} = \frac{P(D|H_1)P(H_1)}{P(D|H_0)P(H_0)} = B_{10}\frac{P(H_1)}{P(H_0)}.$$

In words, the ratio of the posterior probabilities of each hypothesis is equal to the ratio of their prior probabilities multiplied by the Bayes factor. If the prior probabilities of the two hypotheses are equal, then $P(H_1)/P(H_0) = 1$ and the ratio of the posterior probabilities[17] is equal to the Bayes factor.

To compute Bayes factors, one has to average the likelihood associated with each hypothesis, weighted by the prior distribution that the hypothesis places on any parameters.[18] More formally,

$$B_{10} = \frac{P(D|H_1)}{P(D|H_0)} = \frac{\int f_1(D|\theta_1)g_1(\theta_1)\, d\theta_1}{\int f_0(D|\theta_0)g_0(\theta_0)\, d\theta_0}, \tag{10.12}$$

where f_k is the likelihood function associated with hypothesis k, θ_1 represents the parameter(s) included in model k, and g_k is the prior distribution of the parameters under hypothesis k.[19] Usually the Bayes factor is computed by approximating the integrals numerically.

Let's complete one example to clarify the moving pieces. Suppose that in the cereal yield example, we have two hypotheses. H_0 asserts that cereal yield Y is governed by

$$Y_i = \alpha + \epsilon_i,$$

where the ϵ_i are independently drawn from a Normal$(0, \sigma^2)$ distribution. There are two parameters in the model, α and σ^2, and let's say that under H_0, α has a Normal$(0, 100)$ prior distribution and σ^2 has an InverseGamma$(0.0005, 0.0005)$ prior distribution. The other hypothesis, H_1, asserts that cereal yield is governed by

$$Y_i = \alpha + \beta x_i + \epsilon_i,$$

where x indicates fertilizer use. The prior distributions of α and σ^2 are the same as under H_0, and β, like α, has a Normal$(0, 100)$ prior distribution. The difference between the two

---

[17] This ratio is also called the posterior *odds* of hypothesis 1, because if an event has probability p, then it has odds $p/(1-p)$.

[18] The prior probability that the hypothesis is true, which was invoked in the interpretation of Bayes factors given in the previous paragraph, is completely distinct from the prior distributions on the parameters postulated *by* the hypothesis. The prior on the hypothesis being true is sometimes called a *hyperprior*.

[19] Unlike in the general case of computing a posterior distribution, the prior distributions used in computing Bayes factors need to be proper probability density functions, meaning that they need to integrate to 1.

hypotheses is thus the presence of a slope associated with x in the model of H_1. If MCMCpack is loaded, then we can fit the two models using MCMCregress()[20] with[21]

```
y <- anscombe$y1
x <- anscombe$x1
reg0 <- MCMCregress(y~1, b0 = 0, B0 = 1/100, marginal.likelihood = "Laplace")
reg1 <- MCMCregress(y~x, b0 = 0, B0 = 1/100, marginal.likelihood = "Laplace")
```

We compute the Bayes factor using the function BayesFactor(), which reports

```
> summary(BayesFactor(reg1, reg0))
The matrix of Bayes Factors is:
     reg1 reg0
reg1 1.000 3.26
reg0 0.307 1.00
```

The Bayes factor comparing H_1 with H_0 is $B_{10} = 3.26$. (The 0.307 shown in the table is the Bayes factor computed with the roles of the two hypotheses reversed, or B_{01}, which is equal to $1/B_{10}$.) One implication is that if the prior probability of H_1 was $P(H_1) = 0.5$, then the posterior probability is $P(H_1|D) = 0.76$—the data render H_1 somewhat more plausible than it was before the data were collected. A Bayes factor of 3.26 would generally be considered moderate evidence in favor of H_1 over H_0. There are several sets of guidelines available for qualitative interpretations of Bayes factors. One set of influential guidelines comes from Kass & Raftery, who consider Bayes factors of 1 to 3 "not worth more than a bare mention," Bayes factors of 3 to 20 as "positive" evidence for H_1, Bayes factors of 20 to 150 as "strong," and Bayes factors of more than 150 as "very strong." (Of course, these cutoffs are arbitrary, like the $p < 0.05$ criterion in frequentist hypothesis testing.)

Some statisticians prefer Bayes factors over p values. Bayes factors are connected to posterior probabilities, which many people find intuitively appealing. They have a reputation for being more conservative than p values, as well as for avoiding the problem many frequentist tests have of rejecting the null hypothesis when there are many data, even if deviations from the null hypothesis are small. Bayes factors are also applicable in some situations in which p values are hard or impossible to compute, such as in the comparison of non-nested models.

Nevertheless, Bayes factors are controversial. They are dependent on features of prior distributions in ways that Bayesian point and interval estimates are not. Prior distributions used in computing Bayes factors need to be proper—to have density functions that integrate to 1. And whereas Bayesian point and interval estimates often converge on similar answers as the sample size increases, Bayes factors exhibit stronger dependence on the prior distributions associated with each competing hypothesis. You will see an example of this in the following exercise.

---

[20] The inverse gamma parameters we have chosen correspond to the default settings, so they are not specified here.

[21] The "marginal likelihood" of model k is $\int f_k(D|\theta_k)g_k(\theta_k)\,d\theta_k$, in the notation of Equation 10.12. Setting marginal.likelihood = "Laplace" causes the function to approximate the marginal likelihoods necessary for computing the Bayes factor using Laplace's method. Laplace's method works well in large samples and in simple models. We have a small sample here, but we are going through the exercise anyway.

Exercise Set 10-4

(1) Fit three alternative versions of the no-slope (H_0) and simple regression (H_1) models in the main text. The alternative versions will differ only in terms of the prior precision assumed for the coefficients. In each case, make note of the posterior means and 95% quantile-based credible intervals for the linear model coefficients, as well as the Bayes factor comparing H_1 with H_0.

(a) Fit the models used in the main text.
(b) Change the prior precision on the linear model coefficients (B0) to 1/16. (This changes the variance of the prior distribution to 16.)
(c) Change the prior precision on the linear model coefficients (B0) to 1/10000. (This changes the variance of the prior distribution to 10,000.)
(d) What do these results suggest about the sensitivity of the posterior mean, quantile-based credible interval, and Bayes factor to the prior distribution?

10.6 Conclusion: Bayesian vs. frequentist methods

Bayesian approaches suggest a philosophy that is in some ways opposed to that expressed by the frequentist methods developed in Chapters 8 and 9. One comfort is that there are many cases where Bayesian and frequentist methods agree.[22] If the number of data is large compared with both the amount of prior information available and the number of parameters, then posterior means, medians, and modes tend to be similar to maximum-likelihood estimates, and credible intervals tend to be similar to confidence intervals.

Why choose Bayesian methods over frequentist ones? Sometimes, what Bayesian analysis offers is just what is needed. That is, in some cases, posterior probabilities are really what one wants, there is a reasonable way to set priors, and there is a good model for the data-generating process. A separate concern is whether Bayesian or frequentist methods "work" better, which is hard to answer generically. As mentioned before, there is a wide class of problems where they both work well, at least for large datasets. Disagreements about conclusions can arise from small datasets and from high-dimensional problems—research applications in which the number of parameters to be considered is very large. Many data analysts working in high-dimensional settings have embraced Bayesian methods even though they may have little interest in prior or posterior distributions per se—they do it because some aspect of the mechanics of Bayesian statistics helps them solve a problem. As one example, the introduction of prior distributions on parameters is a natural way to *regularize* parameter estimates. "Regularization" describes attempts to prevent estimation procedures from overreacting to chance features of a dataset, and it can be crucial in complicated problems. There are other ways to regularize, but many people feel that Bayesian methods shine in such settings.[23]

How, then, to choose? One obvious answer is to look at problems related to the one you are working on and ask whether frequentist and Bayesian methods both have strong track records. Beyond this, you can ask yourself what criteria define success in your application. If consistency and known coverage probabilities are important, then one might favor

---

[22] Of course, this sword has two edges—one might ask why one would bother with extra Bayesian machinery if the reward at the end is the same.

[23] These applications are usually of the "objective Bayesian" flavor.

frequentist methods; if it is more important to update beliefs coherently, then one might favor Bayesian methods. Another consideration is the level of confidence you place in your field's models. Workers in fields with well-established, reliable mathematical models might be more likely to favor Bayesian methods that can take full advantage of their structure. In fields with less developed models, semiparametric or nonparametric frequentist approaches that require fewer assumptions might be more attractive—Bayesian methods like those in this chapter rely as completely on the likelihood function as do likelihood-based frequentist approaches.[24]

To end this chapter, I'd like to point out one more commonality of Bayesian and frequentist approaches. I said above that in favorable situations, either kind of approach can work. The other side of the denarius is that either approach will produce shoddy results if they are used poorly—for example with incorrect methods, bad data, or wild assumptions. At this writing, attention is focused on a "replication crisis" in many areas of empirical research, and especially in psychology.[25] The replication crisis refers to the growing awareness that many phenomena documented in the academic literature are not replicable in repeated studies—many may be mere flukes that appeared in a single sample but that are not genuine patterns in the world. One of the proposals to address the replication crisis is to switch to Bayesian methods. This proposal has had the salutary effect of leading many researchers to pay closer attention to their methods—the mere realization that there is another way is a jolt to wakefulness. But Bayesian methods will not solve the fundamental issues on their own—among other issues, academic researchers in many fields have strong career incentives to misapply statistical procedures in an effort to gain more publications, and only weak institutional incentives to ensure that that the phenomena they measure are stable. An academic research system that recapitulates the current one but substitutes Bayesian methods for frequentist ones will be gamed just as easily.

10.7 Chapter summary

Bayesian methods allow researchers to combine precise descriptions of prior uncertainty with new data in a principled way. The main object of interest in Bayesian statistics is the posterior distribution, which describes the uncertainty associated with parameters given a prior distribution and the observed data. The posterior can be difficult to compute mathematically, but computational methods can give arbitrarily good approximations in many or most cases. Bayesian point and interval estimates are features of the posterior, such as measures of its central tendency or intervals into which the parameter falls with specified probability. One tool for Bayesian hypothesis testing is the Bayes factor, which compares the probability of observing the data under a pair of distinct hypotheses.

10.8 Further reading

Gelman, A., Carlin, J. B., Stern, H. S., Dunson, D. B., Vehtari, A., & Rubin, D. B. (2014). *Bayesian Data Analysis*. CRC Press, Boca Raton, FL.

---

[24] If you do use a Bayesian method in conjunction with a probabilistic model that is not obviously the "right" model, then it is useful to include some assessment of model fit in your analysis. Many possible suggestions are included in Gelman et al. (2014), in their chapter on model checking and improvement.

[25] This is not necessarily to suggest that the replication crisis is especially severe in psychology in comparison with similar fields, only that it has received more attention.

An authoritative textbook with wide coverage and practical advice. The third edition is integrated with Stan, a flexible piece of software for drawing samples from posterior distributions.

Gigerenzer, G., & Marewski, J. N. (2015). Surrogate science: the idol of a universal method for scientific inference. *Journal of Management*, 41, 421–440.

This article gives a historical perspective on the use of *p* values in psychology and criticizes current practice in ways that overlap with many of the sources recommended at the end of Chapter 7. I include it here rather than there because it makes an additional point—proposals simply to replace frequentist practices with Bayesian ones run the risk of setting up Bayes factors as a new "idol" in place of *p* values, which would lead to a set of problems similar to those faced now.

Griffiths, T. L. & Tenenbaum, J. (2006). Statistics and the Bayesian mind. *Significance*, 3, 130–133.

One interesting offshoot of Bayesian statistics is an effort in psychology to understand human cognition as a fundamentally Bayesian process. This piece is an accessible (and short) introduction to this point of view.

Kass, R. E., & Raftery, A. E. (1995). Bayes factors. *Journal of the American Statistical Association*, 90, 773–795.

A thorough (and enthusiastic) introduction to Bayes factors.

McElreath, R. (2015). *Statistical Rethinking*. Chapman & Hall/CRC Press, Boca Raton, FL.

A wonderful introduction to statistics from a Bayesian perspective. Most of the material is illustrated via computational examples implemented in an R package written for the book, which is itself useful for running Bayesian data analyses in R.

Postlude: models and data

Key terms: Confounding, Cross-validation, Generalized linear model, Mixed model, Multiple regression, Overfitting

In the ten chapters that preceded this one, we have toured statistical thinking by engaging with some central ideas from statistical theory. These topics are critical because researchers who do not know them are unlikely to reach their potential as data analysts. It is hard to become a confident user of statistical methods unless one understands what a probabilistic model is, knows how to use a serious statistical software package, and can examine and manipulate simple equations. A person with these skills has a strong statistical foundation for an empirical research career.

At the same time, mastery of the ideas in this book is, on its own, not nearly enough to make an expert data analyst. For that, one needs domain knowledge, a keen grasp of scientific method, and experience with real datasets. One needs the temperament to ask difficult questions about one's own discoveries and to search aggressively for alternative explanations. In short, one needs to connect probabilistic models with data and with substantive research questions.

In this final chapter, I cannot hope to supply the remaining pieces. Instead, the goal will be to provide some evocative pointers. We will start out with a discussion of the role played by assumptions in statistical analyses, then sketch methods for checking the plausibility of one's assumptions. After that, we will consider a few ways in which the simple regression model that we have examined can be extended.

Post.1 Evaluating assumptions

For the conclusion of a statistical analysis to be sound, the assumptions made during the analysis must also be sound. As data analysts, we would often like to argue something like the following[1,2]:

(i) For data generated by model M, procedure P produces a 99% confidence interval for the parameter θ.

(ii) The data were generated by a process that is described by model M.

(iii) Applying procedure P to data D produces an interval C.

---

[1] This argument considers a confidence interval, but that is not essential. Similar arguments could be constructed for other kinds of estimates or hypothesis tests.

[2] In a research setting, this statistical argument would generally be a piece of a larger scientific argument, which might take many forms.

Statistical Thinking from Scratch: A Primer for Scientists. M. D. Edge, Oxford University Press (2019).
© M. D. Edge 2019. DOI: 10.1093/oso/9780198827627.001.0001

(iv) Interval C is a 99% confidence interval for parameter θ in the population from which the data D were sampled.

The argument's conclusion, (iv), follows from its premises, (i)–(iii). We are near the end of a lengthy discussion of ways in which one can clarify and establish statements like (i). Usually (iii) is uncontroversial in the absence of errors of computation. The most dubious premise is thus (ii), which claims that a theoretical model and a real dataset correspond.

Premise (ii)—that the data were generated by the assumed model—is seldom literally true. For example, real-world data are not literally samples from a normal distribution. If premise (ii) is false and the data are not generated by the assumed model, then the conclusion in (iv) is not guaranteed to be true. The best available response is to argue that the data are *consistent with* the assumed model. Such a response cannot, on its own, substitute for premise (ii): an argument in which the second premise is replaced with

(ii-a) The dataset D is similar to datasets generated by model M

does not validly lead to (iv). Nonetheless, we'll continue by considering ways to defend (ii-a), even though it may not be enough to establish the conclusions we would like to reach.

There are two major approaches to evaluating whether a model's assumptions are reasonable: scrutinizing plots of the data, conducting formal tests of model assumptions, and making predictions in new data.

Post.1.1 Plotting

Recall that the fertilizer data we have been using as an example are stored in the R object anscombe. Specifically, the fertilizer consumption values are in anscombe$x1, and the cereal yield values are in anscombe$y1. To apply lm(), which fits a linear regression model, to the fertilizer data and summarize the results, use the following two commands, just as we did in Chapter 1:

```
> mod1 <- lm(anscombe$y1 ~ anscombe$x1)
> summary(mod1)
```

The second command produces output

```
Coefficients:
             Estimate Std. Error t  value  Pr(>|t|)
(Intercept)    3.0001    1.1247    2.667   0.02573 *
anscombe$x1    0.5001    0.1179    4.241   0.00217 **
—
Residual standard error: 1.237 on 9 degrees of freedom
Multiple R-squared: 0.6665,   Adjusted R-squared: 0.6295
F-statistic: 17.99 on 1 and 9 DF, p-value: 0.00217
```

The estimated intercept and slope are the same as the least-squares, method-of-moments, and maximum-likelihood values we obtained in Chapters 3, 8, and 9, respectively. The standard errors follow from the maximum-likelihood approach considered in Chapter 9. The t statistics reported here are the same as the Wald statistics from Chapter 9, and the p values come from comparing the t statistics with a t distribution (rather than a standard normal, as in the Wald test). The F statistic is equal to the likelihood-ratio statistic from Chapter 9.[3] In other words, `lm()` implements all the maximum-likelihood theory we discussed in Chapter 9 (and more).

The `anscombe` dataframe contains more than just the variables we have been examining. There are three other pairs of variables: `anscombe$x2` and `anscombe$y2`, `anscombe$x3` and `anscombe$y3`, and `anscombe$x4` and `anscombe$y4`. Before reading on, please fit and examine simple regression models for each of these pairs of variables. (You've already done some of these, in Chapter 1 and in an exercise in Chapter 3):

```
> mod2 <- lm(anscombe$y2 ~ anscombe$x2)
> summary(mod2)
> mod3 <- lm(anscombe$y3 ~ anscombe$x3)
> summary(mod3)
> mod4 <- lm(anscombe$y4 ~ anscombe$x4)
> summary(mod4)
```

All three of these additional model estimates produce output nearly identical to the output from the dataset we have been examining throughout the book. That includes coefficient estimates, standard errors, t (Wald) statistics, p values, F statistics, and R^2 estimates. If we were to estimate correlation coefficients for each pair of variables, we would see the same thing.

One might be tempted to claim that because the output is the same for all four datasets, the datasets warrant precisely the same conclusions. But such an inference would be incorrect, as was hinted in Chapter 1. Maximum-likelihood estimates (and their standard errors) depend on model assumptions. If the model assumptions are implausible, then the estimates may be invalid or meaningless. Recall that in Chapter 9, to justify a maximum-likelihood interpretation of the least-squares line, we made the following assumptions:

Assumption A1: linearity $E(\epsilon|X = x) = 0$ *for all x.*

Assumption A2: homoscedasticity $Var(\epsilon|X = x) = \sigma^2$.

Assumption A3-P: independence of units *For all i and all j≠i, ϵ_i is independent of ϵ_j.*

Assumption A4-P: distribution *For all i, ϵ_i is drawn from a normal distribution.*

The guarantees provided by maximum-likelihood theory only apply to the `lm()` output if these four assumptions are met. It is usually impossible to say whether they are "really" true of data, but we can often assess whether they are plausible descriptions of data.

As mentioned in Chapter 1, the data we have been referring to as the "fertilizer data" are not fertilizer data or even real data. They were constructed by Francis Anscombe to make the point I am going to reiterate now: numerical calculations cannot substitute for data visualization.

Look at Figure Post-1, which shows scatterplots of the four pairs of variables with their (identical) least-squares lines. The four datasets are strikingly different, and only one—the one we have been working with throughout the book—could be reasonably analyzed with a simple linear regression. In the top right panel, we have a violation of the linearity assumption—the points are much better described by a curve than by a line. In the bottom left, we do not see a roughly symmetric "cloud" of points as required by the normality assumption. Instead, most of the points fall in a line, and one deviates from the pattern

---

[3] In multiple regression with k independent variables (sketched below), the F statistic reported by `lm()` is the likelihood-ratio test statistic comparing the full model with a model with only an intercept term, divided by k.

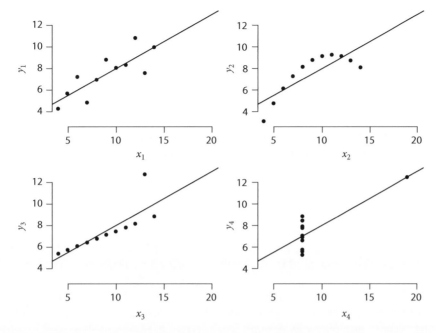

Figure Post-1 The four pairs of variables in the `anscombe` dataframe, plotted as scatterplots with their associated least-squares lines. The least-squares coefficient estimates, as well as their associated standard errors and other statistics, are identical. Only the data in the top-left panel appear consistent with standard linear regression assumptions. The four datasets are known as Anscombe's quartet, after Francis Anscombe, who constructed them.

Modified from a figure on the Wikipedia page for Anscombe's quartet (Anscombe's quartet, n.d.) and published under the CC BY-SA 3.0 License

followed by the rest. In the bottom right, all the estimates and inferences produced by `lm()` depend solely on the position of a single outlying point.

None of these patterns can be read from the summary of the `lm()` output. Looking at the data is the best way to see them. It's a good rule never to rely on a statistical analysis without looking at a plot that supports the reasonableness of the analysis. If a demon ever forces you to choose a single statistical method as your sole research aid, don't choose linear regression: choose the scatterplot.

Exercise Set Postlude-1

One snag associated with making plots is that exploratory data analysis efforts may contaminate hypothesis testing. For example, if the plot leads a data analyst to observe a pattern, then there may be a temptation to explore the pattern with a hypothesis test, possibly leading to further plots and test that explore what are actually random features of the dataset. It is easy to "discover" spurious results in this way. One solution to this problem is to use plots that specifically display the fit of the data to model assumptions rather than displaying associations among variables. If you call `plot()` on the output of an `lm()` command in R, then you can see four such plots designed for linear regression. A residuals vs. fitted values plot principally comments on the linearity assumption (though other patterns can be seen, too). A Q–Q plot (Q–Q is for "quantile–quantile") shows the fit of the residuals to a normal distribution. A spread vs. level plot informs about homoscedasticity violations, and a residuals vs. leverage plot alerts us to outlying points that unduly influence the analysis. Use plot() to view these four plots for the four models based on Anscombe's quartet, and see whether you can see the same violations observable in Figure Post-1.

Post.1.2 Tests of assumptions

Another way to assess the plausibility of assumptions is to conduct a statistical test. The logic of the procedure is the same as outlined in Chapter 7. Compute a test statistic that has a known distribution if a particular assumption holds, and compare the statistic with its distribution to produce a p value. A small p value means that the test statistic falls in a range that is unlikely if the assumption holds.

Tests of assumptions can be useful, but they are subject to all the caveats associated with hypothesis tests generally. For example, the behavior of the test is usually sensitive to sample size. In small samples, even pronounced violations of assumptions may not be detected by the test. In large samples, small violations that might be safely ignored can lead to small p values. It would be unwise to rely solely on a one-size-fits-all criterion, such as $p < 0.05$, to decide whether to proceed with an analysis. Instead, such tests are best thought of as a supplement to data displays. They can reveal whether a pattern observed in a plot is more pronounced than one would typically expect in data produced by the assumed model.

In R, one package for testing linear regression assumptions is gvlma (for "global validation of linear model assumptions": Peña & Slate, 2012). Once the package is installed and loaded, the gvlma() function conducts four specific tests of linear model assumptions, along with an overall p value based on the combination of the four specific tests. After the package is installed and loaded, calling gvlma() on the linear regression model output relating anscombe$x1 to anscombe$y1 produces the following table:

```
> gvlma(mod1)
                    Value p-value                 Decision
Global Stat        1.24763  0.8702  Assumptions acceptable.
Skewness           0.02736  0.8686  Assumptions acceptable.
Kurtosis           0.26208  0.6087  Assumptions acceptable.
Link Function      0.68565  0.4076  Assumptions acceptable.
Heteroscedasticity 0.27255  0.6016  Assumptions acceptable.
```

The tests of "skewness" and "kurtosis" assess the apparent normality of the residuals.[4] The skewness test produces low p values if the residuals are asymmetrically distributed, and the kurtosis test produces low p values if the distribution of the residuals has tails that are too heavy or too light. The "link function" test assesses the plausibility of the linearity assumption. The "heteroscedasticity" test assesses the homoscedasticity assumption. If the data are

---

[4] The assumption made in standard linear regression is that the *disturbances* are drawn from a normal distribution, not that the residuals are. (Recall that the disturbances are vertical distances between y values and the true regression line, $\alpha + \beta x$, whereas residuals are vertical distances between y values and the *estimated* regression line, $\hat{\alpha} + \hat{\beta}x$.) We do not observe the disturbances, so we cannot assess the assumption directly. But if the linearity assumption holds and there are enough data, then the residuals are good estimates of the disturbances.

in fact generated by a model meeting all the standard linear regression assumptions, then each of these four test statistics has a $\chi^2(1)$ distribution, and their sum—displayed as the "global stat"—has a $\chi^2(4)$ distribution.

The gvlma() function does not apply any test of the independence of units assumption. Usually independence is assessed by considering the way in which data were collected and then looking for specific evidence of hypothesized sources of dependence. For example, if data are collected from students in an elementary school, one might look for evidence of dependence of observations from students in the same classroom.

In the following exercise, you will apply the gvlma() tests to the other linear models built from the other three pairs of variables in Anscombe's quartet.

Exercise Set Postlude-2

Apply gvlma() to the lm() objects fitted using the other three pairs of variables in Anscombe's quartet. What do you notice in each case?

Post.1.3 Out-of-sample prediction

A third approach to assessing model appropriateness is to use the model to make predictions about new data—data that were *not* among those used to fit the model.[5] This is the best way to be sure that a model describes something about the world beyond the initial sample, rather than just fitting the data in hand. It is harder to build a model that makes good predictions in new data than it is to build one that fits the data to which it is exposed.

One doesn't need to collect new data to gain many of the benefits of out-of-sample predictions. Instead, one can hold aside a fraction of the dataset for testing model predictions. A variant of this strategy is k-fold **cross-validation**, in which the dataset is broken into k equally sized pieces, and the model is fit k times, each time holding aside one of the pieces. For each version of the model, the quality of the predictions is assessed in the excluded portion of the data.

Out-of-sample prediction is especially useful for assessing the severity of **overfitting**. Models vary in their complexity, and more complex models are capable of identifying more complicated patterns in data. The trouble with complex models is that they can mistake chance fluctuations for genuine patterns in the data—this is overfitting.[6] An example is shown in Figure Post-2. The figure includes the same "fertilizer" data we have used throughout the book, but instead of a line, a higher-order polynomial has been fit to the data. The polynomial fits the data we have perfectly, but it would likely fail spectacularly with new data. Among its nonsensical predictions are (1) Countries that apply less than 4 kg/hectare *or* more than 14 kg/hectare of fertilizer are promised lands

---

[5] Rant: In some areas, the word "predict" is used as a shorthand for something like "is associated with." For example, a data analyst working with the fertilizer data might write that fertilizer application "predicts" cereal yield, when she means only that fertilizer application is associated with cereal yield. This jargon is confusing to outsiders (see, e.g., Abraham, 2010). It also has a pernicious effect on insiders, who can come to conflate the easier task of identifying a model that fits a particular dataset well with the harder task of building a model that generalizes. In fields that rely on this language, the harder task is often not attempted.

[6] Overfitting can also be viewed in terms of what is called the "bias–variance tradeoff." Overly complicated models can fit complex patterns and thus don't induce bias by imposing a simplistic structure. The cost is that they can require a great deal of data to fit and can produce highly variable estimates and predictions. Regularization strategies (see Chapter 10, Section 10.6) attempt to balance problems caused by bias and variance.

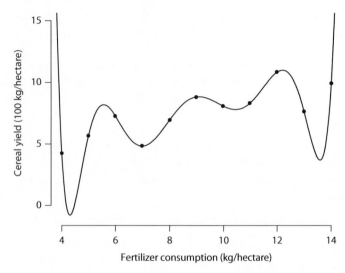

Figure Post-2 The "fertilizer" data, (over)fit with a 10th-degree polynomial rather than a line.

awash with grain and (2) countries that apply a little more than 4 kg/hectare manage to achieve negative cereal yield.

Out-of-sample prediction is the surest cure for overfitting. A model that latches onto noise rather than genuine structure will be exposed in a new dataset, the random aspects of which will differ from the original dataset.

Simple linear regression, which we have considered throughout the book, does not usually pose a great risk of overfitting—the model is too simple. Overfitting can become a major consideration in the extensions of simple regression we consider next.

Post.2 Some extensions of simple linear regression

What does one do when the data provide strong evidence that the assumptions of simple linear regression are violated? One option is to retreat to an interpretation that does not require the violated assumption(s). For example, bootstrap methods do not require the disturbances to be normally distributed, and so one can use bootstrap-based confidence intervals in some situations in which maximum-likelihood interpretations of the least-squares line would be inappropriate. At the extreme, one can always present the least-squares line simply as the line that minimizes squared errors. No probabilistic assumptions are required for this interpretation. Such model-free summaries can sometimes be useful.

Another approach is to work with a model that explicitly accommodates key features of the research question or the data. The literature of applied statistics—and of the fields where statistical thinking is applied—is full of innumerable such modeling efforts. Here we will sketch three ways in which the simple linear regression model can be extended. These methods can be motivated as ways of accounting for specific violations of the model we have studied. Each of them is the topic of many books; we will spare only a few paragraphs. The goal is only to hint at some of the ways in which the ideas you have learned can be applied.

Post.2.1 Multiple regression

Multiple regression—also "multiple linear regression" or just "regression"—extends the simple linear regression model by including more than one independent variable. One way to write the model is

$$Y = \alpha + \beta_1 x_1 + \beta_2 x_2 + \dots + \beta_k x_k + \epsilon. \tag{P.1}$$

Y is a dependent variable, the βs are regression coefficients, the xs are values of k different independent variables, and ϵ is a random disturbance, which is usually assumed to be Normal$(0, \sigma^2)$. Each β in $\beta_1, \dots, \beta_k$ is interpreted as the change in E$(Y|x)$ associated with a one-unit change in the corresponding x, conditional on all the other variables being held constant. The notation has changed a little: the subscripts serve to distinguish among different independent variables rather than different observations. It's implicit that the formula is used to compute one observation's value of Y on the basis of its values of the independent variables $x_1, \dots, x_k$. Whereas simple linear regression is concerned with drawing a line through a two-dimensional scatterplot, you can view multiple regression as concerned with drawing a plane through a higher-dimensional cloud of points. Figure Post-3 shows an example with two independent variables.

The mathematics of multiple regression is most easily explained in matrix notation, and the matrix algebra that is necessary here is a little beyond our scope. Nonetheless, many of the most important facts are straightforward extensions of claims we have already explored. Under the assumptions used in Chapter 8,[7] the least-squares estimates of the coefficients are consistent, and interval estimates can be obtained by bootstrapping. Under the assumptions used in Chapter 9,[8] least-squares estimates are maximum-likelihood estimators, and the likelihood-ratio statistic has a $\chi^2(m)$ distribution under the assumption that m of the β terms are equal to 0.

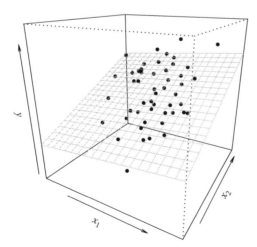

Figure Post-3 Multiple regression output implies a plane that "best" fits (in the least-squares sense) a cloud of points.

---

[7] That is, independent disturbances with expectation 0.
[8] Independent, homoscedastic, normally distributed disturbances with expectation 0.

The multiple-regression least-squares coefficients have formulas similar to those we computed in the simple linear regression case, with adjustments to account for correlations among independent variables. There is even a way to obtain the least-squares estimates from multiple regression by repeated application of simple linear regression, as you can explore in an optional exercise. And many of the methods most people learn in their first course in data analysis—including t-tests, correlation analysis, analysis of variance (ANOVA), and analysis of covariance (ANCOVA)—are special cases of multiple regression, as you will verify in an exercise.

Despite the similarities between simple and multiple regressions, there are several new considerations that appear in the multiple-regression setting. The thorniest issues are related to causality, which we have mostly sidestepped until now. One interpretation of regression, which we could call the *predictive* interpretation, is that regression coefficients are meant merely to assess the patterns of association between independent variables and a dependent variable. Another interpretation, common in economics and allied fields, is *causal*. Under the causal interpretation, regression coefficients measure the strength of the cause-and-effect relationships between independent variables and a dependent variable.

Here is an illustration of the two perspectives. Suppose that a researcher posits a model for reading speed in school-age children,

$$Y = \alpha + \beta_1 * \text{age} + \beta_2 * \text{st} + \epsilon.$$

Y is taken to be reading speed, perhaps measured in words per minute in a standardized text. The variable st is for "school type"—it equals 1 for children in private school and 0 for children in public school. The model is suspect for several reasons, but we will use it anyway. Imagine that we are interested in the difference in reading speed between 8-year-olds in public and private school. By the linearity assumption, $E(\epsilon) = 0$ for all combinations of values of the independent variables, so the expected difference under the model is

$$E(Y|\text{age} = 8, \text{st} = 1) - E(Y|\text{age} = 8, \text{st} = 0) = (\alpha + \beta_1 * 8 + \beta_2 * 1) - (\alpha + \beta_1 * 8 + \beta_2 * 0)$$
$$= \beta_2.$$

Under either the predictive or causal interpretation, the expected difference between private and public school children is β_2. Under the predictive interpretation, β_2 is an expectation that applies to an observational study,[9] where we do not influence which children are sent to which school.

It is not necessary that private schooling *causes* this difference. In contrast, the causal interpretation *does* hold that private schooling causes the difference. Imagine that we take a student who otherwise would enroll in a public school and enroll her in a private school, without changing anything else about her or her circumstances. Under the causal interpretation, this student's reading speed will change from what it otherwise would have been by β_2 words per minute. The predictive interpretation does not make a claim about what would happen in this case.

Of course one can imagine many regression models to which the causal interpretation would not apply—regression coefficients are estimated on the basis of patterns of association among variables and not on their causal relationships. For example, imagine a dataset in which reading speed and shoe size are recorded in school-age children. In a regression

---

[9] An "observational study" is a study in which the researcher is not able to manipulate the putative causal variable(s). The contrast is experimental data, where the researcher changes the putative causal variable, usually by random assignment. If it is possible to manipulate the putative causal variables, then it is much easier to make causal claims based on experimental data. Either observational or experimental data can be analyzed by multiple regression.

with reading speed as a dependent variable and shoe size as an independent variable, the slope estimate will be positive, because shoe size and reading speed will both be correlated with age. But dressing a child in clown shoes does not make a speed reader. Everyone knows this. Advocates of the causal interpretation do not claim that regression—or any related method—is a machine for automatically revealing causal relationships. Instead, they argue that one ought to understand the conditions under which regression coefficients do reveal causal information, and that it is desirable to design studies and data analyses that meet these conditions, allowing causal interpretations.

One important reason that regression coefficient estimates do not generally reflect causal relationships is unmeasured **confounding**. Unmeasured confounds are variables excluded from a regression equation that cause associations between an independent variable and the dependent variable. For example, in a regression associating shoe size with reading speed among school children, age is a confound. Children grow as they age, and their reading skills also improve.[10] To assess the causal effect of shoe size on reading speed, the multiple regression

$$Y = \alpha + \beta_1 * \text{shoe.size} + \beta_2 * \text{age} + \epsilon$$

is an improvement on the simple regression model

$$Y = \alpha + \beta_1 * \text{shoe.size} + \epsilon.$$

If it is possible to include exactly the right variables, including all confounds, and to measure them without error—*and* if the model assumptions hold—then it is defensible to interpret regression estimates causally.

In practice, it is hard to justify causal claims in observational data using regression analysis. Usually it is impossible to know whether all the relevant confounds have been measured. Further, it can be hard to distinguish confounds from *intermediate variables* or *mediators*—variables that are causally affected by an independent variable and that in turn causally affect the dependent variable.[11] Such variables can look like confounds—they are associated with both the independent variable and the outcome—but if they are included in a regression, part of the influence of the independent variable will be falsely attributed to the intermediate variable.[12] Thus the goal is not just to include all the confounds, but to include all the confounds and exclude the intermediate variables—which is much harder. Finally, even if exactly the right variables are included in the regression, if the measurement of those variables is imperfect, then the estimates of causal effects that come from regression will be biased (for some examples, see Westfall & Yarkoni, 2016).

In short, multiple regression is a useful model in many settings, but interpreting regression results is not easy when cause and effect are at issue.

---

[10] This is still too simple—measuring age lumps together cognitive development, amount of practice, and instructional level in school. But it's clear that the shoes aren't in charge.

[11] For example, when a driver depresses a gasoline-powered car's accelerator, a process is initiated that results in fuel being fed into the engine, which is then combusted, causing the car to accelerate. The fuel fed into the engine is an intermediate variable. Intermediate variables are also called *endogenous* variables.

[12] Sometimes the change in the estimated regression coefficients that occurs after an intermediate variable is included is informative. In the gas pedal case (see the previous footnote), the estimated regression coefficient for the gas pedal would be near zero after the fuel injected into the engine is included in the model. In this case, we learn that the gas pedal has its effects via the fuel injection. This conclusion is safe because we already know that the fuel is causally intermediate—causally intermediate variables are also called "mediators"—and not a confound. If we didn't know this, and we thought that fuel injection was a confound—that fuel causes depression of the gas pedal—then we might wrongly conclude that depressing the gas pedal would not cause the car to accelerate.

Exercise Set Postlude-3

(1) [Optional] Least-squares estimates for multiple regression coefficients can be obtained by repeated application of simple linear regression. You will verify this fact using the built-in dataset mtcars, which includes data on various features of cars tested in 1974 by *Motor Trends* magazine.

   (a) Use lm() to fit the model $MPG = \beta_0 + \beta_1 * AM + \beta_2 * HP + \epsilon$ using least-squares, where MPG indicates miles per gallon, AM indicates transmission type (automatic vs. manual), and HP indicates horsepower. The R command to fit the model and save the output to an object called mod.fit is mod.fit <- lm(mpg ~ am + hp, data = mtcars). Call summary() on the result.

   (b) To obtain the same estimated slope for weight as in step 1, but only using simple linear regression, complete the following steps:

     (i) Fit a simple linear regression with transmission type as the dependent variable and horsepower as the independent variable, and save the residuals. E.g., run

```
resid.am.hp <- lm(am ~ hp, data = mtcars)$residuals
```

     (ii) Fit another simple linear regression with miles per gallon as the dependent variable and horsepower as the independent variable, and save the residuals.

     (iii) Fit a third simple linear regression, with the residuals from (ii) as the dependent variable and the residuals from (i) as the independent variable.

   Compare the slope estimate from (iii) with the slope for transmission type in (a). What do you think is going on?

   (c) Modify the procedure in part (b) to obtain the estimated slope for horsepower from part (a) using only simple linear regression.

   (d) In multiple regression, the least-squares regression plane—because it's in more than two dimensions, the regression function is a plane rather than a line—is equal to the sample mean of the dependent variable at the point where all the independent variables are equal to their sample means. This is analogous to the result of Exercise Set 3-2, Problem 1 (b). Use this fact to obtain the intercept from part (a) using the estimated slopes and the sample means of miles per gallon, weight, and horsepower.

(2) [Optional] Several other methods can be viewed as special cases of multiple regression. These methods are sometimes referred to as instances of the *general linear model*. (This is not the same as *generalized* linear models, the topic of the next subsection.) It is clear that simple linear regression is a special case of multiple regression, a multiple regression with only one independent variable. In this problem, we will verify in an example that multiple regression gives the same results as a standard *t*-test, Pearson correlation analysis, and one-way ANOVA. (Don't worry if you have never heard of these methods before.)

   (a) The classic version of Student's independent-samples *t*-test tests the null hypothesis that two samples are drawn independently from the same normal distribution. Compute a *t*-test comparing the gas mileage of cars with automatic vs. manual transmissions in the mtcars dataset, with

```
t.test(mtcars$mpg ~ mtcars$am, var.equal = TRUE)
```

(The specification var.equal = TRUE gives the "classic" *t*-test, which assumes equal variances. There are modifications to account for different variances in the two populations from which the samples are drawn, which are applied by default in R.) Fit a simple linear regression using

```
mod <- lm(mtcars$mpg ~ mtcars$am)
summary(mod)
```

What do you notice about the test statistics and p values reported for the two analyses?

(b) The Pearson correlation coefficient is an estimator of the population correlation coefficient. It is also the basis of a standard hypothesis test of the null hypothesis that the population correlation coefficient for two random variables is equal to zero. Conduct the test in the fertilizer data with

```
cor.test(anscombe$x1, anscombe$y1)
```

Compare with the results from simple linear regression.

(c) Many forms of ANOVA (analysis of variance) can be viewed in terms of linear regression. In so called "one-way" ANOVA, the null hypothesis is that observations divided into k groups are all drawn independently from the same normal distribution. The variable that includes the observations' memberships in k groups is called a "factor." In the built-in `PlantGrowth` data, weights of plants are recorded along with the treatment they received in an experiment. To fit an ANOVA to these data, use

```
anova.fit <- aov(weight ~ group, data = PlantGrowth)
summary(anova.fit)
```

Now fit a linear regression model with the same formula and summarize it. What do you notice?

Post.2.2 Generalized linear models

Multiple regression responds to the need to incorporate more than one explanatory variable in a model. Another limitation of simple linear regression is that it requires a specific type of dependent variable—one that is measured on a continuous scale and that responds linearly to changes in the independent variable. **Generalized linear models** can accommodate a wider range of dependent variables.

The most widely used generalized linear models are models for dichotomous outcomes—dependent variables that can take only two values. In 1987, psychologists Cowles and Davis asked people to complete a personality test and to state whether they would be willing to volunteer to participate in further research. They considered extraversion and neuroticism, as assessed by the personality test, as independent variables, along with gender. The dependent variable was whether the participant would volunteer for future research. We'll represent the dependent variable as and say that $Y = 1$ if the participant says they will participate and $Y = 0$ otherwise.

One possible model for Cowles and Davis's data is

$$Y = \alpha + \beta_1 * Ex + \beta_2 * Ne + \beta_3 * M + \epsilon,$$

where Ex is extraversion, Ne is neuroticism, and $M = 1$ if the participant is a man and otherwise zero. This is a multiple regression model. It is a poor candidate. Y *must* equal 0 or 1, and so this model puts intense constraints on the disturbance term, violating normality, linearity, and homoscedasticity assumptions. An improvement is to replace the decision to volunteer itself with a probability that one will decide to volunteer, as in

$$P(Y = 1) = \alpha + \beta_1 * Ex + \beta_2 * Ne + \beta_3 * M.$$

This can still lead to nonsense, such as probabilities greater than 1 or less than 0.

It is a good idea to model $P(Y = 1)$, but modeling it as a linear function of the independent variables may cause trouble.[13] Assuming that $P(Y = 1)$ is a function of the independent variables that is bounded by 0 and 1 is one solution. There are many such bounded functions—every cumulative distribution function is a possible choice. For example, we could say

$$P(Y = 1) = \Phi(\alpha + \beta_1 * Ex + \beta_2 * Ne + \beta_3 * M), \qquad (P.2)$$

where Φ is the cumulative distribution function of the Normal$(0, 1)$ distribution. Equation P.2 is a case of the probit[14] model, which is a type of generalized linear model.

The most popular choice for dichotomous outcome data is logistic regression, which holds that

$$P(Y = 1) = \frac{e^{\alpha + \beta_1 x_1 + \beta_2 x_2 + \dots + \beta_k x_k}}{1 + e^{\alpha + \beta_1 x_1 + \beta_2 x_2 + \dots + \beta_k x_k}}.$$

Logistic regression coefficients have a natural interpretation in terms of ratios of odds of the outcome occurring, which probit models lack.

The coefficients of probit and logistic models are both typically estimated by maximum likelihood. The likelihood usually cannot be maximized analytically, so numerical approximations are used. Figure Post-4 shows least-squares, probit, and logistic fits to some binary

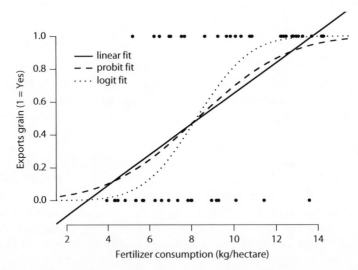

Figure Post-4 Hypothetical data inspired by the running example. The outcome is dichotomous, an index of whether a country exports grain or not. The data are shown along with fits from three different models—a least-squares (linear) model, a probit model, and a logistic (logit) model—each of which could be interpreted as an estimated probability of exporting grain given the fertilizer consumption level shown on the horizontal axis. The fits are similar in some ways, but one salient difference is that the linear model produces estimated probabilities less than zero and greater than 1 for some fertilizer consumption values.

---

[13] Some people do use such approach, which are sometimes called "linear probability models" and can be useful in certain circumstances. See Horrace & Oaxaca (2006) for some references and a recent technical evaluation.

[14] "Probit" is a portmanteau for "probability unit." Don't ask me; I don't know.

outcome data. Interval estimates come from bootstrapping or observed Fisher information. In a Bayesian data analysis, posterior distributions for the coefficients can be estimated using computational methods like Markov chain Monte Carlo.

In the 1970s, it was noticed that these and several other models—importantly including models for count data—shared a common framework. Models described by this common structure are called generalized linear models. A full description is beyond our scope, but the most striking feature is that generalized linear models posit that there is a specific function g that satisfies

$$E(Y|x_1, x_2, ..., x_k) = g(\alpha + \beta_1 x_1 + \beta_2 x_2 + ... + \beta_k x_k).$$

Linear regression satisfies this relationship with the so-called identity function, $g(z) = z$. In probit regression, $g(z) = \Phi(z)$, and in logistic regression, $g(z) = e^z/(1 + e^z)$. In general, the inverse of g, or g^{-1}, is called the *link function*, and generalized linear models are characterized in part by the link functions they use.

Generalized linear models are versatile, but the added versatility sometimes comes at the price of assumptions that are hard to check and that, in some cases, matter more than in the linear regression case. For example, one way to view the probit model is called the *latent variable* formulation. We assume that each unit is associated with a latent—that is, unobserved—variable Z. We don't observe Z itself; we only see whether $Z > 0$, in which case Y—still the dichotomous outcome—is equal to 1. This model is equivalent to the one in Equation P.2 if

$$Z = \alpha + \beta_1 x_1 + \beta_2 x_2 + ... + \beta_k x_k + \epsilon, \tag{P.3}$$

where $\epsilon \sim \text{Normal}(0, 1)$. That is, the unobserved variable Z is assumed to follow exactly the standard multiple regression assumptions—linearity, as well as independence, normality, and homoscedasticity of disturbances. In standard linear regression, the fact that we don't directly observe ϵ creates some difficulties in assessing the plausibility of the assumptions. Now, we don't get to see Z either, so things are even harder. Further, whereas in linear regression, least-squares estimates are consistent even if the disturbances are heteroscedastic, in probit regression, heteroscedastic disturbances lead to inconsistent estimators, as you will see in an optional exercise.

Exercise Set Postlude-4

(1) [Optional] Cowles and Davis's data are available in the `car` package (Fox & Weisberg, 2011), in the dataframe called `Cowles`. Estimate the model in P.2 using the `glm()` function.

(2) [Optional] a) In linear regression, the least-squares estimators are consistent even in the presence of heteroscedasticity. To confirm this, repeat Exercise Set 6-9, Problem 1(a), but introduce heteroscedasticity by assigning the `het.coef` argument a nonzero value. To see the heteroscedasticity in a plot, use, for example,

```
plot(sim.lm(5000, 3, 1/2, het.coef = .25))
```

Then, to estimate the mean and variance of the least-squares estimates with samples of size 100, use

```
ests <- sim.lm.ests(n = 100, nsim = 1000, a = 3, b = 1/2, het.coef = .25)
colMeans(ests)
apply(ests, 2, var)
```

Repeat this for sample sizes of 10, 50, and 1,000.

(b) Simulate data using the latent-variable formulation of the probit model in Equation P.3. Fit the model using the `glm()` function. Here is one way to do it with 100 observations, $\alpha = -1$, and $\beta_1 = 0.2$:

```
#Simulate data
n <- 100
a <- -1
b <- 0.2
x <- rnorm(n, 0, 2)
eps <- rnorm(n, 0, 1)
z <- a + b*x + eps
y <- as.numeric(z > 0)

#Fit the model
prob.fit <- glm(y ~ x, family = binomial("probit"))
summary(prob.fit)
```

Increase the number of observations and see what happens to the estimates. Do you think they are consistent?

(c) Introduce heteroscedasticity to the error terms of the latent variables. One way to do this with the above parameters is to generate the disturbances using

```
eps <- rnorm(n, 0, 1 + max(-1, 0.1*x))
```

Increase the number of observations and see what happens to the estimates. Do you think they are consistent? Try changing the severity of the heteroscedasticity by changing 0.1 in the above function call to a different value.

Post.2.3 Mixed models

Mixed models—also called hierarchical models, multilevel models, or random-effects models—provide a method for dealing with violations of the independence assumption. As an example, imagine that in the fertilizer example, data were collected from the same sub-Saharan African countries, but, rather than just one measurement per country, there are fertilizer use and cereal yield data for each of ten consecutive years. Figure Post-5 shows a simulation.

Fitting a simple linear regression model to the data in Figure Post-5 gives the following output:

```
Coefficients:
              Estimate  Std. Error t  value   Pr(>|t|)
(Intercept)   3.11366     0.34591    9.001    8.63e-15 ***
x             0.52696     0.03626   14.532    < 2e-16  ***
```

The estimated intercept and slope are similar to what they were with one observation per country—this is because the simulation was designed to give similar numbers. The standard errors, however, are much smaller, which makes t statistics larger and p values smaller. The standard errors are smaller because the data set is 10 times as large, and under the independence assumption, the standard error estimates ought to be smaller by a factor of about $\sqrt{10}$, other things being equal.

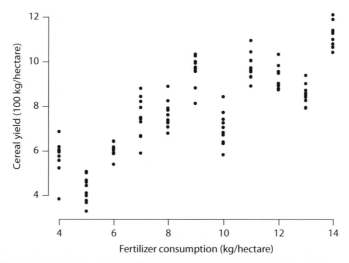

Figure Post-5 Simulated data from 10 years of sampling the same countries as in the fertilizer data. In the simulation, each country uses the same amount of fertilizer every year. Cereal yield observations from the same country tend to be more similar than observations from different countries.

If the observations are not independent, then the reduction in the standard error is too optimistic—dependent observations bring less new information than independent ones. For an illustration, imagine that we are interested in the weights of students at a university and that I have the resources to measure 100 weights. It would be a poor strategy to weigh a single student 100 times. The 100 measurements would be massively dependent—close to identical—and they would reveal far less information about the heights of students at the university than would weight measurements from a sample of 100 randomly selected students. Counting the ten observations from the same country as independent observations is similar—we are crediting dependent observations as independent sources of information, leading to estimates of uncertainty that are too small.

To obtain better standard error estimates, we need to modify the linear regression model to account for the similarity of observations from the same country. One way to do this is to introduce an additional random term to the model that leads observations from the same country to be more similar to each other than are observations from different countries.[15] The standard linear regression model is

$$Y_i = \alpha + \beta x_i + \epsilon_i.$$

Notice that we have revived the subscript i to refer to the ith observation in the dataset, which we had dropped for compactness in the previous two sections. The only random term on the right side of the equation is ϵ_i, the disturbance term associated with observation i. The new model we will use—a *mixed* model—is

$$Y_{ij} = \alpha + \beta x_{ij} + \mu_i + \epsilon_{ij}. \tag{P.4}$$

[15] The approach presented here is not the only option for modifying regression to account for dependence between observations. One reference for many other possible options is Diggle et al. (2002). One less technical resource is Singer & Willett (2003). As is often the case, economists have their own language for these kinds of data, covered in Baltagi (2008).

There are some new notational conventions in Equation P.4—now we have two subscripts, i and j. In the running example, i still refers to the ith country; j now refers to the jth year. So, for example, x_{23} is the value of x associated with the second country in the third year. We are still assuming that the disturbances—now written as ϵ_{ij}—are independent observations drawn from a Normal$(0, \sigma^2)$ distribution. In addition to the disturbances, the μ term is also random. Each country has its own value of μ—for example, every cereal yield observation from country 1 includes μ_1 as one of its terms. If it is easier to grow grain in Angola than in dry Namibia, then the μ for Angola will be larger than the μ for Namibia.

The μ_i are assumed to be drawn independently of each other and of the disturbances, with a Normal$(0, \tau^2)$ distribution. The μ_i term is called a "random effect" because it differs randomly from country to country, whereas α and β do not; they are "fixed effects." The model is "mixed" because it includes both fixed and random effects.

In the model in Equation P.4, the random effects create dependence among observations from the same country. As you will have the chance to show in an optional exercise, the random components of observations from the same country are positively correlated, with correlation coefficient $\tau^2/(\tau^2 + \sigma^2)$. The model can therefore account for some patterns of similarity among observations that violate the standard linear regression model.

Estimation entails estimating both the fixed effects and the parameters of the distribution(s) from which the random effects are drawn. In Equation P.4, this means estimating α, β, τ^2, and σ^2. In frequentist approaches, estimates come from maximizing the likelihood, or, more often, from a related method called restricted maximum likelihood that solves some technical problems that arise in random-effects models. The likelihood generally has to be maximized numerically. Interval estimates come from observed Fisher information or bootstrapping. Bayesian approaches place posterior distributions on the parameters, given data and prior distributions. Using restricted maximum likelihood to fit the model in Equation P.4 to the data in Figure Post-5 gives the following fixed-effects estimates and standard errors:

```
Fixed effects:
              Estimate Std. Error t value
(Intercept)    3.1137    1.0327    3.015
x              0.5270    0.1083    4.868
```

The estimates themselves are almost identical to those produced by linear regression, but the standard error estimates are much larger. You will see in an optional simulation exercise below that these larger standard error estimates are more appropriate in this case.

It is possible to include more fixed and random effects in the model. Adding random effects leads to more elaborate patterns of correlation among the observations. For example, one can add a random effect that gives each country a different slope associated with fertilizer use, as in

$$Y_{ij} = \alpha + (\beta + \beta_i)x_{ij} + \mu_i + \epsilon_{ij},$$

where β_i is a random effect. Or each year could have its own random effect, as in

$$Y_{ij} = \alpha + \beta x_{ij} + \mu_i + \omega_j + \epsilon_{ij},$$

where ω_j is a random effect taking a different value each year. Or there might be several levels of nested random effects, perhaps with equatorial and non-equatorial countries sharing distinct random intercepts, as in

$$Y_{ijk} = \alpha + \beta x_{ijk} + \mu_i + \mu_{ij} + \epsilon_{ijk},$$

where i now indicates broad region (equatorial vs. not), j indicates country, and k indicates year.

All this flexibility can be put to its best use if there is sufficient background information about the problem to support specific models. Sometimes it is hard to justify the modeling assumptions, which can be important. As with generalized linear models, it is harder to identify bad assumptions in mixed models than in linear regression—more of the crucial components are unobserved.

Exercise Set Postlude-5

(1) [Optional] Show that under the model in Equation P.4, two observations of the dependent variable that share the same random effect (in the example, two observations from the same country) are correlated with correlation coefficient $\tau^2/(\tau^2 + \sigma^2)$. (Hint: It is equivalent to show that just the random parts of the observations are correlated at this strength. This is because, for b a constant and X, Y random variables,

$$\text{Cor}(b + X, Y) = \frac{E([b + X]Y) - E(b + X)E(Y)}{\sqrt{\text{Var}(b + X)}\sqrt{\text{Var}(Y)}} = \frac{bE(Y) + E(XY) - bE(Y) - E(X)E(Y)}{\sqrt{\text{Var}(X)}\sqrt{\text{Var}(Y)}}$$
$$= \text{Cor}(X, Y).$$

End hint.)
(2) [Optional] The data in Figure Post-5 were simulated using the following R code:

```
set.seed(8675309)
alpha <- 3
beta <- 1/2
eps.sd <- sqrt(1/2)
re.sd <- 1
yrs <- 10
x <- rep(anscombe$x1, 10)
rand.ints <- rnorm(length(anscombe$x1), 0, re.sd)
y <- alpha + beta*x + rep(rand.ints, 10) + rnorm(length(x), 0, eps.sd)
```

Simulate many more datasets and compute the least-squares estimates of the intercept and slope estimates. What are the estimated expectation and standard deviation of the distribution of the estimators? How do the standard deviations relate to the estimated standard errors using linear regression vs. the mixed model? (Hint: Don't reset the seed each time— that will give you many copies of exactly the same dataset.)

Post.3 Conclusion

We have nearly reached the end of our time together. Thank you for coming all this way. We conclude by revisiting three themes that run through this book: (1) The functions and limits of models, (2) methodological pluralism, and (3) the value of elementary statistical thinking.

Probabilistic models—particularly versions of the simple linear regression model first introduced in Chapter 5—have been the main characters of this book. In Chapter 3, before we had access to probabilistic models, we could visualize and summarize data, and we could even make informal predictions. But there was little more we could say. It was difficult to defend particular summaries and displays as reasonable or to say what they meant. The introduction of model assumptions—which were then combined with data—allowed much richer claims.

Models provide a framework for stating assumptions explicitly, for uncovering the consequences of those assumptions, and for interrogating data with respect to specific research questions. They encode scientific knowledge, or at least scientific hypotheses.

Reliance on models and their associated assumptions also exposes us to risk. If the model is a poor description of the process that generated a dataset, then the substantive conclusions may be wrong—sometimes badly wrong. Assumptions may be grounded in prior knowledge, new data, or good theory. But if they are in doubt—and they almost always are, to some degree—then it is wise to scrutinize their consistency with the data and to consider the consequences that would obtain if the assumptions were false.

A second central theme of the book is pluralism—there are a variety of ways to address any research question. Many of us know this is true in terms of study design and data collection, but it applies equally to modeling and data analysis. We have considered three different views of the same simple linear regression problem, each principled and internally consistent, yet distinct. There are more possible approaches even to this simple problem, and the possibilities multiply in more complicated scenarios. Each approach entails its own strengths, weaknesses, and limits. Though some approaches are clearly better than others in specific situations, there will often be reasonable arguments for multiple ways of addressing a given problem.

The plurality of approaches available—and the opportunity they present for creativity— makes statistics exciting. At the same time, the space of possibilities can be disorienting. One way to shield the student from disorientation is to provide only one recipe for each possible data type. Two continuous variables? Compute a Pearson correlation. One categorical and one continuous? Run a one-way ANOVA. Time for breakfast? Boil the oats. Lunch? Spread the peanut butter.

The premise of this book is that to reason confidently and creatively about data, one needs more than superficial exposure to a range of models for data analysis. In devoting so much effort to the study of simple linear regression, you have become familiar with the language of models, with the questions statisticians consider, and with several important ways of thinking about the relationships between data, models, and parameters. You will find that these foundational ideas are useful as you expand your skills—which I hope you will be motivated to do. The readings suggested at the end of this chapter and the previous ones are possible springboards.

We part as we met, with words from David Hume: "In our reasonings concerning matter of fact, there are all imaginable degrees of assurance, from the highest certainty to the lowest species of moral evidence. A wise man, therefore, proportions his belief to the evidence."

Post.4 Further reading

Statistics, data, and graphical displays

Anscombe, F. J. (1973). Graphs in statistical analysis. *American Statistician*, 27, 17–21.
   A demonstration of the importance of data displays in statistical analyses.

Cleveland, W. S. (1993). *Visualizing Data*. Hobart Press, Summit, NJ.
   An excellent book on data displays.

Donoho, D. (2017). 50 Years of data science. *Journal of Computational and Graphical Statistics*, 26, 745–766.
   A penetrating essay on the relationship of statistics to the emerging field of data science.

Nolan, D., & Lang, D. T. (2015). *Data Science in R: A Case Studies Approach to Computational Reasoning and Problem Solving*. CRC Press, Boca Raton, FL.

This book was recommended at the end of Chapter 2 but warrants a second mention here. Nolan & Lang provide a hands-on tour of several meaty case studies, showing how the concepts you have learned apply to real-world problems.

Tufte, E. R. (1983). *The Visual Display of Quantitative Information*. Graphics Press, Cheshire, CT.

Full of thoughtful, creative ideas for the design of informative data displays. (Incidentally, the term "Anscombe's quartet" was coined here.)

Extending simple linear regression

Angrist, J. D., & Pischke, J. S. (2009). *Mostly Harmless Econometrics: An Empiricist's Companion*. Princeton University Press.

There are many situations in which one might want to understand the causal effects of several variables on an outcome of interest, but it is not possible to do a randomized experiment. Social scientists—especially economists—have thought hard about how one might be able to estimate causal effects in such a scenario. This book is an engaging introduction to frameworks that have resulted from this effort. (There are other frameworks for approaching causality—for overlapping but distinct traditions, see Imbens & Rubin (2015) or Pearl, Glymour, & Jewell (2016).) The mathematical level is similar to this book, but some familiarity with matrices and multiple regression is required.

Dobson, A. J., & Barnett, A. (2008). *An Introduction to Generalized Linear Models*. CRC Press, Boca Raton, FL.

An authoritative introduction to generalized linear models.

Freedman, D. A. (2009). *Statistical Models: Theory and Practice*. Cambridge University Press.

A clear and entertaining exposition of multiple regression and related methods, with an emphasis on assumptions necessary for making causal claims on the basis of regression results.

James, G., Witten, D., Hastie, T., & Tibshirani, R. (2013). *An Introduction to Statistical Learning*. Springer, New York.

An introduction to methods for prediction in large datasets. Many of the methods can be viewed as modifications and generalizations of the regression methods we have covered. An overlapping set of authors (Hastie, Tibshirani, & Friedman, 2009) wrote *The Elements of Statistical Learning*, which covers similar material at a higher mathematical level and is a great reference. The chapter on Model Assessment and Selection contains an expanded discussion of overfitting and methods to address it.

Rabe-Hesketh, S., & Skrondal, A. (2008). *Multilevel and Longitudinal Modeling Using Stata*. STATA Press, College Station, TX.

Though this book (actually two volumes) relies on proprietary Stata software, it provides great explanations of random-effects modeling (also called "multilevel" modeling). The first chapter, which reviews content similar to this postlude, is also excellent.

Cosma Shalizi (to be published), *Advanced Data Analysis from an Elementary Point of View*.

At this writing, a draft of this book is available on Cosma Shalizi's website, but it is not yet in print. One of its strengths is its emphasis on approaches that do not require the regression function to be linear in the independent variables.

APPENDIX A

A tour of calculus

> **Key terms**: Derivative, Differentiation, Function (mathematical), Fundamental theorem of calculus, Integral, Limit, Optimization, Polynomial

Statistics is in part a mathematical discipline, and to understand it, one needs to engage with mathematical language. To understand the arguments in this book, it is necessary to know a little about calculus. You do *not* need to know how to work most calculus problems, but you need to understand a few central calculus concepts and to recognize problems to which they apply. Specifically, you need to be able to give at least informal answers to the following questions:

(1) What is a mathematical function?
(2) What is a limit?
(3) What is a derivative?
(4) How can derivatives be used for optimizing (that is, maximizing or minimizing) functions?
(5) What is an indefinite integral?
(6) What is a definite integral?
(7) How are derivatives and integrals related?
(8) How does one compute derivatives, indefinite integrals, and definite integrals of polynomial functions?

For the level of discussion in this book, you do not need to know the many ingenious methods for computing specific derivatives and integrals taught in introductory calculus courses. You also do not need to have experience with precise definitions of terms like "limit." If you can give accurate—even if somewhat sketchy—answers to the terms in (1)–(8), then you can safely skip this appendix. If you have never seen any of the ideas in (1)–(8), then you may need more support than is provided by this appendix. Some ideas are discussed under "Further reading" at the end.

In this appendix, we will informally discuss the basic concepts and goals of calculus using a simple example. Any math beyond algebra will be introduced as it is required.

A.1 Mathematical functions

The objects of study in calculus are **mathematical functions**. (I specify "mathematical" functions because when we talk about computing in R, we refer to a distinct type of function. In this appendix, "function" always means "mathematical function.") You can think of functions as mathematical objects that receive the values of one or more variables as input and produce another value as output. We will call the input variables "arguments" and the output value the "output."

For example, we might have a function that takes one argument, which we can call x, and squares the argument to produce the output x^2. We would write this as

$$f(x) = x^2.$$

The "$f(x)$" on the left side of the equation is pronounced "f of x." If we pass the argument $x = 2$ to $f(x)$, then we get the output $f(2) = 2^2 = 4$. If we pass 3 as input to $f(x)$, we get $f(3) = 3^2 = 9$. We can plot the relationship between a function's arguments and output, drawing a curve in which the horizontal position represents the argument to a function, and the vertical position represents the output of the function. The resulting plots are called graphs of functions. For example, Figure A-1 shows the graph of the function $f(x) = x^2$.

We can also have functions of more than one variable. For example,

$$f(x, y) = x^2 - y^3$$

is a function of both x and y. When we use functions of more than one variable, we will approach them by working with one variable at a time.

For any combination of values of the function's arguments, exactly one output is produced. In our example, a graphical way of understanding this property is by noting that if one were to draw a vertical line through Figure A-1 at any point along the horizontal axis, it would intersect the curve representing $f(x)$ no more than once.[1]

There are many types of questions we could ask about functions. In this discussion, we will frame the two main types of questions with which calculus is concerned as questions about graphs of functions, the meanings of which questions will be made more precise later:

(1) How quickly does the output of $f(x)$ change as x changes? (See Figure A-2.)
(2) What is the area that falls underneath the graph of $f(x)$, above the x axis, and between the x values a and b? (See Figure A-5.)

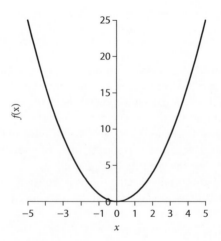

Figure A-1 $f(x) = x^2$.

---

[1] However, several distinct arguments to the same function might produce the same output. In our example, $f(x) = x^2$, choosing either $x = 2$ or $x = -2$ produces $f(x) = 4$. Thus, a horizontal line can intersect a graph of a function more than once.

The first question is answered by differential calculus, and the second question by integral calculus. Remarkably, these two questions are inverses of each other, as described later.

A.2 Differential calculus

Differential calculus is the mathematical study of rates of change. Suppose that the function $f(x) = x^2$ describes the position of a runner with respect to a starting point. We'll assume that in this example, x cannot take on negative values. At time x, the runner is x^2 distance units away from the point from which the measurements are being made. How can we determine the rate at which the runner's position changes? That is, how can we determine the runner's speed?

One way to answer this question is to consider average speeds over defined windows of time. For example, assuming time is measured in seconds and distance is measured in feet, we could determine the runner's average speed in feet per second between seconds two and three. After two seconds, the runner is $f(2) = 2^2 = 4$ feet away, and after three seconds, the runner is $f(3) = 3^2 = 9$ feet away. One second has elapsed, and the runner has traveled five feet, so the runner's average speed in the interval is 5 feet per second. Mathematically, we express this reasoning as

$$\frac{\text{change in } f(x)}{\text{change in } x} = \frac{f(3) - f(2)}{3 - 2} = \frac{3^2 - 2^2}{1} = 5.$$

We can use similar reasoning to determine the runner's average speed over any interval. Note, however, that the runner did not run 5 ft/sec over the whole interval—the runner's speed changed during the interval, and 5 ft/sec is just an average speed.

In differential calculus, we are not concerned with average rates of change over intervals. Rather, we are concerned with *instantaneous* rates of change. In the example of the runner, we might want to know the runner's speed at a particular instant in time. To find the instantaneous rate of change, we need a new concept—the concept of a limit.

Let us start by assuming we want to measure the average rate of change in $f(x)$ over an interval of width Δx (pronounced "delta x") that starts at x. When we found the average rate of change between time $x = 2$ and time $x = 3$, we determined the average rate of change over an interval of width $\Delta x = 1$ that started at time $x = 2$. We saw that in that case, the formula for the average rate of change over the interval was given by

$$\frac{f(3) - f(2)}{3 - 2} = \frac{f(x + \Delta x) - f(x)}{\Delta x}.$$

However, now suppose that we want to know the *instantaneous* rate of change exactly when $x = 3$. One might think of the instantaneous rate of change as the rate of change for an interval of width zero, that is, with $\Delta x = 0$. But we cannot compute the instantaneous rate of change by plugging $\Delta x = 0$ into the expression for the average rate of change (that is, $[f(x + \Delta x) - f(x)]/\Delta x$). If we tried to do that, we would have to divide by zero, which is disallowed by the rules of arithmetic. One thing we *can* do is examine what happens when Δx gets closer and closer to 0. Table A-1 shows what happens when $x = 3$ and Δx is decreased from 2 to 0.001.

As Δx approaches 0, the average rate of change over the interval ranging from $x = 3$ to $x = 3 + \Delta x$ approaches 6. The formal definition of a limit is beyond our scope. Informally, a **limit** of a function is the output value that the function approaches when one of its arguments approaches a particular quantity. For example, the limit, as Δx approaches zero, of the average rate of change in $f(x)$ over the interval ranging from $x = 3$ to $x = 3 + \Delta x$ is 6. We write this as

Table A-1 The approach
of Δx to 0 when $x = 3$

Δx	$\frac{f(3+\Delta x)-f(3)}{\Delta x}$
2	8
1	7
0.5	6.5
0.1	6.1
0.01	6.01
0.001	6.001

$$\lim_{\Delta x \to 0} \frac{f(3 + \Delta x) - f(3)}{\Delta x} = 6.$$

The "lim" stands for limit, and in this case, we are taking a limit as Δx approaches zero, which is indicated by the "$\Delta x \to 0$" below "lim." This limit is the instantaneous rate of change of $f(x)$ when $x = 3$.

To have a general answer to the question "How quickly is $f(x)$ changing as x changes?" we would like to find a function $f'(x)$ (pronounced "f prime of x") that gives the instantaneous rate of change in $f(x)$ for any value of x. We define this function as

$$f'(x) = \lim_{\Delta x \to 0} \frac{f(x + \Delta x) - f(x)}{\Delta x}. \tag{A.1}$$

We call $f'(x)$ the **derivative** of the function $f(x)$, and we call the process of finding the derivative of $f(x)$ **differentiation**. To differentiate $f(x) = x^2$, we start by plugging $f(x)$ into the general expression for the derivative:[2]

$$f'(x) = \lim_{\Delta x \to 0} \frac{f(x + \Delta x) - f(x)}{\Delta x} = \lim_{\Delta x \to 0} \frac{(x + \Delta x)^2 - x^2}{\Delta x}.$$

On the right, we have merely replaced every instance of $f()$ on the left with the square of the argument of f—that's just what the function $f(x) = x^2$ does. Expanding the expression $(x + \Delta x)^2$ and then cancelling the positive and negative x^2 terms gives

$$f'(x) = \lim_{\Delta x \to 0} \frac{(x + \Delta x)^2 - x^2}{\Delta x} = \lim_{\Delta x \to 0} \frac{x^2 + 2x(\Delta x) + (\Delta x)^2 - x^2}{\Delta x} = \lim_{\Delta x \to 0} \frac{2x(\Delta x) + (\Delta x)^2}{\Delta x}.$$

Dividing out the Δx term in the denominator yields

$$f'(x) = \lim_{\Delta x \to 0} 2x + \Delta x$$

---

[2] A note on the use of parentheses: in general, parentheses mean, "perform the operation to the entire quantity represented inside the parentheses." So $f(x)$ means "apply the function to x," and $f(x + \Delta x)$ means "apply the function to the quantity $x + \Delta x$." There is some ambiguity that appears when we don't know whether the letter placed in front of parentheses represents a function or just a number. If parentheses are preceded by a letter representing a function, then we apply the function to the quantity in parentheses. If parentheses are preceded by a letter representing a number, then we just multiply the quantity in parentheses by that number. So in the equations here, $f(x + \Delta x)$ means "apply the function to the quantity $x + \Delta x$" but $2x(x + \Delta x)$ means "multiply $2x$ by $(x + \Delta x)$ " because x is a number and not a function. The information you need to resolve the ambiguity will come from the context. It is also a common practice, and one that I'll (mostly) follow—though there is an exception in Chapter 7—to reserve the letter f for functions and not to use it to represent a number.

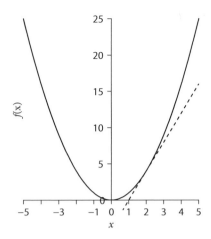

Figure A-2 The dashed line is tangent to $f(x)$ at $x = 2$.

In this form, it is easy to see that as Δx approaches zero, the quantity $(2x + \Delta x)$ approaches $2x$. Thus, when $f(x) = x^2$, differentiation reveals that the derivative of $f(x)$ is

$$f'(x) = 2x.$$

In fact, this is a special case of a more general result. Whenever we have a function of the form $f(x) = ax^n$, called a **polynomial** function of x, the derivative is

$$f'(x) = nax^{n-1}. \tag{A.2}$$

A proof for positive integers n is in an optional exercise below, but Equation A.2 applies to any real number n. The phrase, "out in front and down by one" describes what happens to the exponent during differentiation functions of the form $f(x) = ax^n$. That is, in the derivative, the n in the exponent multiplies the function ("out in front"), and the exponent of x decreases by one ("down by one").

Graphically, the derivative can be interpreted as giving the slope of a line tangent to the graph of the function $f(x)$. The dashed line in Figure A-2 is an example of such a tangent line.

In statistics, differential calculus is used to find arguments that maximize or minimize the output of a function. The pursuit of such arguments is called **optimization**. For example, we might want to find parameter values that maximize quality of a fit of a statistical model to a dataset. When a function reaches its maximum or minimum value, the tangent line to the graph of the function has a slope of zero (Figure A-3).[3] It follows that one way to find candidate maxima and minima is to find all of the values of the arguments that produce derivatives equal to zero.[4] In the case of our function, $f(x) = x^2$, we already know that

---

[3] As a metaphor, imagine throwing a ball straight up in the air. The ball will move upward and decelerate until it reaches its maximum height, where it will stop for an instant before accelerating downward. That is, there is an instant at which the ball's upward movement stops, and that instant occurs exactly when the ball is at its maximum height. Similarly, for functions that have a few basic properties (such as not having discontinuous jumps), when the function reaches a local maximum or minimum, there will be an instant at which the function's rate of change is zero.

[4] This is not guaranteed to find all maxima and minima if the function has points where the derivative is undefined. For example, the absolute value function $f(x) = |x|$ has a minimum at $x = 0$, but the derivative at 0, or $f'(0)$, is undefined in this case. Most of the functions in this book—at least most of those whose derivatives are relevant—are differentiable everywhere, which means they have a defined derivative for all possible values of their arguments.

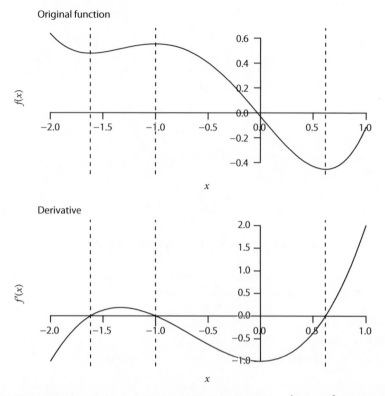

Figure A-3 The upper panel shows a graph of the function $f(x) = x^4/4 + 2\,x^3/3 - x$, with its local minima and maxima marked by dashed lines. The derivative of the function is shown in the lower panel, and we see that the derivative is equal to 0 wherever the original function has a local minimum or maximum.

$f'(x) = 2x$. The only solution to the equation $2x = 0$ is to set $x = 0$. It works—we can see in the graph of $f(x)$ that $f(x)$ is minimized when $x = 0$.[5]

Because we considered the rate of change in $f(x)$ as the argument x changes, we say that we took the derivative "with respect to x." When we consider functions with multiple arguments, it becomes important to state with respect to which argument we are differentiating. Any functions of variables other than the one with respect to which we are differentiating can be treated as constants. What, then, to do with constants? The derivative of a function multiplied by a constant is just that constant multiplied by the original function. That is, if $g(x) = af(x)$ and a does not depend on x, then the derivative of $g(x)$ is just a times the derivative of $f(x)$, or

$$g'(x) = af'(x). \tag{A.3}$$

---

[5] If it were not possible to examine the graph, then to be sure that this is a minimum, we would have to take the additional step of confirming that the second derivative—that is, the derivative of the derivative—is positive, as you will see in Problem 2 of Exercise Set A-1. In the main text, we will be optimizing some functions of multiple variables, and in that case, even a little more work is required—one needs to examine some properties a matrix of second derivatives, which is more technical but conceptually straightforward. Usually we will treat these extra steps as beyond our scope and be content with setting the derivatives equal to zero, but the extra steps are important if the right answers are not known beforehand.

As an example, we know that the derivative of x^2 is $2x$. Equation A.3 implies that the derivative of $2x^2$ is $4x$ and that the derivative of $3x^2$ is $6x$.

To finish bookkeeping, the derivative of the sum of two functions is the sum of the derivatives of those two functions. So, if $h(x) = f(x) + g(x)$, then the derivative of $h(x)$ is

$$h'(x) = f'(x) + g'(x) \tag{A.4}$$

for any functions $f(x)$ and $g(x)$. As an example, we know from Equation A.2 that the derivative of x^2 is $2x$ and that the derivative of x^3 is $3x^2$. Equation A.4 implies that the derivative of the sum $h(x) = x^2 + x^3$ is $h'(x) = 2x + 3x^2$.

As long as you understand how the derivative gives information about rates of change and allows us to maximize and minimize functions, you will be able to follow the arguments that use differentiation in this book.

Exercise Set A-1[6]

(1) For each of the functions below, complete the following steps:
   (i) Sketch the graph of the function for values of x between -5 and 5.
   (ii) Using Equations A.2–A.4, take the derivative. (Remember that $x^0 = 1$ for all x, so you can think of any constant a that is not multiplied by x as ax^0. This is illustrated in part (a) below.)
   (iii) Find the value of x for which $f'(x) = 0$. (In part (c), there are two values of x for which $f'(x) = 0$.)
   (a) $f(x) = x^2 - 2x + 1 (= x^2 - 2x + 1x^0)$;
   (b) $f(x) = -3x^2 + 12x - 5$;
   (c) $f(x) = x^3 - 3x^2$.

(2) For the functions in the previous problem, you will have noticed that the x coordinate at which the derivative is equal to zero corresponds to a maximum or a minimum of the original function. For the functions in parts (a)–(c) of problem 1, take the derivative of the derivative—called the "second derivative" of the original function and denoted $f''(x)$—and evaluate it at the x coordinate or x coordinates for which $f'(x) = 0$. What difference do you notice between values of the second derivatives at minima of the original function and the values of second derivatives at maxima of the original function?

(3) [Optional] Prove the statements in Equations A.3 and A.4.

(4) [Optional, hard] Prove that the derivative of the function $f(x) = ax^n$ is $f'(x) = nax^{n-1}$ for any a and any positive integer[7] n. You will need to use the binomial theorem, which states that

$$(x+y)^n = \sum_{i=0}^{n} \binom{n}{i} x^{n-i} y^i = x^n + nx^{n-1}y + \binom{n}{2}x^{n-2}y^2 + \binom{n}{3}x^{n-3}y^3 + \ldots + nxy^{n-1} + y^n.$$

$\binom{n}{i}$ is the binomial coefficient, which is equal to $n!/[i!(n-i)!]$. The "!" operator is pronounced "factorial," and it tells us to take the product of all positive integers up to and including the number we're interested in. So $5! = 5*4*3*2*1$, and, in general, $k! = k*(k-1)*(k-2)*\cdots*3*2*1$ for positive integers k. A special consideration is that $0! = 1$ by definition. (Hint: You will only need to consider the first few terms of the binomial expansion.)

A.3 Integral calculus

Integral calculus is concerned with what might seem a different question from that of instantaneous rates of change. Suppose that we start with a function, say $f(x) = 2x$, and we

---

[6] See the book's GitHub repository at github.com/mdedge/stfs/ for detailed solutions and R scripts.

[7] The rule in Equation A.2 holds for any real number n, but the approach suggested in this problem proves it for positive integers.

want another function that will give us the area underneath the graph of $f(x)$ and above the x axis. Figure A-4 shows the graph of $f(x) = 2x$.

More specifically, we seek a function $F(x)$ that we can use to give us the area between the x axis and $f(x)$ and between any two points on the x axis. We call the points on the $x$axis in which we're interested a and b. We will use a convention: when $f(x)$ is larger than zero and thus above the x axis in the graph, we will count the area between the x axis and $f(x)$ as positive, and when $f(x)$ is less than zero, the area between the x axis and $f(x)$ will count as negative. We want to construct a function $F(x)$ —which we will call the indefinite **integral** of $f(x)$—that satisfies $F(b) - F(a)$ = area between vertical line $x = a$, vertical line $x = b$, the x axis, and $f(x)$. For $f(x) = 2x$, $a = 1$, and $b = 2$, the area represented by $F(b) - F(a)$ is shown in Figure A-5.

We need some language. The function $F(x)$ is the "indefinite integral" of $f(x)$ with respect to x. The quantity $F(b) - F(a)$ is a "definite integral" and we use the phrase "the definite integral from a to b." In other words, the indefinite integral is a *function* we can use to find areas. The definite integral from a to b is an actual area—or in problems with several variables, a volume. We write the indefinite integral of $f(x)$ with respect to x as

$$\int f(x)\, dx = F(x).$$

The "d" is called a differential operator. Quoting Silvanus Thompson (Thompson & Gardner, 1998), you can take "d" to mean "a little bit of," and think of $\int$ as "merely a long S" meaning "the sum of"—we will see several times in the main text that integrals are analogous to sums. Then, dx tells us that we are integrating "with respect to" x. Thus, the integral will tell us about the area between the function and some segment of the x axis. To indicate the *definite* integral of $f(x)$ with respect to x from a to b, we add the a and b to the bottom and top of the integral symbol, respectively, as in

$$\int_b^a f(x)dx = F(b) - F(a).$$

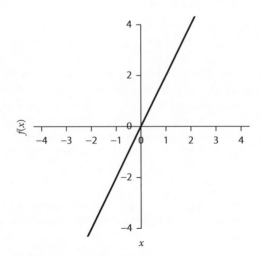

Figure A-4 $f(x) = 2x$.

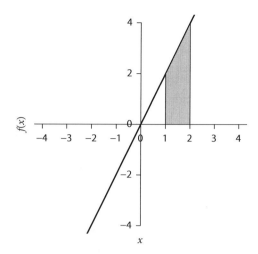

Figure A-5 The gray area represents the definite integral of $f(x)$ from $a = 1$ to $b = 2$.

In the case of $f(x) = 2x$, we have it easy. You can see from Figure A-5 that any definite integral of $f(x) = 2x$ can be represented as either the area of a triangle or the sum of the area of a rectangle and a triangle, and recall that the area of a rectangle is bh, where b is the length of the base of the rectangle and h is the height of the rectangle, and that the area of a triangle is $bh/2$. If you calculate a few strategically chosen definite integrals, you can make a guess about the indefinite integral.

Exercise Set A-2

(1) Deduce the value of the definite integral of $f(x) = 2x$ for the following intervals by computing the relevant areas. Remember that for positive $f(x)$, the area above the x axis and below
$f(x)$ counts as positive in the definite integral and that for negative $f(x)$, the area below the x
axis and above $f(x)$ counts as negative in the definite integral. For each interval, draw a
picture of the area you measure, noting which part (if any) counts as positive and which part
(if any) counts as negative.
   (A) $x = 0$ to $x = 1$;
   (B) $x = 0$ to $x = 2$;
   (C) $x = 1$ to $x = 3$;
   (D) $x = -2$ to $x = 2$;
   (E) $x = -1$ to $x = 3$.
(2) Fill in the following table of the definite integrals of $f(x) = 2x$ from 0 to each integer up to 5.
Do you have a guess about the form the *indefinite* integral might take? (Hint: can you express
the definite integral, in the right column, as a function of b?)

---

b	The definite integral of $f(x) = 2x$ from 0 to b (i.e., $\int_0^b 2x\,dx$)
1	
2	
3	
4	
5	

You may have gathered from the previous two exercises that the indefinite integral of $f(x) = 2x$ is $F(x) = x^2 + C$. (We often write the indefinite integral of a function $f(x)$ with an uppercase $F(x)$.[8]) If you have never seen an integral before, you likely did not add the "C" to the indefinite integral. C represents an arbitrary term that does not depend on x. Why do we add C? Suppose that I told you that the indefinite integral of some function is $F(x) = x + 5$. This means that you can get definite integrals for this function by taking $F(b) - F(a) = (a + 5) - (b + 5)$. But notice that the added 5 makes no difference: $(a + 5) - (b + 5) = a - b$. We could replace the 5 with any constant we'd like—such as 6, -7, 938, or π—and we would get the same definite integrals. In this book, the added C will never be crucial.[9]

In the example of $f(x) = 2x$, we could calculate definite integrals by breaking the area we wanted into simple shapes, and we could guess the indefinite integral using the resulting definite integrals. For most functions, however, the area under the graph of the function cannot be broken into simple shapes. How can we find the indefinite integral—and thus, any particular definite integral we might want—when the method we just used is inapplicable? We will rely on the relationship between integration and differentiation.

Recall that in the previous section, we took the derivative of $f(x) = x^2$ and found that $f'(x) = 2x$. Now, we see that the indefinite integral of $f(x) = 2x$ is $F(x) = x^2 + C$. That is, in this case, differentiation and integration are inverse procedures: if we had started with $f(x) = 2x$, taken an integral, and then taken the derivative of the integral obtained, we would have ended up with our original function. Remarkably, this is not a fortuitous property of this one example; it is true in general. As long as $f(x)$ has a few basic properties,

$$\int f'(x)dx = f(x) + C. \tag{A.5}$$

That is, the indefinite integral of the derivative of $f(x)$ is $f(x)$ itself (plus a constant). Equivalently, if $F(x)$ is the indefinite integral of $f(x)$, then

$$F'(x) = f(x).$$

That is, the derivative of the indefinite integral of $f(x)$ is $f(x)$ itself. The indefinite integrals that one gets by inverting the process of differentiation can be used to find definite integrals, or areas under curves, even when it is not possible to break the area into simple shapes. This relationship, stated more formally, is also known as the **fundamental theorem of calculus**.

Because of the inverse relationship between integrals and derivatives (Figure A-6), versions of Equations A.3 and A.4 have to be true for integrals as well. Indeed, the integral of a function times a constant is equal to that constant times the integral of the original function. That is, if $g(x) = af(x)$ and a does not depend on x, then

$$\int g(x)\,dx = \int af(x)\,dx = a \int f(x)\,dx. \tag{A.6}$$

---

[8] Starting in Chapter 4, we use the lowercase/uppercase notation to distinguish two classes of functions that describe probability distributions. There is some sense to using the same notation—as you'll see, cumulative distribution functions of continuous random variables, which are denoted in uppercase, are indefinite integrals of probability density functions, which are denoted in lowercase.

[9] Returning to the example, when $f(x) = 2x$, the definite integral is still $\int_a^b 2x\,dx = b^2 - a^2$, regardless of the value of C.

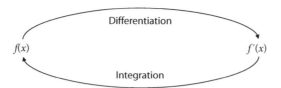

Figure A-6 The inverse relationship of differentiation and integration.

And, parallel to Equation A.4, the integral of a sum of two functions is the sum of the integrals of the individual functions. So if $h(x) = f(x) + g(x)$, then

$$\int h(x)\,dx = \int f(x) + g(x)\,dx = \int f(x)\,dx + \int g(x)\,dx. \tag{A.7}$$

Like differentiation, integration is crucial in probability and statistics. In particular, integration is essential when summaries of the behavior of random variables are required. The role of integration becomes clearer in Chapters 4 and 5.

Exercise Set A-3

What is the indefinite integral with respect to x of $f(x) = ax^n$ for $n \neq -1$? To state the same thing differently, evaluate $\int ax^n dx$. (Hint: Recall that the derivative of ax^n with respect to x is nax^{n-1} (Equation A.2), and remember that $\int f'(x)dx = f(x) + C$ (Equation A.5). Second hint: What function of x would have ax^n as its derivative?)

A.4 [Optional section] An explanation of the fundamental theorem of calculus

The fact that differentiation and integration are inverse procedures is jarring at first. The two operations seem to have nothing to do with each other. The easiest way to understand is with a picture.

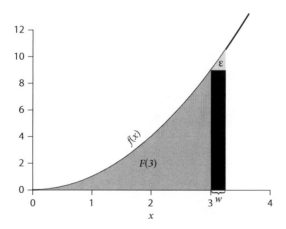

Figure A-7 The graph of $f(x) = x^2$ (black curve), with annotations to explain the fundamental theorem of calculus.

Modified from a plot on the Wikipedia page for the fundamental theorem of calculus (Fundamental theorem of calculus, n.d.). Image by Kabel and published under the CC BY-SA 4.0 License.

Consider Figure A-7, which shows the graph $f(x) = x^2$ as a black curve. Imagine that there is a function $F(x)$ that gives the area under $f(x)$ between 0 and x—$F(x)$ is an indefinite integral of $f(x)$. The value of $F(3)$ is the area of the region shaded in dark gray. The fundamental theorem of calculus holds that the derivative of $F(x)$—the function giving the area under $f(x)$—is $f(x)$ itself. Let's see whether the picture bears this claim out. Recall that the derivative of $F(x)$ is, by the definition in Equation A.1,

$$F'(x) = \lim_{w \to 0} \frac{F(x+w) - F(x)}{w}.$$

This expression is identical to Equation A.1, except that we are now using $F(x)$ instead of $f(x)$, and I have changed the name of Δx to w. Now look again at the picture. The value of $F(3 + w)$ is the area under $f(x)$ between 0 and $3 + w$—that is, it's the summed area of the gray region ($F(3)$), the black rectangle, and the small region shaded in light gray and labeled ϵ. The black rectangle's area is equal to the length of its base (w) times its height ($f(3)$), or $wf(3)$. Thus,

$$F(3 + w) = F(3) + wf(3) + \epsilon.$$

There's nothing special about 3—it just happens to be the value used in the picture. Let's replace it with x, since we'll always be able to break $F(x + w)$ into the sum of $F(x)$, a rectangle, and a term that corrects for overshoot or undershoot by the rectangle, which we can label ϵ. Then we have

$$F(x + w) = F(x) + wf(x) + \epsilon.$$

Now we can substitute this expression for $F(x + w)$ into the expression for $F'(x)$, giving

$$F'(x) = \lim_{w \to 0} \frac{F(x) + wf(x) + \epsilon - F(x)}{w} = \lim_{w \to 0} \frac{wf(x) + \epsilon}{w}.$$

Breaking this into two pieces and cancelling w in the first term, we have

$$F'(x) = f(x) + \lim_{w \to 0} \frac{\epsilon}{w}.$$

We are almost there—the fundamental theorem of calculus holds that $F'(x) = f(x)$, which is true if $\lim_{w \to 0} \epsilon/w = 0$. To sketch the end of the argument, notice that if w is very small and $f(x)$ has does not behave too erratically around x, then the overshoot term ϵ is approximately the area of a triangle. Remember that the area of a triangle is the product of the length of its base and its height divided by two. Look at Figure A-7 again. The triangle that approximates ϵ has height $f(x + w) - f(x)$, and the base has length w, so $\epsilon \approx [f(x + w) - f(x)]w/2$. Plugging this into the expression for $F'(x)$ gives

$$F'(x) \approx f(x) + \lim_{w \to 0} \frac{[f(x+w) - f(x)]w}{2w} = f(x) + \lim_{w \to 0} \frac{f(x+w) - f(x)}{2} = f(x).$$

The last step relies on the assumption that $f(x)$ is a "nice" function—smooth and continuous—such that $f(x)$ will not change much between x and $x + w$ if w is small. If $f(x)$ meets the requirements, then the derivative of $F(x)$—where $F(x)$ is an indefinite integral of $f(x)$, because of the way we constructed the problem—is equal to $f(x)$ itself. This is the fundamental theorem of calculus.

A.5 Summary

One way to view the discipline of statistics is as a way to set up mathematical structures that allow us to reason about data. In this book, our goal is to use accessible mathematics and computation to understand statistical reasoning.

You will need to know a few basic facts about calculus to appreciate the arguments in the rest of the book. Namely, you ought to know that the derivative of $f(x)$ is a function that gives the instantaneous rate of change in $f(x)$ and that the integral of $f(x)$ can be used to find the area between the graph of $f(x)$ and the x axis.

Here are brief answers to the questions posed at the beginning of the appendix:

(1) What is a mathematical function?

A mathematical function takes as input one or more arguments and produces an output. For example, $f(x) = x^3 + 2$ takes a number x as input and produces the cube of that number plus two as output.

(2) What is a limit?

A limit is the value that a function's output approaches as its argument approaches a specific value. For example, if $\lim_{x \to 0} f(x) = 2$, then as one feeds the function argument values that approach 0, the function produces output that approaches 2.

(3) What is a derivative?

Suppose there is a function $f(x)$. The derivative of $f(x)$, often written $f'(x)$, returns the instantaneous rate of change of $f(x)$ at the value x. Graphically, the instantaneous rate of change is interpretable as the slope of a line tangent to the graph of $f(x)$ at x.

(4) How can derivatives be used for optimizing functions?

If a function $f(x)$ has a defined derivative for all values of x, then the local maxima and minima are located at the values of x that satisfy $f'(x) = 0$.

(5) What is an indefinite integral?

Suppose there is a function $f(x)$. An indefinite integral of $f(x)$, often written with a capital $F(x)$, is a function that has $f(x)$ as its derivative. $F(x)$ can also be used to find areas under graphs of functions. (See the next point.)

(6) What is a definite integral?

Suppose there is a function $f(x)$, and that a and b are constants with $a < b$. If $f(x)$ does not take negative values between a and b, then the definite integral of $f(x)$ from a to b is the area that lies under the graph of the function $f(x)$, above the x axis, and between the vertical lines $x = a$ and $x = b$. If the indefinite integral $F(x)$ is known, then the definite integral can be calculated as $F(b) - F(a)$.

(7) How are derivatives, indefinite integrals, and definite integrals related?

Differentiation and integration are inverse procedures. That is, if $\int f(x)dx = F(x)$, then $F'(x) = f(x)$. The derivative of the indefinite integral of a function is equal to the original function.
Definite integrals are computed from indefinite integrals. The definite integral is computed as $\int_a^b f(x)\, dx = F(b) - F(a)$, where $F(x)$ is an indefinite integral of $f(x)$.

(8) How does one compute derivatives and indefinite integrals of polynomial functions?

If $f(x) = ax^n$, then the derivative with respect to x is $f'(x) = nax^{n-1}$, and indefinite integrals are given by $F(x) = ax^{n+1}/(n+1) + C$, where C is a constant.

A.6 Further reading

There are many good calculus books available, including

Edwards, C. H., & Penney, D. E. (1982). *Calculus with Analytic Geometry*. Prentice-Hall, Upper Saddle River, NJ.

This is the book I learned from. I liked it.

Fox, J. (2009). *A Mathematical Primer for Social Statistics*. Sage Publications, Newbury Park, CA.

Fox's treatment of calculus includes the right topics for people who are focused specifically on statistics but want a more thorough introduction to calculus than this appendix provides.

Thompson, S. P., & Gardner, M. (1998). *Calculus Made Easy*. Macmillan, London.

An entertaining, friendly introduction originally published in 1910. The original text is in the public domain and so can be accessed freely.

Another excellent resource is Khan Academy, which provides short, free video lectures on all these topics.

If you want more support and practice opportunities before continuing, you can obtain any calculus book and look specifically at the sections on limits, definitions of derivatives and integrals, the fundamental theorem of calculus, maximization/minimization, and calculus operations on polynomials. You may find yourself inspired to learn about other topics, and that would be wonderful, but you will not need most of them for this book.

APPENDIX B

More R details

In Chapter 2, I either passed over or sped through some features of R. In this appendix, I will more explicitly cover the data types available in R, as well as some of the most important functions. There is much more to be said. Some books and online resources are listed in the further reading at the end of Chapter 2.

B.1 Modes

In Chapter 2, we saved x as a numeric vector of length 1, which is one type of R object. "Numeric" refers to the category of data stored, and "vector" is the category of data structure. Categories of data are called "modes" in R. There are several modes of data that R can use.

Three important R modes are *numeric*, *logical*, and *character*. Most of the data you will use in this book will be numeric. The numeric mode includes both decimals and integers. The assignments x <- 7, x <- 0.479, and x <- sqrt(7) will all produce numeric values. Categorical data are stored using factors, which are a distinct type of numeric data. If you have a set of fruits, and you want to analyze differences between apples, oranges, and pineapples, you can store the fruit types as a factor. We will cover factors later in this appendix.

Logical data can take only two values—TRUE and FALSE. (The capitalization is important.) It is possible to assign TRUE and FALSE values directly, as in x <- TRUE. However, we usually use logical data when we want to check that other data meet certain conditions. For example, suppose I assign a numeric value to the variable p:

```
> p <- pnorm(-2)
> p
[1] 0.02275013
```

I might be interested in whether p is less than some other value, like 0.05. I could check with the following command, which returns a logical value:

```
> p < 0.05
[1] TRUE
```

I could also save this logical value and refer to it later:

```
> sig <- (p < 0.05)
> sig
[1] TRUE
```

In R, "less than" and "greater than" are expressed by the familiar signs "<" and ">". It is less obvious that "less than or equal to" is "<=", "greater than or equal to" is ">=" and "equals" is "==". The double equal is used in part because the single equal is used for assignment.

That is, "x = 5" assigns the variable x to have the value 5, but "x == 5" returns a logical vector that equals TRUE if the variable x is equal to 5 and FALSE if x is not equal to 5.

The third mode, character, will be less important in this book. Strings of text are stored as character data. In R's command line, character data and variable names are distinguished by the presence of quotation marks. Character data have quotations around them:

```
> this.is.an.object.name <- "This, in contrast, is a datum."
> this.is.an.object.name
[1] "This, in contrast, is a datum."
```

Be careful to use quotes when your intention is to specify a character datum. When there are no quotes around text, R assumes that you are either trying to name an object or trying to refer to an object has already been created.

One more point—R stores missing data with the value NA (capitalized, no quotes). NA values can appear in numeric, logical, or character vectors. Because we will use either data sets stored within R or simulated data, we will not encounter missing data in this book.

B.2 Data structures and data extraction

Data structures refer to ways in which data of any mode can be organized. We will discuss four data structures—vectors, matrices, data frames, and lists—and ways to extract specific subsets from them.

B.2.1 *Vectors*

The most basic data structure in R is the vector. In Chapter 2, it was mentioned that within R, individual numbers are treated as numeric vectors of length 1. Vectors can hold data of any mode, though any one vector can hold data of only one mode. In principle, vectors can be of any positive integer length. (In practice, R's memory cannot hold arbitrarily long vectors.) You can combine individual data into a vector using the c() (for "concatenate") function. For example:

```
> x <- c(1,2,4,8)
> y <- c("So What", "Freddie Freeloader", "Blue in Green")
> z <- c(TRUE, TRUE, FALSE, FALSE)
```

These commands make x a numeric vector of length 4, y a character vector of length 3, and z a logical vector of length 4.

As you analyze data, you will frequently need a way to examine specified subsets of the data. The most straightforward way to extract a specific entry from a vector is to provide the index of the entry you desire. For example, if I want the 3rd entry in x, I can type

```
> x[3]
[1] 4
```

I can also get all the entries in x except for one that I do not want. If I don't want the 4th entry in x, I can type

```
> x[-4]
[1] 1 2 4
```

It is possible to select a set of entries as well. To get the 1st and 3rd entries, you can make a vector including the numbers 1 and 3 and use that to extract data:

```
> x[c(1,3)]
[1] 1 4
```

Perhaps the most powerful way to extract information from vectors is to select only the entries in the vector that meet a certain condition. To see only the entries in x that are greater than 3, try

```
> x[x > 3]
[1] 4 8
```

This is important enough to unpack. The "x > 3" inside the brackets actually produces a logical vector:

```
> x > 3
[1] FALSE FALSE TRUE TRUE
```

In words, then, when you type "x[x > 3]," you are asking for the entries in x that correspond to (that is, are in the same position as) entries in the logical vector x > 3 that are TRUE. The logical statement inside the brackets need not be about the vector from which you are trying to extract data—it could be a statement about another vector of the same length.

You can also combine logical statements using the & ("and") and | ("or") operators. For example:

```
> x <- c(1,2,4,8)
> w <- c(1,3,9,27)
> x[x > 3 & w > 20]
[1] 8
```

This statement returns a vector of length one with the value 8. To see why, notice that the two logical vectors combined in the "and" statement are:

```
> x > 3
[1] FALSE FALSE TRUE TRUE
> w > 20
[1] FALSE FALSE FALSE TRUE
```

The first statement is true of the 3rd and 4th entries, and the second statement is true of the 4th entry only. The only entry that is true in both vectors is the 4th, so combining the two statements gives

```
> x > 3 & w > 20
[1] FALSE FALSE FALSE TRUE
```

And thus x[x > 3 & w > 20] returns the 4th entry of x, which is 8. You might also be interested in all the entries in x that are either greater than 3 *or* that are in the same place as entries in w that are greater than 20. To access these, you would use the "or" operator:

```
> x[x > 3 | w > 20]
[1] 4 8
```

Table B-1 Two logical vectors x and y, along with the results of applying the & ("and"), | ("or"), == ("equals"), != ("not equal"), and ! ("not") operators to them

x	y	x & y	x | y	x == y	x != y	! (x&y)
FALSE	FALSE	FALSE	FALSE	TRUE	FALSE	TRUE
FALSE	TRUE	FALSE	TRUE	FALSE	TRUE	TRUE
TRUE	FALSE	FALSE	TRUE	FALSE	TRUE	TRUE
TRUE	TRUE	TRUE	TRUE	TRUE	FALSE	FALSE

In R and other programming languages, we use "or" a little differently than we often do in our everyday language. In everyday language, we often use "or" to mean "one or the other, but not both." The plate is in the cabinet or it is in the dish rack. But in programming, an "or" statement is true if *either or both* of the statements it connects are true. So "two plus two is four or whales are mammals" is true, as is "I am riding a unicorn or two plus two is four."

Table B-1 shows two logical vectors and the results of applying several common logical operators to them.

B.2.2 *Matrices*

The second important data structure in R is the matrix. Whereas vectors can be thought of as data arranged in a single ordered row or column, matrices have both rows and columns. One way to create a matrix is from a vector. Suppose we have a vector of the integers from 1 to 12:

```
> vec <- 1:12
> vec
[1]  1  2  3  4  5  6  7  8  9  10  11  12
```

(Notice that use of the colon ":" creates a vector of integers in sequence. This can be handy when making for() loops.) We can reorganize the vector into a matrix of 3 rows and 4 columns.

```
> mat <- matrix(vec, nrow = 3, ncol = 4)
> mat
     [,1] [,2] [,3] [,4]
[1,]    1    4    7   10
[2,]    2    5    8   11
[3,]    3    6    9   12
```

We have created an object with rows and columns. This particular matrix has 3 rows and 4 columns, which we refer to as a 3 × 4 ("three-by-four") matrix. When analyzing data, one often uses columns to separate different types of measurements and rows to separate different units being measured (for example, different people, different animals, different schools). To extract data from a matrix, we can use the same tactics we used for vectors— either providing numeric vectors of indices to extract or logical vectors in which every desired element corresponds to a TRUE entry in the vector. However, now we have to specify both the row and the column we want to extract. The row information comes first, followed by a comma, and then the column information. If we leave the row information blank,

R will give us all the rows, and if we leave the column information blank, R gives us all the columns. Some examples will help.

To get the entry in the 2nd row and 3rd column:

```
> mat [2,3]
[1] 8
```

To get the entire 1st row:

```
> mat [1,]
[1]  1  4  7  10
```

To get the whole 4th column:

```
> mat [,4]
[1]  10  11  12
```

To get the entries in the 3rd row that are greater than 7:

```
mat [3, mat [3,] > 7]
[1]  9  12
```

To get the entry in the 1st row from the column whose 3rd-row entry is 6:

```
> mat [1, mat [3,] == 6]
[1] 4
```

To unpack that last one, look at the matrix again with annotations:

```
> mat
     [,1]  [,2]  [,3]  [,4]
[1,]   1    4     7      10   #4 is the entry in column 2 in row 1
[2,]   2    5     8      11
[3,]   3    6     9      12   #The entry in column 2 of row 3 is 6
```

The two things to remember when extracting data from matrices are (1) always include the comma separating row information and column information and (2) rows come first.

B.2.3 *Data frames*

A third data structure we will use is the data frame. Data frames are like matrices in that they have rows and columns, and one can extract data from them in the same way one extracts data from a matrix. Data frames have two important features that matrices lack. First, they can store data of different modes in different columns. Second, it is possible to extract specific columns from data frames using the operator $, as we will see below.[1]

We can learn about data frames by examining one of the built-in data frames included in R. We will look at the data frame iris, which is also examined in the tutorial in Chapter 2.

---

[1] One catch, for readers who wish to use their data in matrix-algebraic operations, is that data frames cannot be multiplied or inverted as matrices can.

We can see the whole data frame by typing `iris` at the command line, or we can see the first six lines using the `head()` function:

```
> head(iris)
  Sepal.Length Sepal.Width Petal.Length Petal.Width Species
1          5.1         3.5          1.4         0.2  setosa
2          4.9         3.0          1.4         0.2  setosa
3          4.7         3.2          1.3         0.2  setosa
4          4.6         3.1          1.5         0.2  setosa
5          5.0         3.6          1.4         0.2  setosa
6          5.4         3.9          1.7         0.4  setosa
```

Notice that the columns have names. We can use the $ operator to extract one column from a data frame using the column name:

```
> iris$Sepal.Length[1:6]
[1]  5.1  4.9  4.7  4.6  5.0  5.4
```

Here, I used `[1:6]` to extract just the first six entries. We also appear to have multiple data types in the data frame—the `Species` column is different from the others. A quick glance would suggest that the `Species` column contains character data because every entry consists of a string of letters. That's not correct. Examining the `Species` column directly shows us something we have not yet seen:

```
> iris$Species[1:6]
[1] setosa setosa setosa setosa setosa setosa
Levels: setosa versicolor virginica
```

The "`Levels`" line is a tip-off that the species column is actually a factor, a special type of numeric data used for encoding categories. Factors can have text labels associated with each value, and the labels are what we see here. The "`Levels`" line shows all the unique category labels included in this factor. In this case, the three levels correspond to the three species of iris included in the dataset.

Though the `Species` column turned out to include a special type of numeric data masquerading as text, there is nothing stopping us from having a data frame that includes both numeric and character data:

```
> iris.char <- iris
> iris.char$Species <- as.character(iris.char$Species)
> head(iris.char)
  Sepal.Length Sepal.Width Petal.Length Petal.Width Species
1          5.1         3.5          1.4         0.2  setosa
2          4.9         3.0          1.4         0.2  setosa
3          4.7         3.2          1.3         0.2  setosa
4          4.6         3.1          1.5         0.2  setosa
5          5.0         3.6          1.4         0.2  setosa
6          5.4         3.9          1.7         0.4  setosa
> iris.char$Species[1:6]
[1] "setosa" "setosa" "setosa" "setosa" "setosa" "setosa"
```

In the first step, we made a new data frame called `iris.char` by copying the `iris` data frame. In the second step, we replaced the `Species` column in the new data frame with a version of itself that was coerced to character data using the `as.character()` function. The resulting data frame looks just like the old one, but now, when we look at the `Species` column by itself, we see that it includes character data. Thus, the new data frame includes numeric and character data together.

B.2.4 *Lists*

The final data structure we will consider—albeit briefly—is the list. You have already seen some lists: data frames are actually a subcategory of lists. Lists are a way to organize disparate R objects within a single data structure. For example, here is a command to make a list the first entry of which is a character vector, the second entry of which is a numeric matrix, and the third entry of which is the `iris` data frame:

```
my.list <- list(c("a","b","c"), matrix(1:12, nrow = 3), iris)
```

Entries in a list are indexed with double brackets. For example, to extract the numeric matrix that is the second entry in `my.list`, we would use

```
> my.list[[2]]
     [,1]  [,2]  [,3]  [,4]
[1,]    1     4     7    10
[2,]    2     5     8    11
[3,]    3     6     9    12
```

Extraction can continue from here. To extract the entry in the third row and first column of this matrix, we could use

```
> my.list[[2]][3,1]
[1] 3
```

It is also possible to assign names to the entries in a list and to use these names to extract data. For example, we could remake `my.list` so that its entries have names as follows:

```
my.list <- list(char.vec = c("a","b","c"), num.mat = matrix(1:12, nrow = 3),
iris.dat = iris)
```

We could then extract the second entry either using `my.list[[2]]`, as before, or with

```
> my.list$num.mat
     [,1]  [,2]  [,3]  [,4]
[1,]    1     4     7    10
[2,]    2     5     8    11
[3,]    3     6     9    12
```

B.3 R functions

R functions, like mathematical functions, take arguments and return an output. In the case of an R function, the arguments go in parentheses after the function name, and the output returned is an R object, like a vector or matrix. Many R functions are built-in, many more can be accessed using the thousands of R packages available for free download, and there is no limit to the number of R functions you can write yourself. No one has a complete knowledge of R functions—the trick is knowing how to find the right function at the right time.

The most important R functions are listed in Tom Short's R Reference Card (Short, 2004). You can find the R Reference Card here: http://cran.r-project.org/doc/contrib/Short-refcard.pdf or simply by typing "R Reference Card" into a search engine. I keep a printed copy at my desk. In this book, we will need only a subset of the commands listed. I will list the most important commands for our purposes according to the heading under which they appear in the R Reference Card.

B.3.1 *Getting help*

The most important step in learning to use R is learning how to get help when you are stuck. A web search is often best, especially when you are beginning or doing something new, but here are some useful built-in functions for getting help.

help() The help() function allows you to find out about the usage of any function about which you might be curious. For example, help(mean) brings up a page of information about the mean() function, which calculates the mean of a numeric vector. The page includes the arguments that the function takes, general information about the function, and examples of how to use the function. The trick is that you have to know the function name.

help.search() If you don't know the name of a function that will do what you need, you can use help.search(). help.search() takes a character string (remember to use quotation marks) and searches all the help pages to find functions that might be applicable. For example, if I didn't know that mean() computes the mean of a numeric vector, I could try using help.search("mean"). This would bring up a list of functions that involve taking means, along with the packages in which those functions are included. In this specific case, help.search() is not very helpful. A lot of functions come up, and the one we want—base::mean()—is buried. (We see "base::mean()" because mean() is in the base package, which loads automatically whenever R is started.) A web search is often more effective than help.search().

example() The R help pages have examples at the bottom, and the example() function will show you what happens when you run the examples. The argument to example() is the name of another R function. You can try example(plot).

str() str() shows the structure of an R object. Try str(iris) to see that iris is a data frame with 150 observations of 5 variables, four of which are simple numeric variables and one of which is a factor.

ls() Returns the names of all the objects you have created and not yet removed from the current environment. Try entering ls() after working through the commands in the last two paragraphs to see the names of the variables you have created.

B.3.2 *Data creation*

c() - You have already seen c(), for "concatenate," and you have used it to assemble vectors from individual elements. You can also use c() to combine vectors into a longer vector:

```
> x <- c(1,2)
> y <- c(3,4)
> z <- c(x,y)
> z
[1] 1 2 3 4
```

from:to This is a quick way to make a sequence of integers. For example:

```
> x <- 1:10
> x
[1] 1  2  3  4  5  6  7  8  9  10
```

matrix() You have already seen the use of matrix() to structure a vector into a matrix. For example:

```
> x <- 1:12
> matrix(x, nrow = 3, ncol = 4)
     [,1]  [,2]  [,3]  [,4]
[1,]   1     4     7    10
[2,]   2     5     8    11
[3,]   3     6     9    12
```

You can also change the order in which the entries are added into the matrix:

```
> matrix(x, nrow = 3, ncol = 4, byrow = TRUE)
     [,1]  [,2]  [,3]  [,4]
[1,]   1     2     3     4
[2,]   5     6     7     8
[3,]   9    10    11    12
```

Or you can omit the "x" argument to make a matrix of blank entries to fill in later:

```
> matrix(nrow = 3, ncol = 4)
     [,1]  [,2]  [,3]  [,4]
[1,]   NA    NA    NA    NA
[2,]   NA    NA    NA    NA
[3,]   NA    NA    NA    NA
```

rbind() and **cbind()** These two commands combine vectors by rows or columns into a matrix or data frame:

```
> x <- c(1,2,3)
> y <- c(4,5,6)
> z <- c(7,8,9)
> rbind(x,y,z)
     [,1]  [,2]  [,3]
x      1     2     3
y      4     5     6
z      7     8     9
> cbind(x,y,z)
     x y z
[1,] 1 4 7
[2,] 2 5 8
[3,] 3 6 9
```

B.3.3 *Variable information*

length() Returns the length of a vector. For example:

```
> x <- c(1,2,3,4,5)
> length(x)
[1] 5
```

dim() Similar to `length()`, but for matrices and data frames. Returns the number of rows and the number of columns:

```
> dim(iris)
[1] 150 5
```

B.3.4 *Math*

sum(), mean(), median(), var(), sd() These functions return the sum, mean, median, sample variance, and sample standard deviation of a numeric vector, respectively. All of these functions return NA if the vector in the argument has any missing values, unless the argument na.rm is set to equal TRUE.

```
> x <- iris$Sepal.Length
> sum(x)
[1] 876.5
> mean(x)
[1] 5.843333
> median(x)
[1] 5.8
> sd(x)
[1] 0.8280661
> var(x)
[1] 0.6856935
```

B.3.5 *Plotting*

There is a lot to learn about plotting in R, but two of the most important commands are `plot()` and `hist()`.

plot(x, y) Make a scatterplot of two numeric vectors x and y. Try

```
> plot(anscombe$x1, anscombe$y1)
```

and compare the result with Figure 1-1. It is possible to provide more arguments to `plot()` to alter the title, axis labels, appearance of the points, and many other aspects of the plot. The code used to generate the plots in the book is available at github.com/mdedge/stfs.

hist(x) make a histogram of a numeric vector. Try

```
> hist(iris$Sepal.Width)
```

B.3.6 *Optimization and model fitting*

lm() `lm()` computes estimates and inferential statistics related to linear regression and returns the results as an R object. Linear regression is the central statistical procedure in this book. `lm()` uses formula notation, where the dependent variable is to the left of a tilde ~ and a formula with independent variables is to the right of the tilde. For example, you could write

```
> lm(anscombe$y1 ~ anscombe$x1)
```

to fit a linear model. One way in which R is different from other statistical software is that in R, one typically saves the analyses as objects and then probes those objects later. For example, the command above produces little useful output, but the following gives a lot of information:

```
> my.lm <- lm(anscombe$y1 ~ anscombe$x1)
> summary(my.lm)
```

B.3.7 *Distributions*

rnorm(), **rbinom()**, **rpois()**, etc. These functions generate pseudorandom data from the normal, binomial, and Poisson distributions. It is possible to generate data from many other distributions within R. We say more about pseudorandom number generation in an exercise in Chapter 4. The random number generation functions in R take as arguments the number of random numbers requested and the parameters to use for the probability distribution. (Don't worry if you don't know what this means yet.) For example,

```
> rnorm(10, mean = 0, sd = 1)
```

produces 10 pseudorandom draws from the normal distribution with an expectation of 0 and a standard deviation of 1.

B.3.8 *Programming*

We will write a few small programs to simulate and analyze data. If you have never programmed before, do not worry. The main functions we will use for programming are function(), return(), if(), and for(). Their use will become clearer through the examples included in the tutorial in Chapter 2 and in later chapters.

function() function() allows you to create new R functions that you can call again later in the same session. If you ever find yourself copying and pasting large chunks of code to be re-run, you probably ought to write a function instead. The essential parts of a function definition are the name of the function, the arguments the function takes, the operations the function performs, and the object the function returns. These parts are located as follows, with italics indicating parts that are not actual code:

```
name_of_function <- function(comma-separated arguments){
    operations that the function performs
    return(object the function returns)
}
```

You would then call this function as

```
name_of_function(arguments, comma-separated)
```

In the following paragraphs, mysum() and slowsum() are examples of user-defined functions.

return()   When one writes an R function, one sometimes includes a return() statement at the end, including the object that will be the function's output in the parentheses. For example, you could write the simple function

```
mysum <- function(x, y){
    z <- x + y
    return(z)
}
```

This function simply adds two supplied numbers (or vectors of the same length). We add the two numbers, store the sum as z, and return z. Now if we ran

```
> sum.val <- mysum(2, 3)
```

The variable sum.val would be a numeric vector of length 1 with value 5 (i.e. 2 + 3). We could also just define mysum() as

```
mysum <- function(x, y) {
    return(x + y)
}
```

with the same result. Here's a secret: return() is usually unnecessary. We could just write

```
mysum <- function(x, y) {
    x + y
}
```

and get the same result. R returns the last statement evaluated in the function. Sometimes using return() makes code clearer to human readers, especially beginners, and I use it at times in the book when it's not necessary. return() is useful in cases where a function may need to return a value early for some reason—for example, return() can be deployed inside an if() statement to return a value if some condition is met without evaluating the rest of the code in the function.

After it is defined, we can use mysum() by typing, for example,

```
> mysum(2,2)
[1] 4
```

if() If the logical condition in parentheses is met, R performs the actions that follow in curly brackets. For example, we could modify the mysum() function we wrote above to include an option that lets us double the output if we want:

```
mysum <- function(x, y, doubleit = FALSE) {
    z <- x + y
    if (doubleit == TRUE) {
        z <- z*2
    }
    return(z)
}
```

We have set doubleit to be FALSE by default. If doubleit is FALSE, then the statement in curly brackets is ignored, and we return z. If the user sets doubleit to be TRUE, though, we multiply z by 2 before returning it. So we could do the following:

```
> mysum(2,2)
[1] 4
> mysum(2, 2, doubleit = TRUE)
[1] 8
```

for() for() allows you to repeat a set of commands for every element in a vector. For example, suppose we want to write a function slowsum() that sums all the entries in a vector using a for() loop. (You'll see why it's "slow" sum shortly):

```
slowsum <- function(x) {
  s <- 0
  for (i in x) {
    s <- s + i
  }
  return(s)
}
```

This function starts the value of the sum at 0, then cycles through each entry in the argument and adds it to the sum. Once we have finished, we return the sum. R has a built-in function for doing the same thing called `sum()`. We can check whether our function works by comparing it with `sum()`:

```
> sum(1:100)
[1] 5050
> slowsum(1:100)
[1] 5050
```

This seems to check out. However, using `for()` is actually a bad move in this case, and it's often a controversial choice in R. Using `for()` loops isn't a good fit with R's philosophy, which encourages *vectorized* operations—operations performed on a whole vector rather than on each element in the vector in sequence. Many programmers feel that R code that isn't vectorized is harder to read, and non-vectorized code is sometimes slower than it needs to be, particularly when the `for()` loop is replacing a built-in function that has been optimized. To see the speed difference in this case, we'll make a vector 10 million entries long using `rnorm()` and use the `system.time()` function to see how long `sum()` and `slowsum()` take to run:

```
> vec <- rnorm(10000000, mean = 0, sd = 1)
> system.time(sum(vec))
   user system elapsed
   0.02 0.00 0.02
> system.time(slowsum(vec))
   user system elapsed
   5.33 0.00 5.33
```

You can see that in this case, using `for()` to take the sum takes about 266 times as long as using the built-in `sum()` function. Five seconds may not seem like much, but time adds up quickly when you use a script to perform lots of operations.

Despite the criticism of `for()`, I use it in the text and in some of the answers to exercises. For our purposes, `for()` is fine, and often, solutions that use `for()` are more intuitive for beginners to R to write, read, and understand. Most expert R users will encourage you to use the `apply()` family of functions (`apply()`, `lapply()`, `sapply()`, `tapply()`, etc.) instead of `for()`. I also encourage you to graduate to these functions when you feel comfortable. Once you have started thinking like an R programmer, these functions will often let you do the same thing with fewer lines of code.[2]

apply() `apply()` lets you compute a vectorized function using each row or column in a matrix. For example, if you wanted to know the median of every column in a matrix, you

---

[2] You may run across the claim that `apply()` and its relatives are faster than `for()`. It's more accurate to say that it's sometimes easier to write `for()` loops in ways that will end up running slowly.

could use apply() to get a vector of medians. It's often possible to use apply() instead of for(). Let's make a 10 × 10 matrix called mat:

```
> mat <- matrix(1:100, nrow = 10, ncol = 10)
> mat
      [,1] [,2] [,3] [,4] [,5] [,6] [,7] [,8] [,9] [,10]
 [1,]    1   11   21   31   41   51   61   71   81    91
 [2,]    2   12   22   32   42   52   62   72   82    92
 [3,]    3   13   23   33   43   53   63   73   83    93
 [4,]    4   14   24   34   44   54   64   74   84    94
 [5,]    5   15   25   35   45   55   65   75   85    95
 [6,]    6   16   26   36   46   56   66   76   86    96
 [7,]    7   17   27   37   47   57   67   77   87    97
 [8,]    8   18   28   38   48   58   68   78   88    98
 [9,]    9   19   29   39   49   59   69   79   89    99
[10,]   10   20   30   40   50   60   70   80   90   100
```

If you want to know the mean of every column in mat, one way to find out would be to use a for() loop. For example, you could write

```
mymeans <- numeric(10)
for(i in 1:10){
  mymeans[i] <- mean(mat[,i])
}
```

Printing mymeans would then reveal the means you want:

```
> mymeans
 [1] 5.5 15.5 25.5 35.5 45.5 55.5 65.5 75.5 85.5 95.5
```

This will work just fine, but the R gods might smile on you if you used apply() instead:

```
> apply(mat, 2, mean)
 [1] 5.5 15.5 25.5 35.5 45.5 55.5 65.5 75.5 85.5 95.5
```

To use apply(), you need to specify a matrix, a "margin"—which indicates whether you want to consider the rows or the columns of the matrix—and the function you want to use. In our case, the matrix is mat, the margin is 2 for columns—it would have been 1 for rows—and the function to apply is mean. Many R users feel that apply is easier to read and more "R-like" than for(). In some cases, but not this one,[3] apply() is also faster.

B.3.9 *Data input and output (I/O)*

In this book, we work with datasets that are either generated randomly or built into R. In real applications, you will want to import datasets into R for analysis. Some of the most important functions for data input and output are getwd(), setwd(), write.table(), and

---

[3] In this case, testing on a matrix of 200,000 columns and 5 rows showed that the for() loop above took 23 seconds to run and the apply() solution took 32 seconds. apply() and its relatives are not always faster. In this case, R has a built-in function to take means of columns in matrices, colMeans(), which is faster than Achilles, finishing in 0.06 seconds on my computer. In R, there are many ways to do things, and you have to consider both the time you save in writing code that is easier for you to conceive and the time you would save by optimizing to code to run as quickly as possible.

`read.table()`. For producing image files with graphics, there are several functions depending on the file format you'd like to save—I will give an example using the `tiff()` command.

getwd() and **setwd()** These two functions let you see and change R's working directory. Whenever R is running, you can see the working directory using `getwd()`,

```
> getwd()
[1] "/mnt/Data"
```

Unless you direct R to do otherwise, the working directory is the place in which R will look for files, and it is also the place to which it will write files by default. You can change the working directory with the `setwd()` function. On my machine, I can run

```
> setwd("/mnt/Data/mdedge")
> getwd()
[1] "/mnt/Data/mdedge"
```

R can work with both absolute and relative specifications of directory names. The system for specifying directory names differs by operating system. If you are using RStudio, you can change directories in several ways using options from the pull-down "Session" menu.

write.table() and **read.table()** These two functions allow you to write data frames as text files and read text files into data frames. For example, we can write the `iris` data frame to a text file called iris.txt in the working directory using

```
> write.table(iris, "iris.txt")
```

Doing so produces a text file in the working directory, the first few lines of which look like this:

```
"Sepal.Length" "Sepal.Width" "Petal.Length" "Petal.Width" "Species"
"1" 5.1 3.5 1.4 0.2 "setosa"
"2" 4.9 3 1.4 0.2 "setosa"
"3" 4.7 3.2 1.3 0.2 "setosa"
"4" 4.6 3.1 1.5 0.2 "setosa"
"5" 5 3.6 1.4 0.2 "setosa"
```

By default, names of the columns and rows (in this case, the row names are numbers) are written. Row/column names, character vectors, and factor names are also written with quotation marks by default. Distinct columns are separated by spaces. All these can be changed. For example, to leave off the quotation marks and row names, and to separate data entries with commas rather than spaces, we could use

```
> write.table(iris, "iris.csv", quote = FALSE, sep = ",", row.names = FALSE)
```

which produces a text file that looks somewhat different, with the first few rows

```
Sepal.Length,Sepal.Width,Petal.Length,Petal.Width,Species
5.1,3.5,1.4,0.2,setosa
4.9,3,1.4,0.2,setosa
4.7,3.2,1.3,0.2,setosa
4.6,3.1,1.5,0.2,setosa
```

This file, in .csv format, could be opened by collaborators using Microsoft Excel.

To read data from a text file into an R data frame, use `read.table()`. For example, after running the `write.table()` commands above, we can produce data by reading in the iris.txt file:

```
> iris.dat <- read.table("iris.txt")
```

This command works fine, but it's important that the arguments to `read.table()` be set to match the conventions of the text file you're reading. For example, if we use iris.csv in this way, we get a bad result:

```
> iris.dat <- read.table("iris.csv")
> head(iris.dat)
V1
1 Sepal.Length,Sepal.Width,Petal.Length,Petal.Width,Species
2                                  5.1,3.5,1.4,0.2,setosa
3                                  4.9,3,1.4,0.2,setosa
4                                  4.7,3.2,1.3,0.2,setosa
5                                  4.6,3.1,1.5,0.2,setosa
6                                  5,3.6,1.4,0.2,setosa
```

The resulting dataframe has only one column because the default behavior of `read.table()` is to look for blank spaces as column separators, but iris.csv doesn't contain any blank spaces. It also did not recognize the header row as column names. We can fix these problems by changing the defaults:

```
> iris.dat <- read.table("iris.csv", sep = ",", header = TRUE)
> head(iris.dat)
  Sepal.Length Sepal.Width Petal.Length Petal.Width Species
1          5.1         3.5          1.4         0.2  setosa
2          4.9         3.0          1.4         0.2  setosa
3          4.7         3.2          1.3         0.2  setosa
4          4.6         3.1          1.5         0.2  setosa
5          5.0         3.6          1.4         0.2  setosa
6          5.4         3.9          1.7         0.4  setosa
```

That looks better.

If a collaborator gives you data in a format used by another data analysis package, such as SAS, SPSS, or Stata, then you can use the functions in the R package `foreign` (R Core Team, 2017) to read it in.

It is also possible to use the `save()` function to save data frames or other R objects in a .RData file, which is unreadable by standard text editors but easily loadable into R using the `load()` function. The `save()` and `load()` functions are especially useful when you need to load extremely large data files or complex data structures, such as lists or arrays, into R.

Saving graphics with `tiff()` and its relatives The best way to get figures and graphics from R and into another software package is to save your graphics directly to a file. For example, the following sequence of three commands produces a .tif image file in your working directory that is similar to Figure 1-1:

```
> tiff("fig1_1.tif", width = 6, height = 4.5, units = "in", res = 600)
> plot(anscombe$x1, anscombe$y1, xlab = "Fertilizer Consumption (kg/hectare)",
ylab = "Cereal Yield (100 kg/hectare)", pch = 19, bty = "n")
> dev.off()
```

The first command, `tiff()`, initiates a .tif file with the specified name, size, and resolution. The second file prints the plot to the .tif file. The third command, `dev.off()`, closes the graphics object and saves it to a .tif file. There are similar utilities for producing BMP, JPEG, PNG, PDF, and PostScript graphics files. The advantages of using any of these utilities include the ability to specify the size and resolution of your graphics as well as to reproduce them exactly should you ever have need.

APPENDIX C

Answers to selected exercises

Exercise Set 2-2

(1) See github.com/mdedge/stfs for code. You'll see that petal length and width are strongly associated and that all three species can be distinguished on the basis of petal length and width.
(2) Code is in the question. The gpairs() plot includes scatterplots for every possible pair of variables in the iris dataset.

Exercise Set 3-1

(1) See github.com/mdedge/stfs for code. The intercept is 3.00, and the slope is 0.500.
(2) See github.com/mdedge/stfs for the full derivation.

Exercise Set 3-2

(1) (a) Suppose that we already know $\widetilde{b}$. Then the sum of squared errors as a function of a is

$$g(a) = \sum_{i=1}^{n}(y_i^2 - 2ay_i - 2\widetilde{b}x_iy_i + a^2 + 2a\widetilde{b}x_i + \widetilde{b}^2 x_i^2).$$

The derivative with respect to a is

$$g'(a) = \sum_{i=1}^{n}(-2y_i + 2a + 2\widetilde{b}x_i).$$

We find $\widetilde{a}$ by finding any values of a for which $g'(a) = 0$. Through a set of steps parallel to those used for $\widetilde{b}$ in the main text (see github.com/mdedge/stfs), we find the sole value of a for which $g'(a) = 0$: $\widetilde{a} = (\sum_{i=1}^{n}y_i - \widetilde{b}\sum_{i=1}^{n}x_i)/n$.
   (b) If $x = \bar{x}$, then $\widetilde{y} = \widetilde{a} + \widetilde{b}\bar{x}$. Replacing $\widetilde{a}$ with $\bar{y} - \widetilde{b}\bar{x}$ gives $\widetilde{y} = (\bar{y} - \widetilde{b}\bar{x}) + \widetilde{b}\bar{x} = \bar{y}$, so when $x = \bar{x}$, the y coordinate of the line is $\bar{y}$.
(2) The slope is $b' = (\sum_{i=1}^{n}x_iy_i)/(\sum_{i=1}^{n}x_i^2)$. One way to get the slope is to set $\widetilde{a} = 0$ in Equation 3.5, which works because we know that Equation 3.5 gives the value of b that minimizes the squared line errors given any provided value of a, including 0.
(3) See github.com/mdedge/stfs.
(4) (a–c) See github.com/mdedge/stfs for code. The intercept is 3.24, and the slope is 0.48. The least-squares and L1 lines are similar in this case.
   (e) The L1 line goes through almost all the points and misses the one that falls off the line. The least-squares line, in contrast, is pulled toward the outlying point. The sum

of the squared line errors is sensitive to large line errors. As such, the least-squares line will be pulled toward individual outlying points in order to avoid making large errors. In contrast, the L1 line weights large line errors less heavily than the least-squares line.

Exercise Set 4-1

See github.com/mdedge/stfs for solutions to this exercise set.

Exercise Set 4-2

(1) If events A and B are independent, then $P(A|B) = P(A)$. Replacing the conditional probability with its definition gives $P(A \cap B)/P(B) = P(A)$. Multiplying both sides by $P(B)$ gives $P(A \cap B) = P(A)P(B)$. To prove the second part, divide both sides of $P(A \cap B) = P(A)P(B)$ by $P(A)$. The left side becomes $P(A \cap B)/P(A)$, which, by the definition in Equation 4.1, is $P(B|A)$. Thus, $P(B|A) = P(B)$.
(2) See github.com/mdedge/stfs or the next section for a derivation. The solution is called Bayes' theorem, and we will have more to say about it in the next section.

Exercise Set 4-3

(1) The probability mass function is
$f_X(0) = f_X(3) = \frac{1}{8}, f_X(1) = f_X(2) = \frac{3}{8}$, and $f_X(x) = 0$ for all other x.
(2) The sum would be 1. Axiom (ii) of probability (in optional Section 4.1) tells us that the probability of the event that includes all possible outcomes is 1.
(3) $F_X(b) - F_X(a)$. Remember that $F_X(b) = P(X \leq b)$ and $F_X(a) = P(X \leq a)$. Thus, $F_X(b) - F_X(a) = P(X \leq b) - P(X \leq a)$, or the probability that X is less than or equal to b but not less than or equal to a, $P(a < X \leq b)$.

Exercise Set 4-4

(1) There are three parts to consider. For any $x < 0$, $P(X \leq x) = 0$, so the cumulative distribution function starts with a flat line at height 0 extending from negative infinity to $x = 0$. Similarly, if $x > 1$, then $P(X \leq x) = 1$, so the cumulative distribution function has another flat line, this one at height 1 and extending from $x = 1$ to positive infinity. The remaining interval is between 0 and 1, the values that X can take. The requirement that all equally sized intervals in $[0, 1]$ are equally likely to contain X results in a line of uniform slope connecting (0,0) and (1,1). In this case, the slope required is 1. You can simply draw the function, but see github.com/mdedge/stfs for code for plotting it in R.
(2) Just as was the case for discrete random variables, $P(a \leq X < b) = F_X(b) - F_X(a)$. This means that if the probability of landing in two intervals of equal size differs, then the average slope of F_X in those two intervals must differ. Specifically, the interval with the higher probability must have the higher average slope. See github.com/mdedge/stfs for R code for drawing one possible cumulative distribution function that meets the description in the problem.

Exercise Set 4-5

(1) $\int_{-\infty}^{\infty} f_X(x)dx = 1$. For any cumulative distribution function, $\lim_{x \to \infty} F_X(x) = 1$ and $\lim_{x \to \infty} F_X(x) = \int_{-\infty}^{\infty} f_X(x)dx$. Probability density functions always have a total area under the curve of exactly 1.

(2) See github.com/mdedge/stfs.

Exercise Set 4-6

(1) The probability mass function of the Poisson distribution is $P(X = k) = \lambda^k e^{-\lambda}/k!$. Plugging in the appropriate values for k and λ gives (i) e^{-5}, (ii) $5e^{-5}$, and (iii) $(25/2)e^{-5}$.

(2) Use the probability mass function of the geometric distribution with parameter 1/2. If our first "heads" occurs on the 6th flip, then we have five tails before it. We plug $p = 1/2$ and $k = 5$ into $P(X = k) = (1 - p)^k p$ to get $P(X = 5) = (1/2)^5(1/2) = 1/64$.

(3) (a)

```
> x <- seq(-3, 3, length.out = 1000)
> plot(x, dnorm(x, mean = 0, sd = 1), type = "l")
```

(b)

```
> x <- seq(-3, 3, length.out = 1000)
> plot(x, pnorm(x, mean = 0, sd = 1), type = "l")
```

(c) Use `qnorm(0.975, mean = 0, sd = 1)` to get 1.96.

(4) (a)

```
> normsims <- rnorm(1000, mean = 0, sd = 1)
> hist(normsims)
```

(b) See github.com/mdedge/stfs for code and an explanation.

Exercise Set 5-1

(1) (a) If X is a Bernoulli random variable, then the mass function is $f_X(x) = P(X = x) = p^x(1 - p)^{1-x}$ for $x \in \{0, 1\}$. Because there are only two possible values of X, summing is easy. The expectation is

$$E(X) = \sum_{x=0}^{1} x f_X(x) = \sum_{x=0}^{1} x p^x(1 - p)^{1-x} = 0p^0(1 - p)^1 + 1p^1(1 - p)^0 = p.$$

(b) A binomial random variable is the number of successes out of n independent trials each with probability of success p. We already know that a Bernoulli random variable can model a single trial with probability of success p. Thus, a binomial random variable is the sum of n independent Bernoulli random variables, which we can label $X_1, X_2, \ldots, X_n$. So if $X = \sum_{i=1}^{n} X_i$, where the X_i are independent draws from a Bernoulli distribution with success probability p, then X is distributed as a binomial random variable with parameters n and p. We can now write $E(X) = E(\sum_{i=1}^{n} X_i)$. Because the expectation is linear, the expectation of the sum is equal to the sum of the expectations. This lets us quickly finish the job using the result from part (a): $E(X) = E(\sum_{i=1}^{n} X_i) = \sum_{i=1}^{n} E(X_i) = \sum_{i=1}^{n} p = np$.

(c, d) See github.com/mdedge/stfs for derivations.

(2) (a) When $n = 1$, the histogram ought to resemble the density function for a normal random variable with expectation 0 and standard deviation 1: we are taking individual samples from a normal distribution and plotting them. That is, it ought to be symmetric, centered around 0, and roughly bell-shaped. The great majority of the data should fall between -2 and 2. As we draw larger samples and take their means, the law of large numbers suggests that the sample means should generally be closer to the expectation than the individual observations are. Indeed, as we increase the sample size, we find that fewer samples have means that are far from the expectation. The shape of the distribution continues to look roughly normal.

(b) See github.com/mdedge/stfs for a modified version of the code that uses `rexp()` to simulate exponential random variables. The shape of the exponential density is markedly different from the shape of the normal distribution. When the expected value is set to 1, most of the observations are near 0, and the observations trail off far to the right. Again, when we take means of samples, we find that the sample means get closer to the expectation as the sample size increases. Notice something else: the shape of the distribution also changes. Namely, it starts to look more normal. This is a preview of the central limit theorem, which will appear soon.

Exercise Set 5-2

(1) Our definition says that $Var(X) = E([X - E(X)]^2)$. We expand the expression to get $Var(X) = E(X^2 - 2XE(X) + [E(X)]^2)$. Applying the linearity of expectation then gives $Var(X) = E(X^2) - E(2XE(X)) + E([E(X)]^2)$. $E(X)$ is a constant, as is $[E(X)]^2$. The expectation of a constant is just the constant itself, so we have that $Var(X) = E(X^2) - 2E(X)E(X) + [E(X)]^2$. Finally, notice that $E(X)E(X) = [E(X)]^2$ and collect the $[E(X)]^2$ terms to get $Var(X) = E(X^2) - [E(X)]^2$.

(2) (a) Using the definition, we have $Var(X + c) = E([X + c - E(X + c)]^2)$. Linearity of expectation (Equation 5.4) lets us write this as $Var(X + c) = E([X + c - E(X) - c]^2)$, which is $Var(X + c) = E([X - E(X)]^2) = Var(X)$. Thus, adding a constant to a random variable does not change the variance of the random variable.

(b) This time, we'll start with Equation 5.7, which gives $Var(cX) = E([cX]^2) - [E(cX)]^2$. Linearity of expectation lets us pull the constants out of the expectations, giving $Var(cX) = c^2E(X^2) - [cE(X)]^2 = c^2(E(X^2) - [E(X)]^2) = c^2Var(X)$.

(3) See github.com/mdedge/stfs for a full derivation.

(4) (a) Bernoulli random variables only take two values: 0 and 1. Note that because $0^2 = 0$ and $1^2 = 1$, if X is a Bernoulli random variable, then $X = X^2$ in every case, which implies that $E(X^2) = E(X) = p$. By Equation 5.7, this means that the variance is $Var(X) = p - p^2 = p(1 - p)$.

(b) A binomial random variable with parameters n and p is the sum of n independent Bernoulli random variables with parameter p. Because the variance of the sum of independent random variables is the sum of the variances, the variance of the binomial random variable is $Var(X) = np(1 - p)$. (This is much easier than calculating $E(X^2) = \sum_{x=0}^{n} x^2 \binom{n}{x} p^x (1 - p)^{n-x}$.)

(5) First, using Equation 5.8, notice that $\mathrm{Var}[(1/n)(X_1 + X_2 + \ldots + X_n)] = (1/n^2)$ $\mathrm{Var}(X_1 + X_2 + \ldots + X_n)$ because $1/n$ is a constant. We also know that because the Xs are independent, the variance of the sum is the sum of the individual variances (Equation 5.9), which is $n\sigma^2$. Thus, the variance is $\mathrm{Var}[(1/n)(X_1 + X_2 + \ldots + X_n)] = (1/n^2)n\sigma^2 = \sigma^2/n$. The standard deviation is the square root of the variance, or $\sigma/\sqrt{n}$. This is an important result in statistics. As we add observations to a sample, the sample mean becomes less and less variable, as long as we are taking observations from a distribution with finite variance.

(6) See github.com/mdedge/stfs for the steps.

Exercise Set 5-3

(1) (a) We have to find $\mathrm{E}(XY)$, $\mathrm{E}(X)$, and $\mathrm{E}(Y)$. We do this by summing over the joint mass function. We can do this with a table like the following:

Outcome	Probability	X	Y	XY
$X = -1, Y = 1$	1/3	-1	1	-1
$X = 1, Y = 1$	1/3	1	1	1
$X = 0, Y = -2$	1/3	0	-2	0
		$\mathrm{E}(X) = 0$	$\mathrm{E}(Y) = 0$	$\mathrm{E}(XY) = 0$

In the bottom row, we compute the expectations by taking an average weighted by the probabilities, which, in this case, are all equal. Thus, $\mathrm{Cov}(X, Y) = \mathrm{E}(XY) - \mathrm{E}(X)\mathrm{E}(Y) = 0$.

(b) $\mathrm{Cov}(X, Y) = 0$, but X and Y are not independent. For example, $\mathrm{P}(X = 0 \cap Y = -2) = 1/3 \neq \mathrm{P}(X = 0)\mathrm{P}(Y = -2) = 1/9$. You can compute similar examples for the other outcomes, but only one is needed to show that the two random variables are not independent.

(2) See github.com/mdedge/stfs for a proof that the covariance changes with linear scaling of the random variables being considered, but the correlation does not change with scaling.

(3) See github.com/mdedge/stfs for a proof.

Exercise Set 5-4

(2) See github.com/mdedge/stfs for commands that will let you explore the parameter set (1, 1). In this case, the normal distribution is an acceptable approximation for the distribution of means of samples of size 5, and it's a great approximation for samples of 10 and 50. The other values of the shape parameters suggest that for distributions that are more skewed or more U-shaped, larger samples are required, but for the parameter sets seen here, the means of samples of size 50 are well approximated by a normal distribution.

(3) See github.com/mdedge/stfs for code to run the simulations. With the parameters and sample size requested, the distribution of sample means is a good fit to the normal within about 2 standard deviations of the expectation. Beyond 2 standard deviations, though, the Pareto sample mean distribution has much heavier tails than the normal: extreme observations are much more likely than predicted. For example, there are

about 100 times as many observations beyond 5 standard deviations from the expectation as would be predicted by the normal distribution. Thus, with this distribution and $n = 1,000$, convergence in the center of the distribution is good, but convergence in the tails is poor. If the probability of an extreme event (such as, say, an earthquake of Richter magnitude >8) is important to know, then the central limit theorem can lead to spectacularly poor predictions.

Exercise Set 5-5

(1) From Equation 5.16, the correlation coefficient is $\rho_{X,Y} = \beta \sigma_X / \sigma_Y = \beta \sigma_X / \sqrt{\beta^2 \sigma_X^2 + \sigma_\epsilon^2}$. Squaring the correlation coefficient gives

$$\rho_{X,Y}^2 = \left(\frac{\beta \sigma_X}{\sqrt{\beta^2 \sigma_X^2 + \sigma_\epsilon^2}} \right)^2 = \frac{\beta^2 \sigma_X^2}{\beta^2 \sigma_X^2 + \sigma_\epsilon^2} = 1 - \frac{\sigma_\epsilon^2}{\beta^2 \sigma_X^2 + \sigma_\epsilon^2} = 1 - \frac{\mathrm{Var}(Y|X = x)}{\mathrm{Var}(Y)}.$$

If the relationship between X and Y is linear—as it is in the model here—then the square of the correlation coefficient of X and Y equals one minus the proportion of the variance of Y that remains after conditioning on X. This is why people sometimes refer to the squared correlation coefficient as "the proportion of variance explained." This phrase is justified when the relationship between X and Y has the properties of the model we developed in the previous section.

(2) When one repeatedly simulates with the same parameters, the resulting data vary. Making sigma.disturb larger both decreases the apparent strength of the relationship between x and y and increases the extent to which the results of different simulations vary. Varying a changes the numbers on the y axis but little else. Varying b changes the strength and direction of the relationship between x and y: large absolute values give apparently stronger relationships, and changing the value from positive to negative changes the direction of the relationship. Changing sigma.x changes the spread of the observations on both the x and y axes.

Exercise Set 6-1

(1) The expectation of the sample mean is

$$\mathrm{E}[\hat{\theta}_n(D)] = \mathrm{E}\left(\frac{1}{n} \sum_{i=1}^{n} X_i \right) = \frac{1}{n} \sum_{i=1}^{n} \mathrm{E}(X_i) = \mathrm{E}(X_1) = \theta.$$

The first step follows from applying the expectation operator to the definition in Equation 6.1. The second step comes from the linearity of expectation (Equation 5.4). The third step comes from the fact that all the X_i have the same expectation, and the fourth step comes from the fact that the first parameter of a normal distribution is equal to its expectation (Equation 5.21). Because the expectation of the estimator is equal to the quantity we are trying to estimate, the estimator has a bias of zero—we say that it is "unbiased."

(2) See github.com/mdedge/stfs for code. You should see that the histogram of the sample median is centered around θ and that the mean of the sample medians is very close to θ. The results correctly suggest that the sample median is an unbiased estimate of θ when

the data are independent samples from a Normal(θ, 1) distribution. These results do not constitute proof, but they do suggest what turns out to be the right answer in this case.

Exercise Set 6-2

(1) We have seen this before (Exercise Set 5-2, Problem 5). Because each observation is independent, it follows from Equations 5.8 and 5.9 that $\text{Var}(\widehat{\theta}_n) = 1/n$, remembering that the variance of each X_i is 1 under the model. This argument appeals only to properties of the variance and to the fact that the observations are independent and identically distributed. We did not use the normality assumption. Thus, if the data $X_1, X_2, \ldots, X_n$ are independent and identically distributed with $\text{Var}(X_1) = \text{Var}(X_2) = \ldots = \text{Var}(X_n) = \sigma^2$, then the variance of the sample mean as an estimator of the expectation is $\text{Var}(\widehat{\theta}_n) = \sigma^2/n$.

(2) See github.com/mdedge/stfs for code. When I ran the simulation, I found that for samples of size 25, the variance of the sample mean was 0.04, whereas the variance of the sample median was 0.06. You can also use `boxplot()` to see that the sample median is less precise. Try it with different sample sizes by changing the n argument in the `mat.samps()` command. You should see that though the variance of both the sample mean and the sample median decrease when the size of each sample increases, the sample median has a larger variance—is less precise—than the sample mean.

Exercise Set 6-3

(1) For notational compactness, we write $\widehat{\theta}_n(D)$ as $\widehat{\theta}_n$, remembering implicitly that estimators are functions applied to random data. Starting with the definition of mean squared error given in Equation 6.4, then expanding the squared term in the definition and applying linearity of expectation (Equation 5.4) gives

$$\text{MSE}(\widehat{\theta}_n) = \text{E}[(\widehat{\theta}_n - \theta)^2] = \text{E}(\widehat{\theta}_n{}^2 - 2\theta\widehat{\theta}_n + \theta^2) = \text{E}(\widehat{\theta}_n{}^2) - 2\theta\text{E}(\widehat{\theta}_n) + \theta^2.$$

By the identity for the variance given in Equation 5.7, the first term is $\text{E}(\widehat{\theta}_n{}^2) = \text{Var}(\widehat{\theta}_n{}^2) + [\text{E}(\widehat{\theta}_n)]^2$, letting us write $\text{MSE}(\widehat{\theta}_n) = \text{Var}(\widehat{\theta}_n{}^2) + [\text{E}(\widehat{\theta}_n)]^2 - 2\theta\text{E}(\widehat{\theta}_n) + \theta^2$. Noticing that $[\text{E}(\widehat{\theta}_n)]^2 - 2\theta\text{E}(\widehat{\theta}_n) + \theta^2 = [\text{E}(\widehat{\theta}_n) - \theta]^2 = \text{B}(\widehat{\theta}_n)^2$ completes the proof: $\text{MSE}(\widehat{\theta}_n) = \text{B}(\widehat{\theta}_n)^2 + \text{Var}(\widehat{\theta}_n)$.

(2) We have already seen that the sample mean and sample median are unbiased estimators of the first parameter of a normal distribution (Exercise Set 6-1, Problem 2). Because the bias of each estimator is zero, the mean squared error of each estimator is equal to its variance by Equation 6.5. We saw that the mean of a sample of normally distributed data has lower variance than the median of a sample of normally distributed data (Exercise Set 6-2, Problem 2). Thus, the sample mean has lower mean squared error than the sample median as an estimator of the first parameter of a normal distribution.

Exercise Set 6-4

(1) The sample mean is a consistent estimator of the expectation of a random variable's distribution, regardless of the distribution family of the random variable, assuming that the expectation exists. This follows from the weak law of large numbers (Equation 5.3)—just note that $\overline{X}_n$ is the estimator and μ is the quantity being estimated,

and Equation 5.3 is equivalent to Equation 6.6. The sample mean is also a consistent estimator of the first parameter of a normal distribution because if $X \sim \text{Normal}(\theta, 1)$, then $E(X) = \theta$ (Equation 5.21).

(2) See github.com/mdedge/stfs for code. You should see that the variance of the sample median decreases as the size of the sample increases. We cannot prove it rigorously by simulation, but the variance of the sample median continues to approach zero as the sample size increases.

(3) (a) The sample mean $\bar{X} = (1/n)\sum_{i=1}^{n}X_i$ is both unbiased and consistent. We proved that it is unbiased in Problem 1 of Exercise Set 6-1, and we have just proved that it is consistent in Problem 1 above.

(b) The shifted sample mean is biased. $E[(1/n)\sum_{i=1}^{n}X_i + 1] = \theta + 1$. The shifted sample mean is also inconsistent. The shifted sample mean converges in probability to $\theta + 1$, it does not converge to θ, and therefore it is inconsistent.

(c) The first observation is unbiased; $E(X_1) = \theta$. This follows from Equation 5.21. However, the first observation is not consistent. No matter how large the sample gets, the first observation has variance 1. If we draw a large sample, this estimator still uses only the first observation.

(d) The "shrunk" sample mean is biased; $E\{[1/(n+1)]\sum_{i=1}^{n}X_i\} = [n/(n+1)]\theta$, which implies that $B\{[1/(n+1)]\sum_{i=1}^{n}X_i\} = [n/(n+1)] - 1]\theta = -\theta/(n+1)$. At the same time, the shrunk sample mean is consistent. Invoking the assumption that the true variance of each observation is 1, the variance of the shrunk sample mean is $\text{Var}\{[1/(n+1)]\sum_{i=1}^{n}X_i\} = n/(n+1)^2$. Using Equation 6.5, the mean squared error of the shrunk sample mean is then $(\theta^2 + n)/(n+1)^2$. As the sample size n increases, the denominator becomes much larger than the numerator, and so the mean squared error approaches zero, which, by Equation 6.7, implies that the shrunk sample mean is a consistent estimator of θ. Together, parts (a)–(d) show that it is possible for an estimator to be unbiased and consistent, biased and inconsistent, unbiased and inconsistent, or biased and consistent.

(4) See github.com/mdedge/stfs for a proof.

Exercise Set 6-5

(1) (a) See github.com/mdedge/stfs for some R code that estimates the relative efficiency of the sample mean and median as an estimator of the first parameter of a normal distribution using a sample of five observations. When I ran the code, I obtained a relative efficiency of 1.4. In this situation, the sample mean is a more efficient estimator than the sample median is.

(b) See github.com/mdedge/stfs for code. The relative efficiency appears to level off between 1.5 and 1.6 as the sample size increases. This agrees with theoretical results—a little math (beyond our scope) shows that the true asymptotic relative efficiency is $\pi/2 \approx 1.57$.

(2) See github.com/mdedge/stfs for R code. You'll see that things have reversed: if the data are Laplace distributed, then the median is actually a more efficient/lower-variance estimator than the mean is, particularly for large samples. The point is that efficiency is not a property of a statistic; it is a property of an estimator under a model. If the model changes, then the relative efficiency of estimators may also change.

Exercise Sets 6-6 and 6-7

See github.com/mdedge/stfs for solutions to these exercise sets.

Exercise Set 6-8

(1) See github.com/mdedge/stfs for code. When γ and Λ are large, both the median and the mean are biased upward, and they both increase in variance. The median, however, is much less affected by the aberrant observations than the mean is. This exercise demonstrates the median's robustness against outliers. One way to formalize this property is to define a statistic's *breakdown point*—roughly, the proportion of observations from a contaminating distribution required to make the statistic perform arbitrarily badly. The mean has a breakdown point of zero: in principle, a single observation from a different distribution can mess up the sample mean as much as you want, provided that the contaminating distribution is far enough removed from the target. In contrast, the median has a breakdown point of $1/2$, which is the maximum possible. As long as more than 50% of the data come from the distribution of interest, the sample median will at least be within the range of observations drawn from the correct distribution.

Exercise Set 6-9

(1) (a) In this scenario, the least-squares estimators are unbiased and consistent. (You will have a chance to prove this in some optional exercises in Chapter 9.) At each sample size, the means of the estimates are close to the true values, and the variances of the estimates decrease as the sample size increases.

(b) The results for the least-absolute-errors estimators are similar to those in part (a), though the variances are somewhat larger.

(c) In this scenario (normally distributed disturbances of constant variance), the least-squares estimators are more efficient than the least-absolute-errors estimators: both sets of estimators are unbiased, and the variances of the least-squares estimators are smaller than those of the least-absolute-error estimators at each sample size.

(2) (a) The cloud of observations has a different vertical dispersion pattern when the disturbances are Laplace distributed, but the effect is too subtle to detect reliably just by looking.

(b) Both sets of estimators appear to be approximately unbiased, and the simulations suggest that they may be consistent. However, the relative efficiency is reversed: now the least-squares estimators are less efficient than the least-absolute-errors estimators.

(3) (a) The command shown plots a cloud of points centered on a line (not drawn) with intercept 3 and slope $1/2$. In some trials, there are some points in the lower-right corner that are far removed from the rest of the data. These are outliers both in the sense of being removed from the rest of the data and from being created by a different process.

(b) With outliers, neither set of estimators is unbiased or consistent. Both tend to produce slope estimates that are too low: the line is being "pulled down" by the outlying points in the lower right. However, the least-absolute-errors estimators are much more robust than the least-squares estimators: they are closer to the true values on average and have lower variance.

(4) (a, b) In this case, $E(\widetilde{\beta}) = \beta + \gamma\rho$, where ρ is the correlation of X and Z. This means that if either Z has no effect on Y ($\gamma = 0$) or X and Z are uncorrelated ($\rho = 0$), then $\widetilde{\beta}$ is unbiased. Otherwise, if the data analysis ignores Z, then the estimate of β is biased, and the size and direction of the bias depend on γ and ρ. This problem is *not* specific to least-squares estimators. Econometricians call this "omitted variable bias." In words, the problem is that part of the causal effect of Z on Y is being wrongly attributed to X.

(c) Confounding is certainly a problem if we want to interpret the estimate of β causally —for example, if we would like to make claims like, "increasing X by one causes a value of Y that is β units larger." $\widetilde{\beta}$ retains its interpretation as the slope of the least-squares line, and it may be useful for prediction if the data used to fit the line are representative of the prediction context. But it will mislead us if we want to manipulate X to produce a desired change in Y. This problem remains even in the largest samples. You may have been taught that randomized experiments are the best way to infer causation. The formula $E(\widetilde{\beta}) = \beta + \gamma\rho$ is a way to explain this claim. Consider the running agricultural example. We expect that cereal yield would be a function of fertilizer application, but it might also depend on many other factors: latitude, crops grown, irrigation, availability of labor, etc. These can cause omitted variable bias if we fail to account for them. In an experiment, the treatment (X) is assigned randomly to the units being studied. For example, if we wanted to do a country-level experiment on fertilizer yield, then we would randomly decide how much fertilizer each country ought to apply to its fields. If we assign the levels of fertilizer use randomly, then we ensure that the confounds are each uncorrelated with the level of fertilizer use—that is, we set $\rho = 0$. If $\rho = 0$ for all possible confounds, then $E(\widetilde{\beta}) = \beta$, and we can get unbiased estimates of β—the causal effect of X on Y—without accounting for confounds explicitly. There are approaches for inferring causation from observational data—that is, data that do not result from a randomized experiment—and we will say more about the problem in the Postlude.

Exercise Set 7-1

(1) (a) The standard deviation of the observations is σ. This comes from the fact that the second parameter of the normal distribution is equal to the distribution's variance (Equation 5.22).

(b) The standard error of the estimate is $\text{SE}(\widehat{\theta}_n) = \sigma/\sqrt{n}$. Recall from Exercise Set 6-2, Problem 1, that $\text{Var}(\widehat{\theta}_n) = \sigma^2/n$. The standard error follows immediately from this fact and Equation 7.1.

(c) Plugging in the numbers in the problem gives $\text{SE}(\widehat{\theta}_n) = 1/5 = 0.2$.

(d) Use the `mat.samps()` function to draw a set of samples:

```
> s.mat <- mat.samps(n = 25, nsim =10000)
```

Calculate the median of each sample using `apply()` (or using the `for()` loop given in Exercise Set 6-1, Problem 2):

```
> ests.median <- apply(s.mat, 1, median)
```

Use the sd() function to estimate the standard error of the sample median, as in sd(ests.median). You will get an answer of approximately 0.25, which is larger than the standard error of the sample mean.

(2) (a) We seek the probability $P(\hat{\theta}_n - \omega < \theta < \hat{\theta}_n + \omega)$. θ is fixed, and in this problem so is ω, so this is really a probability statement about $\hat{\theta}_n$, the only random variable in the expression. The statement $\hat{\theta}_n - \omega < \theta < \hat{\theta}_n + \omega$ is equivalent to $\theta - \omega < \hat{\theta}_n < \theta + \omega$, so $P(\hat{\theta}_n - \omega < \theta < \hat{\theta}_n + \omega) = P(\theta - \omega < \hat{\theta}_n < \theta + \omega) = P(\hat{\theta}_n < \theta + \omega) - P(\hat{\theta}_n < \theta - \omega)$. Recalling that normally distributed random variables are continuous, we know that $P(\hat{\theta}_n < \theta + \omega) = P(\hat{\theta}_n \le \theta + \omega)$, and therefore, if the cumulative distribution function of $\hat{\theta}_n$ is written $F_{\hat{\theta}_n}$, then the probability we seek is $F_{\hat{\theta}_n}(\theta + \omega) - F_{\hat{\theta}_n}(\theta - \omega)$. Because $\hat{\theta}_n$ is normally distributed with expectation θ and standard deviation ω, this is the probability that a normally distributed random variable falls within 1 standard deviation of its mean. You can look up this probability in a table or use R's pnorm() function, which evaluates the cumulative distribution function of the normal distribution, as in pnorm(1)-pnorm(-1), which returns 0.683. Thus, the interval $(\hat{\theta}_n - \omega, \hat{\theta}_n + \omega)$ would contain θ about 68% of the time. There is a subtle but important point here, which will be emphasized in the next section: the probability statement is about the *interval*, which is random, and not about θ, which is fixed.

(b) Reasoning along similar lines as in part (a), we evaluate the necessary probability in R with pnorm(2)-pnorm(-2), which returns .954. Thus, the interval $(\hat{\theta}_n - 2\omega, \hat{\theta}_n + 2\omega)$ will contain θ about 95% of the time.

Exercise Set 7-2

(1) (a) $z_{\alpha/2} \approx 0.674$, which you can verify with the R command qnorm(0.75). See github.com/mdedge/stfs for R code that plots an appropriate picture.

(b) By Equation 7.3, the confidence interval was constructed as $(\hat{\theta} - \omega z_{\alpha/2}, \hat{\theta} + \omega z_{\alpha/2})$. Thus, $\hat{\theta}$ is at the midpoint between the boundaries of the interval—in this case, $\hat{\theta} = 3$.

(c) The lower bound of the interval is 2, and $\hat{\theta} = 3$, giving the equation $3 - \omega z_{\alpha/2} = 2$, which implies that $\omega z_{\alpha/2} = 1$. By part a, $z_{\alpha/2} \approx 0.674$, and it follows that the standard error is $\omega = 1/z_{\alpha/2} \approx 1/0.674 \approx 1.48$.

(d) For a 95% confidence interval, we need $z_{0.025}$, which is approximately 1.96. The 95% confidence interval is then $(\hat{\theta} - \omega z_{\alpha/2}, \hat{\theta} + \omega z_{\alpha/2}) \approx (3 - 1.48 * 1.96, 3 + 1.48 * 1.96) \approx (0.09, 5.91)$. This is a larger range than that covered by the 50% confidence interval.

(e) $3/1.48 \approx 2.02$. The estimate was about two standard errors away from zero.

(f) We need the value of α that solves the equation $\hat{\theta} - \omega z_{\alpha/2} = 3 - 1.48 * z_{\alpha/2} = 0$. That is, we need the value of α that gives $z_{\alpha/2} = 3/1.48 \approx 2.02$. The necessary value of α is thus one minus the probability that a random variable drawn from a normal distribution will fall within 2.02 standard deviations of its expectation. The probability that such a random variable will fall more than 2.02 standard deviations *above* its expectation is $1 - \Phi(2.02)$, or in R, 1-pnorm(2.02). Because the normal distribution is symmetric, the probability that such a random variable will fall more than 2.02 standard deviations *above* its expectation *or* more than 2.02 standard deviations *below* its expectation is therefore double this quantity, or approximately 0.043. This is the necessary value of α.

(g) If $\theta = 0$, then the probability of observing $|\hat{\theta}| > 3$ is the probability of observing a normal random variable more than $(\hat{\theta} - \theta)/\omega = 3/1.48 = 2.02$ standard deviations away from its expectation. By the reasoning in the solution to part (f), this probability is 0.043.

Exercise Set 7-3

(1) There are many possible responses. Let's divide the responses into two: first, are there reasons why, if the hypothesis were true, you would not be convinced that the theory is true? Second, are there reasons why, if the hypothesis were false, you would not be convinced that the theory is false? First, suppose that I had found that Arizonan women were shorter than other women from the US. There are many reasons not to take such a finding as evidence in favor of the theory, which posits cause-and-effect relationships between the weather, one's size, and one's decisions about where to live. The most important is *confounding*. Arizona has many properties besides hot weather—for example, it is dry, it borders Mexico, it contains most of the Navajo Nation, it is a popular destination for retirees, and it has been a center of copper mining. When one chooses to study Arizonans, one does not get a set of people who live in a place that is hotter than average and otherwise like everywhere else. One instead gets a set of people who live in a place with a slew of unusual properties. Any difference between Arizonans and other people could in principle—though perhaps not always plausibly—be due to any of these additional properties.

Second, a failure to find a significant difference between the heights of Arizonan women and other women does not kill the theory. For one thing, I only sampled 25 women. As we will see soon, a sample of 25 is not large enough to detect small or moderate differences between population means, and even if the theory were true, the proposed effect on height would probably be subtle. The assumption that Arizona is a good test case for the effects of a hot climate is crude. Though the major population centers are hot, the northern part of the state is at high altitude and experiences reasonably cold winters. Sampling from across the state thus dilutes the effects of heat, possibly masking effects that would be apparent if only the hottest parts of the state were sampled. Using height—an imperfect proxy of body size—as the measured variable has the same effect. Confounding can also mask true effects in addition to generating spurious ones. It may be that smaller people are genuinely attracted to hotter climates but that some other feature of Arizona has attracted tall people, masking the effect. This list of problems is not exhaustive but gives a sense of the many difficulties facing empirical researchers.

(2) (a) The standard error is the standard deviation divided by the square root of the sample size, or, in this case, 1 mm.

  (b) From Equation 7.3, a confidence interval is $(\hat{\theta} - \omega z_{\alpha/2}, \hat{\theta} + \omega z_{\alpha/2})$, where $\hat{\theta}$ is the estimator, ω is the standard error, and $z_{\alpha/2} = \Phi^{-1}(1 - a/2)$, the $(1 - \alpha/2)$th quantile of the standard normal distribution. For a 95% confidence interval, $z_{\alpha/2}$ can be found using the R command qnorm(0.975), and it is equal to 1.96. Thus, because $\omega = 1$, a 95% confidence interval is given by $(\bar{x} - 1.96, \bar{x} + 1.96)$, where $\bar{x}$ is the sample mean.

  (c) We start by finding the values of $\bar{x}$ that would give two-sided $p = 0.05$. The lower of these values is the same as would give a one-sided p of 0.025. Thus, we want to find the value of $\bar{x}$ that satisfies the equation $\Phi_{100,1}(\bar{x}) = 0.025$, where $\Phi_{100,1}$ is the cumulative distribution function of the Normal$(100, 1)$ distribution, or, equivalently, $P(\overline{X} \leq \bar{x}|\mu = 100) = 0.025$. That is, we need $\Phi_{100,1}^{-1}(0.025) = \bar{x}$. In R, we use qnorm

(0.025, mean = 100, sd = 1) to get $\bar{x} = 98.04$, or $\bar{x} = 100 - 1.96$. By symmetry, the two-sided p is 0.05 if $\bar{x} = 100 + 1.96$. Thus, the two-sided p of a test of the null hypothesis that $\mu = 100$ will be less than 0.05 if either $\bar{x} < 98.04$ or $\bar{x} > 101.96$. By the result of part (b), these are exactly the cases in which a 95% confidence interval for μ excludes 100.

(d) See github.com/mdedge/stfs for R code. The distribution of the p values is approximately uniform. The proportion of p values less than 0.05 should be approximately 0.05. Similarly, the proportion of p values less than 0.10 should be about 0.10. This is a good result—it means that when the null hypothesis is true, the test works approximately as advertised.

(e) See github.com/mdedge/stfs for code to simulate normal samples of size 4 from a Normal(101, 4) distribution and test the null hypothesis that $\mu = 100$. The distribution of p values is no longer uniform: it has a concentration of low p values, representing samples that would be unlikely to be drawn if μ were in fact 100. About 17% of the p values are less than 0.05 and about 26% are less than 0.10.

(f) See github.com/mdedge/stfs for code to simulate normal samples of size 4 from a Normal(102, 4) distribution and test the null hypothesis that $\mu = 100$. Again, the distribution of p values shows a concentration of low values, even more pronounced than in part (e). About 51% of the p values are less than 0.05 and about 64% are less than 0.10.

(g) See github.com/mdedge/stfs for code to simulate normal samples of size 16 from a Normal(101, 4) distribution and test the null hypothesis that $\mu = 100$. Again, the distribution of p values shows a concentration of low values, even more pronounced than in part (e). About 51% of the p values are less than 0.05 and about 64% are less than 0.10. These are the same values as in part (f): doubling the difference between the true mean and the mean under the null hypothesis had the same effect on the distribution of p values as quadrupling the sample size. There is a good reason for this: both changes have the effect of doubling the number of standard errors separating the true parameter from the value postulated by the null hypothesis.

Exercise Set 7-4

(1) (a) The proportion of the time that the null hypothesis is true is $\tau = (t_n + f_p)/(t_n + f_p + f_n + t_p) = t_n + f_p$. The simplification comes from the fact that the denominator $t_n + f_p + f_n + t_p = 1$.

(b) The proportion of the time the null hypothesis is false is $\varphi = f_n + t_p$.

(c) The power of the test is $\pi = P(R|H_0^C) = t_p/(t_p + f_n)$. By part (b), this quantity equals t_p/φ.

(d) See github.com/mdedge/stfs for the proof that $P(H_0|R) = \alpha\tau/(\alpha\tau + \pi\varphi)$. Because τ, α, and π are all between 0 and 1, the false discovery rate decreases as the power of the test, π, increases. This is one reason for valuing tests with high power.

(e) In the notation of the table, the negative predictive value is $P(H_0|R^C) = t_n/(t_n + f_n)$. Using reasoning parallel to that used in part (d), the negative predictive value is equal to $P(H_0|R^C) = [(1 - \alpha)\tau]/[(1 - \alpha)\tau + (1 - \pi)\varphi]$. Because τ, α, and π are all between 0 and 1, the negative predictive value increases as the power of the test, π, increases.

Exercise Set 7-5

(1) (a) If the research group adopted the proposed procedure, testing 7 variables and rejecting the null hypothesis if any one of the tests is significant, then they would falsely reject the null hypothesis about 18% of the time, which is more than three times the nominal level of each of their tests. This is true even though each of the individual tests rejects the null hypothesis at the correct rate. See github.com/mdedge/stfs for code.

   (b) Other things being equal, increasing the number of tested variables increases the probability that at least one of the tests leads to an incorrect rejection of the null hypothesis, also called the familywise error rate. Increasing the degree of correlation between the measurements tends to decrease the probability that at least one of the tests leads to a rejection of the null hypothesis. One way to control the familywise error rate is with Bonferroni correction, in which each p value is compared with the value α/k, where α is the desired familywise error rate and k is the number of hypothesis tests being conducted.

(2) Checking as the size of each group grows to 50 leads to a rejection 11–12% of the time, and checking until the size of each group grows to 100 leads to false rejection almost 20% of the time. It grows worse with more repeated testing. See github.com/mdedge/stfs for code.

Exercise Set 7-6

See github.com/mdedge/stfs.

Exercise Set 8-1

(1) (a) As the sample size n increases, the empirical distribution function matches the true cumulative distribution function more closely. For large sample sizes, the empirical distribution function and true cumulative distribution function are not visibly distinguishable.

   (b) See github.com/mdedge/stfs for code to make a similar plot with the exponential distribution and Poisson distribution. The Poisson distribution looks different because it is discrete, so its true cumulative distribution function looks like a step function.

(2) See github.com/mdedge/stfs.

Exercise Set 8-2

(1) (a) There are two expressions. Using the identity in Equation 5.7, the plug-in estimator is

$$\tilde{\sigma}^2 = \frac{1}{n}\sum_{i=1}^{n} X_i^2 + \left(\frac{1}{n}\sum_{i=1}^{n} X_i\right)^2.$$

Equivalently, we can use the definition of the variance (Equation 5.6) directly to get

$$\tilde{\sigma}^2 = \frac{1}{n}\sum_{i=1}^{n}(X_i - \overline{X})^2, \text{ where } \overline{X} = \frac{1}{n}\sum_{i=1}^{n}X_i.$$

These two expressions are equivalent. See github.com/mdedge/stfs for a proof.

(b) The plug-in estimator of the standard deviation is the square root of the plug-in estimator of the variance.

(c) As with the variance, there are two equivalent expressions for the plug-in estimator of the covariance, depending on which of the expressions in Equation 5.15 is used as a basis. The first is

$$\widetilde{\sigma_{X,Y}} = \frac{1}{n}\sum_{i=1}^{n}(X_i - \overline{X})(Y_i - \overline{Y}),$$

where

$$\overline{X} = \frac{1}{n}\sum_{i=1}^{n}X_i \text{ and } \overline{Y} = \frac{1}{n}\sum_{i=1}^{n}Y_i.$$

The second expression is

$$\widetilde{\sigma_{X,Y}} = \frac{1}{n}\sum_{i=1}^{n}X_iY_i - \overline{X}\,\overline{Y}.$$

The proof that these two expressions are equal is similar to the proof that the two expressions in part (a) are equal.

(d) The correlation is the covariance scaled by the product of the standard deviations of the two variables. We can thus construct expressions for the plug-in estimator of the correlation—often denoted r—from the plug-in estimators of the covariance and standard deviation. For example,

$$r_{X,Y} = \frac{\sum_{i=1}^{n}(X_i - \overline{X})(Y_i - \overline{Y})}{\sqrt{\sum_{i=1}^{n}(X_i - \overline{X})^2}\sqrt{\sum_{i=1}^{n}(Y_i - \overline{Y})^2}}.$$

This is one of several equivalent expressions, with other versions using different equivalent forms of the plug-in estimators of the covariance and standard deviation.

(2) (a) See github.com/mdedge/stfs for code, which gives answers very close to 0.8, not 1.

(b) Your simulated answers will be close to those in the following table:

n	2	3	4	5	6	7	8	9	10
Plug-in	1/2	2/3	3/4	4/5	5/6	6/7	7/8	8/9	9/10

If σ^2 is the true variance and $\tilde{\sigma}_n^2$ is the plug-in estimator of the variance using a sample of n independent observations, then $E(\tilde{\sigma}_n^2) = [(n-1)/n]\sigma^2$. Thus, $\tilde{\sigma}_n^2$ is biased downward, especially for small sample sizes. We can obtain an unbiased estimator of σ^2 by multiplying $\tilde{\sigma}_n^2$ by $n/(n-1)$, making it slightly larger to correct its downward bias. This yields what is called the "sample variance," $s^2 = \sum_{i=1}^{n}(X_i - \overline{X})^2/(n-1)$, where $\overline{X}$ is the sample mean. The var() function in R computes the sample variance. The "sample standard deviation" is the square root of the sample variance. Though the sample *variance* is unbiased, the sample *standard deviation* is biased slightly downward, but less biased than the plug-in estimator of the standard deviation.

(And neither version is robust—the sample standard deviation, because it relies on the sum of squared deviations from the mean—is delicately sensitive to outlying points.)

(c) See github.com/mdedge/stfs for one approach to the proof.

(3) See github.com/mdedge/stfs for code that estimates the moments using the sample moments. The true values are $E(X^4) = 3$, $E(X^5) = 0$, $E(X^6) = 15$, $E(X^7) = 0$, and $E(X^8) = 105$.

Exercise Set 8-3

(1) From Exercise Set 5-1, Problem 1(d), the expectation of X if X has a continuous Uniform(a, b) distribution is $E(X) = (a + b)/2$. Here, $a = 0$, so $E(X) = b/2$. Thus, $b = 2E(X)$, and the method-of-moments estimator is $\tilde{b} = (2/n)\sum_{i=1}^{n} X_i$.

(2) See github.com/mdedge/stfs.

Exercise Set 8-4

(1) Running `wrap.bm()` shows that n and B have to be reasonably large in order for the bootstrap distribution of the sample mean to approximate the true distribution of the sample mean. When n is 5, the bootstrap distribution does not look normal, and its mean and standard deviation vary widely around the true value, even when B is large. Similarly, when B is 10, the bootstrap distribution is a poor approximation of the true distribution, even if n is large. For this problem, the bootstrap distribution of the sample mean is a reasonable approximation of the true distribution of the sample mean when $n \geq 20$ and $B \geq 1,000$. The values of n and B that lead to useful answers vary in different settings. This setting, that of estimating the distribution of the mean of a normal sample, is less demanding than many problems one sees in practice.

(2) The bootstrap standard error is imprecise for the midrange with the values of n and B specified in Problem 1. Even increasing n to 1,000 still gives variable standard errors. The midrange is sensitive to small changes in the data because it is a function of two values—the maximum and the minimum of the sample—that can change a lot if the sample changes slightly.

(3) Imagine that we have a distribution function $F_X(x)$. Because it is a distribution function, $F_X(x)$ is monotonically increasing with x, its minimum value is zero, and its maximum value is 1. Graphically, drawing one observation from a distribution function is equivalent to the following procedure:

(i) Sample a Uniform$(0, 1)$ random number $Y = y$

(ii) Draw a horizontal line with height equal to y

(iii) Where the horizontal line with height y intersects $F_X(x)$, draw a vertical line down to the horizontal axis.

(iv) The value on the horizontal axis immediately below the intersection of $F_X(x)$ and the horizontal line with height y is an observation from the distribution function $F_X(x)$.

To get more independent samples from $F_X(x)$, one repeats this procedure with independent Uniform$(0, 1)$ random numbers.

Suppose we construct an empirical distribution function for a set of observations $x_1, x_2, \ldots, x_n$. The empirical distribution function is horizontal everywhere except at points with x coordinates equal to the observations $x_1, x_2, \ldots, x_n$, where there will be vertical line segments of length $1/n$. (If there are sets of observations with equal value among $x_1, x_2, \ldots, x_n$, then some of the vertical sections with have length k/n, where k is the number of observations that share the same value.) This means that, for every uniform random number $Y = y$ drawn in step (i), there is a $1/n$ probability that the horizontal line at height y intersects the empirical distribution function in the vertical strip associated with each observation. If the horizontal line at height y intersects the empirical distribution function in the vertical strip associated with a given observation, then a value equal to that observation is added to the bootstrap sample. Thus, each time we sample from the empirical distribution function, every observation from the original dataset has a $1/n$ probability of being copied into the bootstrap sample, regardless of whether it has been chosen before. This is equivalent to sampling from the original sample with replacement.

Exercise Set 8-5

(1) The first step is to choose a test statistic. One sensible choice is the difference in mean wheat yield between the fields that received substance Z and the fields that did not. We might also choose the difference in median wheat yields, or something else. The next step is to identify a way of permuting the data. Suppose that Y represents a field's wheat yield. Further assume that $Z = 1$ if a field is randomly assigned to receive substance Z and that $Z = 0$ if the field is randomly assigned not to receive it. Then one model for the data is $Y_i = \alpha + \beta Z_i + \epsilon_i$, where α and β are constant Xs and ϵ is a random variable with $E(\epsilon) = 0$. The subscript i identifies a particular field, and all fields are assumed to be independent of each other, meaning that the ϵ_i terms are independent. This is the model we developed in the main text, with Z in place of X. We can therefore test it in a similar way, shuffling the labels Z_i independently of the wheat yields Y_i. If $\beta = 0$, then every permutation leads to a hypothetical dataset that is exactly as probable as the original data. For every permutation we try, we compute the mean difference in wheat yield between fields associated with $Z = 1$ and fields associated with a label $Z = 0$. These differences form a permutation distribution, to which we compare the original mean difference observed.

  As to the hypothesis being tested: one is tempted to claim that the null hypothesis is that substance Z does not affect the expected yield. But we are really testing the null hypothesis that substance Z does not affect the entire *distribution* of the wheat yield— that is, that $\beta = 0$ *and* that nothing about the distribution of ϵ_i depends on Z_i.

(2) To call `sim.perm.B()`, assess the rate at which the null hypothesis is rejected with a significance level of 0.05, and plot a histogram of the permutation p values for $n = 10$ and $\beta = 0$, use the following commands:

```
ps <- sim.perm.B(0, 0, n = 10)
mean(ps < 0.05)
hist(ps)
```

When I ran these simulations, I arrived at the following results. Your exact results may differ slightly.

	$n = 10$	$n = 50$	$n = 100$
$\beta = 0$	.034	0.046	0.052
$\beta = 0.1$	0.086	0.300	0.486
$\beta = 0.2$	0.182	0.760	0.978

The top row of the table, in which $\beta = 0$, is encouraging. When we set the significance level to 0.05, the type I error rate is in fact approximately 0.05. In the next two rows, we can see, unsurprisingly, that as the sample size increases (left to right), and as the effect size increases ($\beta = 0.2$ vs. $\beta = 0.1$), the power of the permutation test increases.

Exercise Set 9-1

(1) This statement is false. In frequentist statistics, θ is not random, so we cannot make probability statements about it. In Bayesian statistics, the value of θ that maximizes $L(\theta)$ is not generally the most probable value of θ given the data, but sometimes it is (see Chapter 10). A better statement is *"The value of θ that maximizes $L(\theta)$ is the one that maximizes the probability of obtaining the observed data."* This statement is correct for discrete random variables; for continuous random variables, it could be modified to *"The value of θ that maximizes $L(\theta)$ is the one that maximizes the joint probability density associated with the observed data."*

(2) (a) The density is $f_X(x) = [1/(\sigma\sqrt{2\pi})]e^{-(x-\mu)^2/(2\sigma^2)}$.

(b) The log-likelihood is the log of the density, $l(\mu) = \ln\{[1/(\sigma\sqrt{2\pi})]e^{-(x-\mu)^2/(2\sigma^2)}\}$. This expression can by simplified. First, notice that it is a product and that the log of a product is the sum of the logs of the terms being multiplied: $l(\mu) = \ln[1/(\sigma\sqrt{2\pi})] + \ln(e^{-(x-\mu)^2/(2\sigma^2)})$. The second term contains the log of an exponent. Raising e to a power and taking a natural log are inverse operations, so

$$l(\mu) = \ln\left(\frac{1}{\sigma\sqrt{2\pi}}\right) - \frac{(x-\mu)^2}{2\sigma^2}.$$

We could simplify the first term further, but this is enough for us.

(c) Because the two random variables are independent, the joint density function is the product of the two marginal density functions: $f_{X_1,X_2}(x_1,x_2) = \frac{1}{\sigma\sqrt{2\pi}}e^{-\frac{(x_1-\mu)^2}{2\sigma^2}} *$
$\frac{1}{\sigma\sqrt{2\pi}}e^{-\frac{(x_2-\mu)^2}{2\sigma^2}} = \frac{1}{\sigma^2 2\pi}e^{-\frac{(x_1-\mu)^2+(x_2-\mu)^2}{2\sigma^2}}$.

(d) The log-likelihood for two observations is the log of the density of two observations. We could either take the log of the expression in part (c) directly, or we could start with part (b), remembering that the log of a product is the sum of the logs of the terms being multiplied. Either way, we obtain

$$l(\mu) = 2\ln\left(\frac{1}{\sigma\sqrt{2\pi}}\right) - \frac{(x_1-\mu)^2}{2\sigma^2} - \frac{(x_2-\mu)^2}{2\sigma^2}.$$

(e) Extending the result in part (d) gives

$$l(\mu) = n\ln\left(\frac{1}{\sigma\sqrt{2\pi}}\right) - \sum_{i=1}^{n}\frac{(x_i-\mu)^2}{2\sigma^2} = n\ln\left(\frac{1}{\sigma\sqrt{2\pi}}\right) - \frac{1}{2\sigma^2}\sum_{i=1}^{n}(x_i-\mu)^2.$$

Exercise Set 9-2

(1) (a) The likelihood function is equal to the joint probability mass function. Because the observations are assumed to be independent, their joint probability mass function is the product of their individual probability mass functions. Thus, it is

$$L(p) = \prod_{i=1}^{n} p^{x_i}(1-p)^{1-x_i}.$$

(b) The log-likelihood is the natural log of the likelihood function, which in this case is

$$l(p) = \ln[L(p)] = \ln\left[\prod_{i=1}^{n} p^{x_i}(1-p)^{1-x_i}\right] = \sum_{i=1}^{n} \ln[p^{x_i}(1-p)^{1-x_i}]$$

$$= \sum_{i=1}^{n} x_i \ln(p) + (1-x_i)\ln(1-p).$$

(c) See github.com/mdedge/stfs for code.

(d) The maximum-likelihood estimate of p is the mean of the sample, the proportion of 1s.

(2) (a) The log-likelihood function is

$$l(\theta) = \ln[L(\theta)] = n \ln\left(\frac{1}{\sigma\sqrt{2\pi}}\right) - \frac{1}{2\sigma^2}\sum_{i=1}^{n}(x_i - \theta)^2$$

$$= n \ln\left(\frac{1}{\sigma\sqrt{2\pi}}\right) - \frac{1}{2\sigma^2}\sum_{i=1}^{n}(x_i^2 - 2x_i\theta + \theta^2).$$

To maximize the log-likelihood, we take the derivative with respect to θ:

$$\frac{\partial l(\theta)}{\partial\theta} = -\frac{1}{\sigma^2}\sum_{i=1}^{n}(\theta - x_i) = -\frac{n}{\sigma^2}(\theta - \bar{x}),$$

where $\bar{x}$ is the sample mean, $(1/n)\sum_{i=1}^{n}x_i$. Because n and σ^2 are both positive, the only value of θ that sets the derivative equal to 0 is $\theta = \bar{x}$. To check that this solution maximizes, rather than minimizes, the log-likelihood function, we confirm that the second derivative is negative (Problem 2 of Exercise Set A-1). The second derivative is equal to $-n/\sigma^2$ and therefore always negative. Setting $\theta = \bar{x}$ maximizes the log-likelihood function, and thus the likelihood function. The maximum-likelihood estimator of θ is therefore $\hat{\theta} = \bar{x}$, the sample mean.

(b) The natural logs of the observations have a normal distribution, $\ln(Y_i) \sim$ Normal(θ, σ^2) for all i. Thus, by part (a), the maximum-likelihood estimator of θ is $\hat{\theta} = (1/n)\sum_{i=1}^{n}\ln(Y_i)$. By the functional invariance property of maximum-likelihood estimators, the maximum likelihood estimator of e^{θ} is $e^{\hat{\theta}}$. This is different from the method-of-moments estimator, which would estimate $e^{\theta} = E(Y)$ as the sample mean, $(1/n)\sum_{i=1}^{n}Y_i$.

Exercise Set 9-3

(1) See github.com/mdedge/stfs for the proof. The unbiasedness of the least-squares estimators is not guaranteed by their status as method-of-moments or maximum-likelihood estimators, so we have to show it directly. However, the proof does not rely on the assumptions of normality, independence, or constant variance of the disturbances. Thus, the least-squares estimator is also unbiased under the weaker assumptions used in Chapter 8.

(2) See github.com/mdedge/stfs for a derivation. The requested variance is Var$(\hat{\beta}) = \sigma^2/\sum_{i=1}^{n}(x_i - \bar{x})^2$. The variance of $\hat{\beta}$ decreases as the number of observations increases. The

proof relies on the independence and constant variance of the disturbances, but does not invoke normality.

(3) See github.com/mdedge/stfs for the derivation. The maximum-likelihood estimator of σ^2 is $\widehat{\sigma^2} = (1/n)\sum_{i=1}^{n}(Y_i - \alpha - \beta x_i)^2$. When α and β are unknown, we replace them with maximum-likelihood estimates, and the maximum-likelihood estimate of σ^2 becomes $\widehat{\sigma^2} = (1/n)\sum_{i=1}^{n}(Y_i - \widehat{\alpha} - \widehat{\beta} x_i)^2$, the average of the squared line errors. However, if α and β must be estimated, then the expectation of the maximum-likelihood estimator of the disturbance variance is $E(\widehat{\sigma^2}) = [(n-2)/n]\sigma^2$. Thus, the maximum-likelihood estimator is biased downward. One way to understand the bias is to notice that the maximum-likelihood estimator of σ^2 is proportional to the sum of the squared line errors, and $\widehat{\alpha}$ and $\widehat{\beta}$ are chosen to make this sum *as small as possible*. Thus, to the extent that estimates of α and β err, they will err in ways that make the squared line errors smaller than they would be if the true values of α and β were known. One unbiased estimator of the variance of the disturbances is $\widecheck{\sigma^2} = [n/(n-2)]\widehat{\sigma^2} = [1/(n-2)]\sum_{i=1}^{n}(Y_i - \widehat{\alpha} - \widehat{\beta} x_i)^2$. When n is large, $n/(n-2) \approx 1$, and the two estimators are nearly identical. In practice, the unbiased estimator is used more often.

(4) See github.com/mdedge/stfs for the proof. In combination, you have proven in Problems 1, 2, and 4 that, given the assumptions in this section, $\widehat{\beta}$ is normally distributed with expectation β and variance $\sigma^2/[\sum_{i=1}^{n}(x_i - \bar{x})^2]$.

Exercise Set 9-4

(1) See github.com/mdedge/stfs for derivations. $\widehat{\theta} = \bar{x}$ is the maximum-likelihood estimator of θ, and $Var(\widehat{\theta}) = \theta/n$, which we would estimate by plugging in $\widehat{\theta}$ for θ. The Fisher information method (part c) only gives the asymptotic variance of $\widehat{\theta}$, but because parts (b) and (c) give the same expression, it is also the small-sample variance.

Exercise Set 9-5

(1) The maximum-likelihood estimate $\widehat{\beta}$ is the least-squares slope, which we have already computed as $\widehat{\beta} = 0.5$. (See, for example, Chapter 3.) Because of the null hypothesis specified in the problem, $\beta_0 = 0$. The variance of $\widehat{\beta}$ appears in Equation 9.12; replace σ^2 with $\widecheck{\sigma^2}$ as given in Equation 9.14b. Computing this value for the example data gives a result of approximately 0.014. See github.com/mdedge/stfs for code to R-function compute the test statistic. The lm() function also computes the Wald statistic, but it labels it "t." The test statistic is $W^* \approx 4.24$, and $p = 2\varphi(-|W^*|) \approx 0.00002$. Comparing the test statistic against the appropriate t distribution gives a p value of 0.002 (using the t distribution with 9 degrees of freedom, because there are 11 data points minus 2 parameters being estimated), which is in close agreement with the permutation test from Chapter 8.

(2) Under the null hypothesis, the Wald test statistic is distributed as Normal(0, 1). The square of the Wald statistic is thus distributed as the square of a Normal(0, 1) random variable—in other words, it is distributed as $\chi^2(1)$.

(3) See github.com/mdedge/stfs for code to compute Wald-test p values from simulated datasets with independent, normally distributed disturbances (by default). When I ran the simulations, I arrived at the following results. Your results may differ slightly.

	$n = 10$	$n = 50$	$n = 100$
$\beta = 0$	0.088	0.060	0.039
$\beta = 0.1$	0.127	0.317	0.530
$\beta = 0.2$	0.257	0.789	0.956

The top row of the Wald test table, in which $\beta = 0$, is unsettling. When $n = 10$, the Wald test produces $p < 0.05$ in 9% of the simulations. This problem arises because the variance of the disturbances has to be estimated. Increasing n allows for better estimation of the unknown variance, and the problem is ameliorated. One solution—the one adopted by R's lm() function—is to compare the Wald statistic with an appropriate t distribution rather than a standard normal distribution. In these simulations, the power of the Wald test is similar to that of the permutation test for $n = 50$ and $n = 100$.

Exercise Set 9-6

(1) (a, b) See github.com/mdedge/stfs for a derivation of an expression for the likelihood-ratio test statistic and code to compute it. Applying the code to the agricultural data in the running example gives $\Lambda^* \approx 17.99$. (I've added the asterisk to indicate that this value of Λ has been calculated using an estimate of σ^2.) Because we held one parameter constant—namely, β—we compare Λ^* with a $\chi^2(1)$ distribution. The p of 0.00002 is found using pchisq(17.99, 1).

(c) The p values from part (b) and from Problem 1 of Exercise Set 9-5 are identical. Moreover, the test statistic from part (b) is the square of the Wald statistic. This relationship explains the identity of the p values. The test statistic in part (b) is compared with a $\chi^2(1)$ distribution, which is the distribution of the square of a single draw from the standard normal distribution, and we compare the test statistic from Problem 1 of Exercise Set 9-5 with a standard normal distribution. The agreement between the two tests is not a coincidence—in the simple linear regression case, the Wald test and the likelihood-ratio test are equivalent, even for small sample sizes.

(2) See github.com/mdedge/stfs for a proof.

Exercise Set 10-1

(1) (a) See github.com/mdedge/stfs for code. The mean of my sample is 2.15, fairly close to the true expectation of 2. With my simulated data, the expectation of the conjugate posterior is 2.047, and the posterior variance is 0.0476. The posterior standard deviation is the square root of the variance, approximately 0.218 here.

(b) After installing and loading the MCMCpack package and defining the variables in part (a), use

```
mn.mod <- MCnormalnormal(z, sigma2 = 1, mu0 = prior.mean, tau20 = prior.sd^2, mc =
10000)
```

Calling summary(mn.mod) reveals results extremely similar to part (a).

(c) Using the reject.samp.norm() function from Problem 2(d) gives results that are extremely similar to those obtained in parts (a) and (b). The small differences are due to stochasticity inherent in MCMC and rejection sampling.

(2) (a) The ratio is equal to the likelihood: $f_D(d|\theta)f_\theta(\theta)/f_\theta(\theta) = f_D(d|\theta) = L(\theta)$.

(b) If c times the unscaled posterior is equal to the prior, then $cL(\theta)f_\theta(\theta) = f_\theta(\theta)$, which implies $c = 1/L(\theta)$. Similarly, if c times the unscaled posterior is less than the prior, then $c < 1/L(\theta)$. If we set $c = 1/\max\big(L(\theta)\big)$, then $cL(\theta)f_\theta(\theta) = f_\theta(\theta)$ for the value(s) of θ that maximize the likelihood, and $cL(\theta)f_\theta(\theta) < f_\theta(\theta)$ for all other values of θ.

(c) The only step that changes from the algorithm given in the text is the second one. Instead of computing m as the likelihood, we can compute m as the likelihood divided by the maximum possible value of the likelihood, which guarantees that m is between 0 and 1 but also that it can be as large as 1. This usually means that a larger proportion of the samples can be accepted, which increases the efficiency of the algorithm.

(d) See github.com/mdedge/stfs for functions that provide one way to do it.

Exercise Set 10-2

(1) (a) Obtain the least-squares estimates of the intercept and slope (3 and ½, respectively) with `summary(lm(anscombe$y1~anscombe$x1))`.

(b) See github.com/mdedge/stfs for code to fit the model with all nine possible priors. When prior precision is low—meaning that prior variance is high—then the prior means are not especially important; these three choices lead to similar conclusions. If the prior precision is higher, then the prior means matter much more, and estimates are generally close to the prior means.

(2) See github.com/mdedge/stfs for proofs.

Exercise Set 10-3

(1) After fitting the models in Problem 1 of Exercise Set 10-2, look at the 2.5th and 97.5th quantiles of the slope parameter in the summary output. For these models, the quantile and highest-posterior-density intervals are similar. When the prior precision is very low, the credible interval largely agrees with the frequentist confidence intervals you have already calculated. If the precision is higher, then the credible interval is pulled toward the prior mean.

Exercise Set 10-4

(1) See github.com/mdedge/stfs for code.

(a) Bayes factor B_{10}: 3.26. By K & R's scale, this is positive evidence for H_1 over H_0.

(b) Bayes factor B_{10}: 26.9. By K & R's scale, this is strong evidence for H_1 over H_0.

(c) Bayes factor B_{10}: 0.26. By K & R's scale, this is positive evidence for H_0 over H_1—notice that we have switched from having the data support H_1 to having them support H_0.

(d) In this case, changing the prior precision/variance had a small effect on point estimates and credible intervals. In contrast, the Bayes factors changed consequentially. With intermediate prior variance, we had relatively weak but "positive" support for the model with the slope included. After decreasing the prior variance, that support became much stronger. Increasing the prior variance caused the Bayes factor to reverse.

Exercise Set Postlude 1

(1) You can see the set of plots for the first dataset in the quartet with `plot(lm(y1 ~ x1, data = anscombe))`. For the other datasets in the quartet, change the variable names accordingly.

Exercise Set Postlude 2

(1) Once the package is installed and loaded, fit the four models and run the diagnostics with `gvlma(lm(y1 ~ x1, data = anscombe))`, replacing `y1` and `x1` with the other variable names as necessary. As expected from the plot, tests using the first dataset reveal no reasons for concern. (But remember that the sample size in this example is very small, which means that power to detect deviations is low.) The second model shows that the "link function" test—which is intended to detect departures from linearity—returns a low p value. This makes sense: the data clearly fit a curve and not a line. The third dataset reveals trouble with normality. This isn't as informative as looking at the plot, but at least the test has returned an alarm when it should. The fourth result is more disquieting—the tests detect no problems with the assumptions, even though the plot suggests something is wrong. In short, the tests are useful in conjunction with the plots, but they do not replace them.

Exercise Set Postlude 3

(1) (a) Code in text.

(b) Here's some code:

```
resid.am.hp <- lm(am ~ hp, data = mtcars)$residuals #1
resid.mpg.hp <- lm(mpg ~ hp, data = mtcars)$residuals #2
resid.mod.hp <- lm(resid.mpg.hp ~ resid.am.hp) #3
summary(resid.mod.hp)
```

The estimated slope for the residuals of transmission type here is the same as the estimated slope for transmission type in part (a). Think of multiple regression coefficient estimates as assessing the association between an independent variable and the dependent variable when the other independent variables are held constant. In this case, by residualizing both weight and miles per gallon on horsepower, we are obtaining versions of these two variables whose association with horsepower has been "removed."

(c)

```
resid.hp.am <- lm(hp ~ am, data = mtcars)$residuals #1
resid.mpg.am <- lm(mpg ~ am, data = mtcars)$residuals #2
resid.mod.am <- lm(resid.mpg.am ~ resid.hp.am) #3
summary(resid.mod.am)
```

(d)

```
mean(mtcars$mpg) - mean(mtcars$am)*5.277 - mean(mtcars$hp)*(-0.05888)
```

(2) In parts (a) and (b), the t-statistics and associated p values from the two pairs of analyses are equal. In part (c), the F statistics and associated p values are equal.

Exercise Set Postlude 4

(1) `glm()` is for "generalized linear model." Once the car package is loaded, you can use

```
probit.fit <- glm(volunteer ~ extraversion + neuroticism + sex, data = Cowles,
family = binomial("probit"))
summary(probit.fit)
```

to get the estimates. To fit the logistic model instead, you would use `family = bino-mial("logit")`

(2) (a) See github.com/mdedge/stfs for code.
   (b) Yes, they are consistent. In line with this, setting the n to 10,000 or 100,000 gives estimates very close to the coefficients specified in the simulation.
   (c) In the presence of heteroscedasticity in the model for the latent variables, the estimators for the model coefficients are inconsistent—they converge on the wrong numbers. The degree to which they're wrong increases with the severity of the heteroscedasticity.

Exercise Set Postlude 5

(1) See github.com/mdedge/stfs for the proof.
(2) See github.com/mdedge/stfs for code. The least-squares estimates are unbiased. With the parameters specified in the code, the standard deviation of the intercept estimates is about 0.93, and the standard deviation of the slope estimates is about 0.10. These standard deviations are roughly in agreement with the mixed-model standard error estimates reported in the main text. They are much larger than the standard error estimates from simple linear regression. Ignoring dependence among the observations causes us to overestimate the amount of information we have, leading to standard error estimates that are too small.

Exercise Set A-1

(1) (a) (ii) $f'(x) = 2x - 2$; (iii) $x = 1$.
   (b) (ii) $f'(x) = -6x + 12$; (iii) $x = 2$.
   (c) (ii) $f'(x) = 3x^2 - 6x = 3x(x - 2)$; (iii) $x = 0$ and $x = 2$.
(2) Notice that the function in Problem 1, part (a) is minimized at $x = 1$, the function in part (b) is maximized at $x = 2$, and the function in part (c) is locally maximized at $x = 0$ and locally minimized at $x = 2$. The second derivatives are (a) $f''(x) = 2$, (b) $f''(x) = -6$, and (c) $f''(x) = 6x - 6$. Plugging in the appropriate x-coordinates suggests that if $f'(c) = 0$ and $f''(c)$ is positive, then $f(x)$ is locally minimized when $x = c$. Similarly, if $f'(c) = 0$ and $f''(c)$ is negative, then $f(x)$ is locally maximized when $x = c$. These conjectures turn out to be true in general. One way to think of this is that when the slope is zero but increasing (as indicated by the positive second derivative), the function is minimized.

(3) (a) By the definition of the derivative, $g'(x) = \lim_{\Delta x \to 0}[g(x + \Delta x) - g(x)]/(\Delta x)$. Because $g(x) = af(x)$,

$$g'(x) = \lim_{\Delta x \to 0} \frac{af(x + \Delta x) - af(x)}{\Delta x} = a \lim_{\Delta x \to 0} \frac{f(x + \Delta x) - f(x)}{\Delta x} = af'(x).$$

The first step comes from the definition of $g(x)$, the second step comes from the distributive property, and the third step comes from the definition of $f'(x)$.

(b) Because $h(x) = f(x) + g(x)$,

$$h'(x) = \lim_{\Delta x \to 0} \frac{f(x + \Delta x) + g(x + \Delta x) - f(x) - g(x)}{\Delta x}$$

$$= \lim_{\Delta x \to 0} \frac{f(x + \Delta x) - f(x)}{\Delta x} + \lim_{\Delta x \to 0} \frac{g(x + \Delta x) - g(x)}{\Delta x} = f'(x) + g'(x).$$

(4) See github.com/mdedge/stfs for a proof.

Exercise Set A-2

(1) (a) 1; (b) 4; (c) 8; (d) 0; (e) 8.

(2)

b	The definite integral of $f(x) = 2x$ from 0 to b (i.e., $\int_0^b 2x\,dx$)
1	1
2	4
3	9
4	16
5	25

Exercise Set A-3

(1) To solve this problem, consider that the derivative of ax^n with respect to x is nax^{n-1} and that differentiation and integration are inverse processes, meaning that $\int f'(x)\,dx = f(x) + C$. Jointly, these two facts imply that $\int nax^{n-1}dx = ax^n + C$. That is, to integrate x raised to an exponent, we raise the power of the exponent by one (so $n - 1$ becomes n) and divide by the new value of the exponent (this gets rid of the n in front). Then we add C to get the indefinite integral. The same thing applies when we start with n in the exponent instead of $n - 1$: we add one to the exponent and divide by the new value of the exponent. Thus, $\int ax^n dx = (ax^{n+1})/(n + 1) + C$. You can say "up and under" as a way to remember this—the exponent increases by one ("up") and then divides the whole expression ("under"). Remember that for differentiation of polynomials, the phrase to remember is "out in front and down by one." When we integrate, we are doing the opposite.

This rule does not apply when $n = -1$. When $n = -1$, the "up and under" rule would force us to divide by zero, which is not allowed. We will use the integral of ax^{-1} in Chapter 9.

Table of mathematical notation

∞	Infinity
$+,-$	Plus, minus
$*,/$	Multiplied by, divided by (multiplication is usually indicated by juxtaposition)
x^k	x raised to the kth power
$\sqrt{}$	Square root
$=,\neq,\approx$	Equals, does not equal, is approximately equal to
$<,>,\leq,\geq$	Less than, greater than, less than or equal to, greater than or equal to
$\in$	Is an element of. Examples: $x \in (1,3)$ is $1 < x < 3$; $x \in [1,3]$ is $1 \leq x \leq 3$, and $x \in \{1,2,3\}$ means that x is 1, 2, or 3. Do not confuse with Greek letter ε.
$\binom{n}{k}$	Binomial coefficient, "n choose k."
$k!$	k factorial, i.e. $k*(k-1)*(k-2)*\ldots*1$. By definition, $0! = 1$.
max, min	Maximum, minimum
argmax, argmin	Argument of the maximum; argument of the minimum. i.e., the value of a function's argument that maximizes or minimizes the function
$\cup, \cap$	Union, intersection
log, ln	Logarithm, natural logarithm (i.e. logarithm with base e, Euler's number)
$f(x)$	A function of x. Typically, the letters f and g are reserved for functions.
$\sum_{i=1}^{n} x_i$	Sum over an index i from 1 to n, i.e. $x_1 + x_2 + \ldots + x_n$
$\lim_{x \to a} f(x)$	Limit of $f(x)$ as x approaches a
$f'(x)$	Derivative of $f(x)$; writing $\frac{d}{dx}f(x)$ clarifies that the derivative is with respect to x.
$\int f(x)dx$	Indefinite integral of $f(x)$ with respect to x
$\int_a^b f(x)dx$	Definite integral of $f(x)$ from a to b
P	Probability
$P(A|B)$	Conditional probability (of A given B). Conditional distribution functions, expectations, variances, etc. are also indicated with a vertical bar.
W, X, Y, Z	Random variables are usually written as capital Latin letters, instances of random variables as corresponding lowercase letters
$\text{Normal}(\mu, \sigma^2)$	Normal distribution with parameters μ and σ^2. A specific distribution may be noted by the name of the family followed by the parameter values in parentheses.
f_X, F_X	Lowercase f_X indicates the probability mass function or density function of a random variable X; capital F_X is for the cumulative distribution function. Joint distributions are indicated by multiple variables in the subscript, separated by commas, and conditional distributions by a vertical bar in the subscript.
E	Expectation, also called expected value
Var, SD	Variance, standard deviation
Cov, Cor	Covariance, correlation
$\theta, \hat{\theta}$	Parameters are often indicated by Greek letters, and estimators by an accent or "hat"
SE	Standard error
H_0, H_a	Null hypothesis, alternative hypothesis
p	p value
L, l	Likelihood function, log-likelihood function

Glossary

Assignment (R)   In R, a command that results in a variable being assigned a value. See Section 2.1.

Asymptotic   In general, "asymptotic" refers to limiting behavior (see "Limit"). In this book, "asymptotic" typically refers to the way a statistical estimator or other procedure behaves as the number of data collected (i.e. sample size) becomes very large. For example, asymptotic normality means that a statistic's distribution converges to a normal distribution as the sample size approaches infinity.

Bayes estimator   In Bayesian statistics, an estimator that minimizes the expected value of a loss function, where the expectation is taken over the posterior distribution. See Section 10.3.

Bayes factor   In Bayesian statistics, a statistic comparing the degree of compatibility of the data with each of two hypotheses. See Section 10.5.

Bayes' theorem   A mathematical statement relating conditional probabilities to each other. In particular, if A and B are events, then Bayes' theorem holds that $P(A|B) = P(B|A)P(A)/P(B)$. See Section 4.3.

Bayesian statistics   An approach to statistics in which probabilistic statements about parameters are updated in the light of data using Bayes' theorem. See Box 4-2, the Interlude, and Chapter 10.

Bean machine   A device, useful for illustrating the central limit theorem, in which balls or other particles are dropped through a triangular grid of evenly spaced pegs. The balls will settle at the bottom in a shape that resembles the probability density function of a normally distributed random variable. Also called a quincunx or Galton board. See Section 5.5.

Best-fit line   A line drawn to pass through points in an "optimal" way. More formally, a line that minimizes the sum of a loss function computed on the points, where the loss function encodes losses incurred when a particular distance separates a point and the line. We often use the least-squares line, which minimizes the sum of the squared vertical distances between the points and the line, but we could choose to minimize another function of the distances (for example, absolute value), or distances that are not vertical. See Chapter 3.

Bias   – An estimator's bias is the difference between the estimator's expected value and the true value of the estimand. An unbiased estimator has the true value of the estimand as its expected value. See Section 6.1.

Bootstrapping   A set of approaches for approximating the sampling distribution of a statistic by simulating new samples using the original data. See Section 8.2.

Causality   Underlying cause-and-effect relationships between sets of variables. Causal inference is the attempt to discern causal relationships between variables being studied.

Central limit theorem (CLT)   There are several versions of the CLT. The one discussed in this book holds that the distribution of means or sums of samples of n independent, identically distributed random variables, each with finite variance, approaches a normal distribution as the sample size n approaches infinity. See Section 5.5.

Conditional distribution   Suppose that there are two random variables X and Y. The conditional distribution of $Y|X = x$ is the distribution of Y under the assumption that X takes the specific value x. The conditional distribution is also associated with all the same descriptions and summaries as other probability distributions. That is, one can define a conditional cumulative distribution function, conditional probability mass function, conditional probability density function, conditional expectation, conditional variance, etc. See Section 5.4.

Conditional probability   The conditional probability $P(A|B)$ is the probability that A occurs if it can be assumed that B has occurred or will occur. It is defined as $P(A \cap B)/P(B)$. See Section 4.2.

Confidence interval   An interval estimate that is constructed using a procedure that will include the true value of an estimand with a specified probability. A confidence set has the same property, but it is not necessarily an uninterrupted interval. See Section 7.2.

Confounding A confound is a variable that causes associations between an independent variable and the dependent variable. Estimates of the independent variable's causal effect size will be biased if the confound is not accounted for. See Section Post.2.1.

Conjugate prior A prior distribution that, when combined with the likelihood using Bayes' theorem, leads to a posterior distribution that is a member of the same distribution family as the prior. For example, if data are drawn from a Poisson distribution, then if the prior density for the parameter λ is a gamma distribution, then the posterior for λ will also be a gamma distribution. When applicable, conjugate priors make Bayesian procedures easy because the posterior distribution can be computed directly.

Consistency A consistent estimator converges to the true value of the estimand as the sample size increases. See Section 6.4.

Continuous random variable A random variable that can take on any of an uncountably infinite number of possible values. (Contrasted with discrete random variable.) A continuous random variable's distribution can be characterized by a cumulative distribution function or a probability density function. See Section 4.5.

Correlation The correlation of two variables is their covariance rescaled so that 1 is the maximum possible value and -1 is the minimum possible value. See Section 5.3.

Cost function See "Loss function."

Covariance A measurement of the degree to which two random variables depart from independence. It is defined as $\mathrm{Cov}(X, Y) = \mathrm{E}([X - \mathrm{E}(X)][Y - \mathrm{E}(Y)])$. If X and Y are independent, then $\mathrm{Cov}(X, Y) = 0$, but the covariance is not sensitive to all forms of dependence—it is possible for two random variables to have zero covariance but not be independent. See Section 5.3. The sample covariance is an estimator of the covariance based on a sample of data.

Coverage probability The probability that a procedure for constructing confidence intervals produces an interval that includes, or *covers*, the true value of the estimand. See Section 7.2.

Credible interval/region In Bayesian statistics, a set of candidate parameter values that are plausible according to the posterior distribution. See Section 10.4.

Cross-validation A strategy for evaluating the ability of a predictive model to generalize beyond the data to which it was fit, that is, to curtail the effects of overfitting. In k-fold cross-validation, the dataset is broken into k equally sized pieces, and the model is fit k times, each time holding aside one of the pieces. For each version of the model, the quality of the predictions is assessed in the excluded portion of the data. See Section Post.1.3.

Cumulative distribution function If X is a random variable, then its cumulative distribution function gives the probability that X takes a value less than or equal to the value of its argument. That is, if F_X is the cumulative distribution function of X, then $F_X(x) = P(X \leq x)$. See Section 4.4.

Data frame (R) A data structure in which data are arranged into a rectangle of rows and columns, but distinct columns can hold distinct data types (i.e., modes). See Appendix B, Section B.2.3.

Data-generating process See "Process, data-generating."

Decision theory Decision theory is a broad field. Statistical decision theory is a branch of decision theory and of statistical theory concerned with evaluating the properties of statistical procedures, viewed as decision rules. See Sections 6.6 and 10.3.

Dependent variable A variable in a research study or probabilistic model that is treated as the outcome of the model or process. For example, in a linear regression model, the dependent variable Y is viewed as a weighted sum of the independent variables plus a disturbance term. The decision as to which variable is the dependent variable might be made on the basis of a presumed causal relationship between the variables under study (where the dependent variable is viewed as an outcome and independent variables are viewed as potential causes) or because the dependent variable is of particular interest in applications. Also called a response variable or outcome variable.

Derivative Suppose we have a mathematical function $f(x)$. The derivative of $f(x)$, written as $f'(x)$ or $\frac{d}{dx}f(x)$, is a function whose output can be viewed as the instantaneous rate of change of $f(x)$. The "second" derivative is the derivative of the derivative. Derivatives are useful in optimization problems. See Box 3-1 and Appendix A, Section A.2.

Differentiation The process of identifying a function's derivative. See Box 3-1 and Appendix A, Section A.2.

Discrete random variable   A random variable with a countable number of possible values, as opposed to a continuous random variable, which can take on any of an uncountably infinite number of possible values. A discrete random variable's distribution can be characterized by a cumulative distribution function or a probability mass function. See Section 4.4.

Distribution family   A set of probability distributions where the functions that describe them share a general mathematical form, but specific numbers, called "parameters," may differ. For example, the $f_X(x) = 2e^{-2x}$ and $g_X(x) = 5e^{-5x}$ (for $x \geq 0$) are probability density functions that share the general form $\lambda e^{-\lambda x}$ but have different values of λ, the parameter. (Distributions described by probability density functions of this form are members of the exponential distribution family, and their associated random variables are exponentially distributed random variables.) See Section 4.7.

Disturbance   In a linear regression model, a random variable representing contributions to the dependent variable that are unrelated to the independent variables. Also called an "error" term. Disturbances are related to residuals or line errors, but whereas residuals measure a vertical distance between an estimated line (or, in multiple regression, an estimated plane) and a data point on a scatterplot, disturbances represent a distance between a data point and the true, unknown regression line (or plane). See Section 5.6.

Effect size   A measurement of the strength of association between one variable and another. (Though the word "effect" suggests a causal relationship, effect sizes are used to measure the strength of associations that are not necessarily causal.) Effect sizes are useful both as estimands of primary interest in many research studies and in power analysis, where they are arguments to the power function of a statistical test. See Section 7.8.

Efficiency   Loosely speaking, an efficient estimator is one that achieves relatively high precision with a given sample size. See Section 6.5.

Empirical distribution function   A consistent estimator of the cumulative distribution function. If the argument to the empirical distribution function is c, then the output is the proportion of values in a dataset that are less than or equal to c. See Chapter 8.

Estimation   The attempt to identify, as nearly accurately as possible, an unknown quantity of interest using data. The target of estimation—that is, the unknown quantity of interest—is called the estimand. The function into which the data are fed is called an estimator, and the resulting quantity is called an estimate. See Chapter 6.

Event   Loosely, something that can happen in a probability model. More formally, a subset of the set of outcomes defined in a probability model. See Section 4.1.

Expectation or Expected value   Loosely, the mean of a random variable's distribution. A weighted average of the values that a random variable can take, where the weights are the probabilities that the random variable takes each value (or the probability density associated with each value). See Section 5.1.

Exploratory data analysis   Data analysis in which the basic properties of data are explored with numerical summaries and data displays. See Chapter 2 and the Interlude.

Expression (R)   An R command that results in output being displayed to the human user rather than stored as a variable. Contrasted with assignment. See Section 2.1.

Exponential distribution   A distribution family often used to model waiting times. See Section 4.7.

False discovery rate   The proportion of rejected null hypotheses that are actually true (meaning that the null hypothesis was rejected in error). See Exercise Set 7-4.

Fisher information   A function of the likelihood that is used to identify the asymptotic standard error of maximum-likelihood estimators. See Section 9.2.2.

for() loop   In computer programming, a scheme for iterating a procedure many times, often once on each member of a set of inputs. See Section 2.2 and Appendix B, Section B.3.8.

Frequentist statistics   An approach to statistics in which procedures are justified in terms of their long-run properties, that is, their behavior when applied to many possible data sets. Contrasted with Bayesian statistics. See Box 4-2, the Interlude, and Chapters 6–9.

Function (mathematical)   A mathematical function takes an input, or argument, and produces an output. For example, if we define the function $f(x) = x^2$, then $f(3) = 9$. In this case, 3 is the argument and 9 is the output. See Box 3-1 and Appendix A.

Function (R)   Similar conceptually to a mathematical function, a predefined procedure in R that can be applied to a user-supplied input. Many functions exist in base R, many more exist in user-written packages, and users can write their own functions. See Section 2.2 and Appendix B, Section B.3.

Functional   A functional is a mathematical function whose argument is another function. See Section 8.1.1.

Functional invariance property of maximum-likelihood estimators   If $\hat{\theta}$ is a maximum-likelihood estimator of a parameter θ, then the maximum-likelihood estimator of a function of θ, $g(\theta)$, is the same function applied to the maximum-likelihood estimator of θ; that is, the maximum-likelihood estimator of $g(\theta)$ is $g(\hat{\theta})$. See Section 9.1.

Fundamental theorem of calculus   The statement that integration and differentiation are inverse procedures. If $f(x)$ is a function and $F(x)$ is its integral, then the derivative of $F(x)$ is $f(x)$. See Appendix A, Section A.4.

Generalized linear model   A modeling strategy for building regression-like models for dependent variables that do not obey regression assumptions, such as dichotomous (0/1) outcomes and counts. Logistic regression and probit regression are examples. See Section Post.2.2.

Heteroscedasticity   In a linear regression model, the disturbance term is heteroscedastic if its variance changes depending on the value(s) of the independent variable(s).

Homoscedasticity   In a linear regression model, the disturbance term is homoscedastic if its variance is constant, regardless of the value(s) of the independent variable(s). See Section 5.6.

Hypothesis test   Loosely, a statistical procedure that entails evaluating the degree to which an observed data set is consistent with a hypothesis about the data-generating process. There are multiple approaches to hypothesis testing in statistics. See Chapter 7 from Section 7.3 onward.

Independence   Two events A and B are independent if conditioning on one of the variables does not influence the probability of the other, i.e. $P(A|B) = P(A)$. If A and B are independent, it is also true that $P(A \cap B) = P(A)P(B)$. Similarly, if two random variables X and Y are independent, then the probability density or mass function of one of the variables is not affected by conditioning on the other, i.e., $f_{X|Y=y}(x|y) = f_X(x)$. Their joint mass or density function is also the product of the individual densities, $f_{X,Y}(x,y) = f_X(x)f_Y(y)$. See Sections 4.2 and 5.3.

Independent variable   A variable in a research study or probabilistic model that is treated as input to a model or process. For example, in a linear regression model, a dependent variable Y is viewed as a weighted sum of the independent variables plus a disturbance term. The decision as to which variables are independent might be made on the basis of a presumed causal relationship between the variables under study (where the dependent variable is viewed as an outcome and independent variables are viewed as potential causes) or because the dependent variable is of particular interest in applications. In regression analyses, independent variables are also called covariates (often "covariate" is a term for independent variables of secondary interest), explanatory variables, predictor variables, or regressors.

Indicator variable   A random variable that takes the values 0 or 1. Indicator variables are useful constructions in some probability problems, where they can be used to "indicate" that an outcome has occurred. For an example of their use, see the description of the empirical distribution function at the beginning of Chapter 8.

Instance   An instance of a random variable is a value that the random variable turns out to take in a given trial or sample. Also called a realization. To repeat an example given in the main text, we can model the outcome of a six-sided die roll with a random variable X, which might take any integer value from one to six. After we roll the die, we observe an *instance* of the random variable. Typically, we write random variables as capital Latin letters and instances of those random variables with corresponding lowercase Latin letters. There are exceptions, including the disturbance term in a linear regression, which is typically written using a Greek letter ϵ. See Section 4.4.

Integral   The *definite* integral of a function $f(x)$ from a to b, denoted $\int_a^b f(x)dx$, is equal to the area that falls vertically underneath the graph of $f(x)$ and above the horizontal axis and horizontally between $x = a$ and $x = b$. Indefinite integrals are functions useful for computing definite integrals— if $F(x)$ is an *indefinite* integral of $f(x)$, then $\int_a^b f(x)dx = F(b) - F(a)$. See Box 4-3 and Appendix A, Section A.3.

Interquartile range   The difference between the 75th percentile and the 25th percentile of a dataset.

Intersection   In set theory, "both and," symbolized with $\cap$. For example, if A and B are sets, then $A \cap B$ is the collection of overlapping elements—i.e., elements that are members of both A and B. In probability, if A and B are events, then $P(A \cap B)$ refers to the probability that both A and B occur. See Box 4-1.

Interval estimation   Loosely, the attempt to identify ranges of possible estimates that are plausible candidate values for an unknown quantity of interest. See Chapter 7 through Section 7.2 and Sections 8.2, 9.2, and 10.4.

Joint distribution   The distribution of two (or more) random variables considered together. Joint distributions can be defined by joint cumulative distribution functions, by joint probability mass functions (if all random variables are discrete), and by joint probability density functions (if all random variables are continuous). See Section 5.3.

Law of large numbers   A law stating that the mean of n independent, identically distributed random variables $X_1, X_2, ..., X_n$ approaches the shared expected value $E(X)$ as the sample size n approaches infinity, assuming that $E(X)$ exists. See Section 5.1.

Law of the unconscious statistician   The expected value of a function g of a random variable X is $E[g(X)] = \int_{-\infty}^{\infty} g(x)f_X(x)\, dx$ if X is continuous (with f_X the probability density function of X), and $E[g(X)] = \sum_x g(x)f_X(x)$ if X is discrete (with f_X the probability mass function of X). See Section 5.1.

Level   The maximum tolerable type I error rate (or size) for a statistical hypothesis test. See Section 7.4.

Likelihood function   Loosely, a function that answers the question, "What is the probability of observing the data in hand given an assumed data-generating process with particular parameter value(s)?" The likelihood function has the same form as the joint probability mass function or joint probability density function of the data, but it is viewed as a function of the parameter(s) rather than as a function of the data. Likelihood functions form the basis of many methods for estimation and inference in frequentist statistics. See Chapter 9.

Likelihood-ratio test   A statistical test in which the maximum values taken by the likelihood function under two competing hypotheses are compared. See Section 9.4.

Limit   The value that a function approaches as its argument approaches a specified value. For example, $\lim_{x \to 0} f(x)$ is the value that $f(x)$ approaches as x approaches zero. See Appendix A, Section A.2.

Line error   A vertical distance between a data point on a scatterplot and a line drawn through the data. Also called a residual. See Section 3.1.

Line values   The vertical coordinates, along a line drawn through a scatterplot, that are immediately above or below data points. Also called "fitted values" or "predicted values."

Linear regression   A statistical procedure in which a dependent variable is modeled as a weighted sum of a set of independent variables plus a random disturbance term. That is, if Y is a dependent variable and $X_1, ..., X_k$ are the independent variables, then a standard linear regression model holds that $Y = \beta_0 + \beta_1 X_1 + ... + \beta_k X_k + \epsilon$, where the β terms are (possibly unknown) constants. The final term, ϵ, is also called the "disturbance" or "error" and is a random variable, typically with expected value 0. (In the version written here, the X terms are treated as random variables, but in practice they are often treated as fixed constants, so the disturbance is the only random variable involved.)

Linearity assumption   In a linear regression model, the linearity assumption is that the disturbance terms have expected value 0 regardless of the value(s) of the independent variable(s). See Section 5.6.

Linearity of expectation   A property of expectation holding that the expected values of linear sums of random variables are corresponding linear sums of the random variables' expectations. In particular, $E(aX + bY + c) = aE(X) + bE(Y) + c$, where X and Y are random variables and a, b, and c are fixed constants. See Section 5.1.

Logarithm, or log   The logarithm, or just "log," is a function that is useful for turning multiplication problems into addition problems. In particular, if a is the base of a logarithm, then the logarithm function satisfies $x = a^{\log(x)}$. In words, the log of x (with base a) is the power to which a must be raised to equal x. For our purposes, the most useful property of the logarithm is that

$\log(a \star b) = \log(a) + \log(b)$. In this book, we use Euler's number $e \approx 2.781$ as a base, which is also called the "natural" logarithm and denoted ln. See Box 9-1.

Log-likelihood   The natural logarithm of the likelihood function.

Loss function   In general, a function that identifies the amount of loss (or profit) resulting from a decision. In this book, there are two distinct but related uses of "loss function." In Chapter 3, the loss function describes the loss associated with drawing a line that "misses" a data point on the scatterplot by a certain amount, i.e., the loss associated with a line error of a given size. In Chapters 6 and 10, we discuss loss functions that quantify the loss associated with estimates that miss the true value of the estimand by a specified amount. Also called a cost function or objective function. See Sections 3.1, 6.6, and 10.3.

Markov chain Monte Carlo (MCMC)   A computational method for drawing samples from posterior distributions. See Section 10.2.

Matrix (R)   A data structure in which entries of a single data type (i.e., mode) are arranged into a rectangle of rows and columns. See Appendix B, Section B.2.2.

Maximum-likelihood estimation   An approach to statistical estimation in which the estimators are candidate parameter values that maximize a likelihood function. In looser English, the parameter values that make the data most probable are selected as the estimates. See Section 9.1.

Mean, sample   The sample mean is the average of a sample. If the sample values are denoted $x_1, ..., x_n$, then the sample mean is $\frac{1}{n}\sum_{i=1}^{n} x_n$. The sample mean is also the first sample moment and a plug-in estimator of the expectation.

Mean squared error (MSE)   The expected value of the squared difference between an estimator and the true value of the estimand. See Section 6.3.

Median   Loosely, a "middle number" of a random variable's distribution or of a sample if the sample data are arranged from least to greatest. More specifically, a median of a random variable X's distribution is a number m that satisfies $P(X \leq m) = 0.5$, or the 0.5 quantile of the distribution. A sample median is a plug-in estimator of a distribution median.

Method of moments   A method for estimating a distribution's parameters. In the method of moments, a distribution's k parameters are expressed in terms of k of its moments. To form method-of-moments estimators for the parameters, one replaces the moments in these expressions with plug-in estimators of the moments, that is, the sample moments. See Section 8.1.2.

Mixed model   A regression model including one or more "random" effects in addition to fixed effects. Mixed models have many uses, including accommodating non-independent data. Also called "hierarchical models" and "random-effects models." See Section Post.2.3.

Mode (R)   In R, a mode is a type of data. We discuss three modes: numeric, logical, and character (see Appendix B, Section B.1).

Mode (statistics)   The sample mode is the value that appears most frequently in a sample of data. Analogously, the mode of a random variable is the possible value associated with the largest value of the probability density function (if the random variable is continuous) or the largest value of the probability mass function (if the random variable is discrete).

Moment   If X is a random variable, then its kth moment is the expected value of X raised to the kth power, that is, $E(X^k)$.

Multiple regression   Linear regression with more than one independent variable. See Section Post.2.1.

Nonparametric   A nonparametric probabilistic model is a probabilistic model that cannot be described by a finite set of parameters. In practice, nonparametric procedures are often used when data cannot be assumed to have been sampled from a known probability distribution. See the Interlude and Chapter 8.

Normal distribution   Normally distributed random variables have symmetric, bell-shaped probability density functions. One reason that normal distributions are important is that many statistics—particularly those that can be viewed as sums or means of independent random variables—may have distributions that are approximately normal. See Sections 4.7 and 5.5.

Null hypothesis   In both Fisherian and Neyman–Pearson hypothesis testing, the null hypothesis is the hypothesis whose degree of consistency (loosely speaking) with the data is evaluated directly. Typically, we check whether a dataset or test statistic is far removed from what we would expect to observe if the null hypothesis were true. If the test statistic or data set is unusually far removed from what we would expect under the null hypothesis, then that is taken as evidence against the null hypothesis, and we may make a decision to "reject" the null hypothesis. See Sections 7.3 and 7.4.

Observational study   A study in which the variables of interest are not controlled by the researchers. Establishing causal claims on the basis of observational data is sometimes possible in theory but difficult in practice. Contrasted with experiment or experimental study, in which some or all of the independent variables are controlled by the researchers.

Optimization   The attempt to find the arguments of a function that maximize it or minimize it. One approach to optimization is to use calculus, finding the arguments that set the derivative of the function to zero. Another approach is to optimize functions numerically, using a computer and algorithms that explore possible optimizing arguments in a reasonable way.

Outlier   An observation that is far removed from the rest of a data distribution and/or that is drawn from a data-generating process that is not the one of interest. See Section 6.7.

Overfitting   A model that mistakes chance patterns in the data—that is, patterns that will not generalize to other datasets—for genuine features is overfit. Overfitting may result from using overly complex models. See Section Post.1.3.

p **hacking**   Any procedure that causes *p* values to be or appear systematically lower than they would be under correct practice. Common examples include conducting many hypothesis tests and only reporting the ones that produce low *p* values, or collecting data and checking the *p* value every so often, stopping data collection only when the *p* value is low. See Section 7.6.

p **value**   A measure of the strength of evidence against a null hypothesis. Loosely, the *p* value is the probability that a sample would produce a result as extremely removed or more extremely removed from what would be expected if the null hypothesis were in fact true. See Sections 7.3 and 7.4.

Parameter   A number that controls the properties of a random variable drawn from a particular probabilistic model. Parameters are often denoted with Greek letters, such as β, θ, or μ. One exception to this pattern is the disturbance term in a linear regression, which is a random variable and not a parameter, but is usually denoted by ϵ. See Section 4.7.

Parametric   A parametric probabilistic model is a probabilistic model whose behavior is completely described by a finite set of parameters.

Percentile   See "Quantile."

Permutation test   A type of statistical test in which some part or feature of the data is randomly shuffled or permuted, and the test statistic is recalculated using the shuffled data. After the data have been permuted and the test statistic has been recalculated many times, the original value of the test statistic is compared with the distribution of test statistics computed under random permutation. See Section 8.3.

Plug-in estimator   Loosely, an estimator for a population-level quantity that is constructed by treating the distribution of data in the sample as if it were the probability distribution associated with the data-generating process. See Section 8.1.1.

Point estimation   An estimation setting in which the goal is to identify a single guess, or point estimate, for an unknown quantity of interest. See Chapter 6.

Polynomial   A polynomial function of *x* is a sum of terms, each of which is formed by multiplying *x* raised to some power by a constant. That is, a polynomial function of *x* has the form $f(x) = \sum_i a_i x^{b_i}$, where the *a* and *b* terms are constants. See Box 3-1 and Appendix A.

Population   One type of data-generating process is to draw a random sample from a population of interest, where the value of some statistic in the entire population is the quantity of interest. See also "Process, data-generating."

Posterior distribution   In Bayesian statistics, a posterior distribution is a distribution that "describes" (in some sense) uncertainty about a parameter of interest that remains after observing data. Posterior distributions are constructed by combining a prior distribution and a likelihood function using Bayes' theorem. Various summaries of the posterior distribution might be used for point estimation (e.g., the posterior distribution's mean, median, or mode) or interval estimation (e.g., credible intervals). See Chapter 10.

Power   The probability that a given statistical test rejects the null hypothesis under a specific alternative hypothesis. See Section 7.8.

Power function A function, associated with a statistical test, whose output is the power of the test in a situation described by the function's arguments. For example, the power function might take as inputs a sample size, the test's level (i.e., type I error rate), and an assumed effect size under the null hypothesis, and return the power of the test. See Section 7.8.

Prior distribution In Bayesian statistics, a probability distribution taken to "describe" (in some sense) possible values of a parameter of interest *before* (i.e., "prior" to) observing the data. There are several approaches to Bayesian statistics that adopt different attitudes about the role of the prior distribution. The prior distribution might be taken to reflect the data analyst's beliefs about a parameter, or expert knowledge about a parameter, or it might simply be chosen because it leads estimates and inferences that have desirable properties. See Chapter 10 through Section 10.1.

Probabilistic model A model designed to reflect a data-generating process that contains at least one random (i.e. probabilistic) component. Also called a statistical model.

Probability There are multiple possible real-world interpretations of probability, including that of a long-run frequency of an event and that of a degree of belief. Regardless of the real-world interpretation, probabilities are mathematical objects that obey the axioms of probability. See Chapter 4 through Section 4.1.

Probability density function If X is a continuous random variable, then its probability density function is the derivative of its cumulative distribution function. Probability density functions behave analogously to probability mass functions in many ways. See Section 4.6.

Probability mass function A function that describes the distribution of a discrete random variable. Specifically, if X is a discrete random variable, then its probability mass function returns the probability that X takes the value specified in its argument. That is, if f_X is X's probability mass function, then $f_X(x) = P(X = x)$. See Section 4.4.

Process, data-generating A process that generates (or that is assumed to generate) observed data. Often, but by no means always, the data-generating process is assumed to be that of selecting and measuring a randomly chosen sample of units from a population. Discussed in several places in the book, including the beginning of Chapter 4, the Interlude, and Footnote 1 of Chapter 6.

Publication bias A situation in which studies associated with hypothesis tests that achieve statistical significance are easier to publish than those that do not. See Section 7.6.

Quantile The pth quantile of the distribution of a random variable X is a number c satisfying $P(X \leq c) = p$. Sample quantiles are analogous; the pth sample quantile is a number c such that p is the proportion of data values in the sample that are less than or equal to c. Percentiles are quantiles where p is expressed as a percentage.

Quantile function The quantile function of a random variable X takes as its argument a number p between 0 and 1 and returns a number c that satisfies $P(X \leq c) = p$. The number c is called the pth quantile of the distribution of X. The quantile function is the inverse of the cumulative distribution function. That is, if the quantile function of X is Q_X and the cumulative distribution function of X is F_X, then $F_X(c) = p$ implies that $Q_X(p) = c$ and vice versa.

R A programming language useful for statistics and data analysis. See Chapter 2 and examples throughout.

Random variable A mathematical object that encodes possible outcomes of a random process. For example, if we are interested in rolls of a six-sided die, we might construct a random variable that can take integer values from 1 to 6. Random variables are not numbers, but their realizations (also called instances) are numbers. Random variables are often written with capital Latin letters, and their instances are written with lowercase Latin letters. One important counterexample is the disturbance term in linear regression, which is a random variable but is usually denoted with a Greek ϵ. See Section 4.4.

Realization See "Instance."

Rejection sampling A method for drawing samples from a posterior distribution. In rejection sampling, simulated data are drawn from a prior distribution and accepted with a probability that depends on how compatible they are with the likelihood function. See Section 10.2.

Resampling methods Statistical methods that involve drawing samples from the original data or reshuffling (i.e., permuting) the original data. Bootstrapping and permutation testing are examples. See Chapter 8, Sections 8.2–8.3.

Residual See "Line error."

Risk   In statistical decision theory, the risk of an estimator (or more generally, of a decision rule), is the expected loss associated with adopting estimates produced by the estimator. (The expectation is taken with respect to the estimator's distribution, and the parameter is treated as fixed.) See Section 6.6.

Robustness   Loosely, a robust statistical procedure is one that gives useful information even if its assumptions are violated to some degree or if some proportion of the data is actually drawn from a data-generating process that is not the one of interest. See Section 6.7.

Sample   A set of data drawn from a population or a data-generating process.

Sample moment   The kth sample moment is the mean of the observations in a sample of data after each observation has been raised to the kth power. Sample moments are plug-in estimators of the distribution's moments. See Section 8.1.1.

Sample size   The number of data included in a sample. Most statistical procedures improve as sample sizes increase—estimators typically become more precise, and other things equal hypothesis tests have greater power at higher sample sizes, assuming that the null hypothesis is false.

Sampling distribution   Suppose a statistic $S(D)$ is computed on the basis of a dataset D. The sampling distribution of S is the probability distribution that describes the behavior of S across repeated samples drawn from the same data-generating process as D. Put another way, the sampling distribution of a statistic computed on the basis of a dataset describes the values the statistic would take if we were to repeat the study many times and collect equivalent—but different because of random sampling—datasets each time. See Chapter 6.

Semiparametric   A semiparametric probabilistic model is a model with a parametric component and a nonparametric component. See Chapter 8.

Set   An unordered collection of objects. See Box 4-1.

Significance, statistical   The outcome of a hypothesis test is said to be significant if it leads to a rejection of the null hypothesis. Statistical significance does not imply practical importance. See Section 7.4.

Simple linear regression   Linear regression with one independent variable. Simple linear regression is the main example of a statistical procedure throughout this book.

Simulation   A procedure in which random numbers are generated to "simulate" datasets that obey a particular probabilistic model. The behavior of statistical procedures can be studied across simulated datasets, with the advantage that the true values of all relevant parameters are known.

Size   Also type I error rate, the probability that a statistical test leads to a rejection of the null hypothesis given that the null hypothesis is actually true. See Section 7.4.

Standard deviation   A measure of the dispersal of a random variable that is equal to the square root of the variance. See Section 5.1. The sample standard deviation is a measure of the dispersal of a dataset and also an estimator of the random variable's standard deviation.

Standard error   The standard deviation of an estimator, where the estimator is conceived as a random variable that would take many values across different repetitions of a study, i.e., that would have a sampling distribution. See Section 7.1.

Standard normal distribution   A Normal$(0, 1)$ distribution—that is, a normal distribution with expectation 0 and variance 1. Because the standard deviation is the square root of the variance, the standard deviation is also 1.

Statistic   A quantity resulting from applying a mathematical function to data.

Test statistic   In a statistical hypothesis test, a test statistic is typically compared with its sampling distribution under the null hypothesis. The test statistic is a function of the data designed to detect departures from the null hypothesis. See Section 7.3.

Type I error   In Neyman–Pearson hypothesis testing, an error in which the null hypothesis is true but the test rejects it. See Section 7.4.

Type II error   In Neyman–Pearson hypothesis testing, an error in which the null hypothesis is false but the test does not reject it. See Section 7.4.

Union   In set theory, an inclusive "or," symbolized with $\cup$. For example, if A and B are sets, then $A \cup B$ is the collection of elements that are members of A or B or both. In probability, if A and B are events, then $P(A \cup B)$ refers to the probability that any of A, B, or both A and B occur. See Box 4-1.

Variance   An index of the dispersal of a random variable. In particular, $\mathrm{Var}(X) = \mathrm{E}[(X - \mathrm{E}[X])^2]$. See Section 5.1. The sample variance is a measure of the dispersal of a dataset and also an estimator of a random variable's variance.

Vector (R)   In R, an ordered, one-dimensional collection of objects of the same mode. See Appendix B, Section B.2.

Wald test   A statistical test in which the difference between a parameter estimate and its value under the null hypothesis is scaled by the estimator's standard error (or, in most applied settings, an estimate of the standard error). The resulting test statistic has an asymptotic normal distribution under the null hypothesis. See Section 9.3.

Wilks' theorem   A theorem stating that for a wide class of probabilistic models and null hypotheses, a likelihood-ratio-based test statistic obeys a χ^2 distribution if the null hypothesis is true. See Section 9.4.

Winner's curse   Suppose we test the null hypothesis that a parameter $\theta = 0$ by comparing an estimator of θ with its sampling distribution under the null hypothesis. Estimates that lead to a rejection of the null hypothesis are likely to be overestimates of θ, particularly in low-power settings. This is the "winner's curse." See Section 7.8.

Bibliography

Abraham, L. (2010). Can You Really Predict the Success of a Marriage in 15 Minutes? An Excerpt from Laurie Abraham's The Husband and Wife's Club. *Slate.* https://slate.com/human-interest/2010/03/a-dissection-of-john-gottman-s-love-lab.html?via=gdpr-consent

Angrist, J. D., & Pischke, J. S. (2009). *Mostly Harmless Econometrics: An Empiricist's Companion.* Princeton University Press.

Anscombe, F. J. (1973). Graphs in statistical analysis. *American Statistician*, 27, 17–21.

Anscombe's quartet (n.d.). *Wikipedia.* https://en.wikipedia.org/wiki/Anscombe%27s_quartet

Baltagi, B. (2008). *Econometric Analysis of Panel Data.* Wiley, Hoboken, NJ.

Bassett R. & Deride, J. (2018). Maximum a posteriori estimators as a limit of Bayes estimators. *Mathematical Programming.* https://doi.org/10.1007/s10107-018-1241-0

Beckerman, A. P., Childs, D. Z., & Petchey, O. L. (2017). *Getting Started with R: An Introduction for Biologists.* Oxford University Press.

Beaumont, M. A. (2010). Approximate Bayesian computation in evolution and ecology. *Annual Review of Ecology, Evolution, and Systematics*, 41, 379–406.

Bem, D. J. (2000). Writing the empirical journal article. In J. M. Darley, M. P. Zanna, & H. L. Roediger (Eds.), *The Compleat Academic: A Career Guide* (pp. 171–201). Washington, DC: American Psychological Association.

Benjamini, Y., & Hochberg, Y. (1995). Controlling the false discovery rate: a practical and powerful approach to multiple testing. *Journal of the Royal Statistical Society, Series B*, 57, 289–300.

Bickel, P. J., & Doksum, K. A. (2007). *Mathematical Statistics: Basic Ideas and Selected Topics*, Volume I, 2nd Edition. Pearson Prentice Hall, Upper Saddle River, NJ.

Blitzstein, J. K., & Hwang, J. (2014). *Introduction to Probability.* CRC Press. Boca Raton, FL.

Canty, A. & Ripley, B. (2017). boot: Bootstrap R (S-Plus) functions. R package version 1.3-20.

Casella, G., & Berger, R. L. (2002). *Statistical Inference*, Volume 2. Duxbury Press, Belmont, CA.

Champely, S. (2009). pwr: Basic functions for power analysis. R package.

Cleveland, W. S. (1993). *Visualizing Data.* Hobart Press, Summit, NJ.

Cohen, J. (1994). The earth is round (p <.05). *American Psychologist*, 49, 997–1003.

Cowles, M., & Davis, C. (1987). The subject matter of psychology: volunteers. *British Journal of Social Psychology*, 26, 97–102.

Dalgaard, P. (2008). *Introductory Statistics with R.* Springer, New York, NY.

Davison, A. C. & Hinkley, D. V. (1997). *Bootstrap Methods and Their Applications.* Cambridge University Press.

Diaconis, P., & Skyrms, B. (2017). *Ten Great Ideas about Chance.* Princeton University Press.

Diggle, P. J., Heagerty, P., Liang, K. Y., & Zeger, S. (2002). *Analysis of Longitudinal Data.* Oxford University Press.

Dobson, A. J., & Barnett, A. (2008). *An Introduction to Generalized Linear Models.* CRC Press, Boca Raton, FL.

Donoho, D. (2017). 50 Years of data science. *Journal of Computational and Graphical Statistics*, 26, 745–766.

Edwards, C. H., & Penney, D. E. (1982). *Calculus with Analytic Geometry.* Prentice-Hall, Upper Saddle River, NJ.

Efron, B. (1979). Bootstrap methods: another look at the jackknife. *Annals of Statistics*, 7, 1–26.

Efron, B. (1987). Better bootstrap confidence intervals. *Journal of the American Statistical Association*, 82, 171–185.

Efron, B. (1998). RA Fisher in the 21st century. *Statistical Science*, 13, 95–114.

Efron, B., & Tibshirani, R. (1986). Bootstrap methods for standard errors, confidence intervals, and other measures of statistical accuracy. *Statistical Science*, 1, 54–75.

Eliason, S. R. (1993). *Maximum Likelihood Estimation: Logic and Practice*. Sage Publications, Newbury Park, CA.

Emerson, J. W., Green, W. A., Schloerke, B., Crowley, J., Cook, D., Hofmann, H., & Wickham, H. (2013). The generalized pairs plot. *Journal of Computational and Graphical Statistics*, 22, 79–91.

Ernst, M. D. (2004). Permutation methods: a basis for exact inference. *Statistical Science*, 19, 676–685.

Fisher, R. A. (1936). The use of multiple measurements in taxonomic problems. *Annals of Eugenics*, 7, 179–188.

Fox, J. (2009). *A Mathematical Primer for Social Statistics*. Sage Publications, Newbury Park, CA.

Fox, J. & Weisberg, S. (2011). *An R Companion to Applied Regression*, 2nd Edition. Sage Publications, Thousand Oaks, CA.

Freedman, D. A. (2009). *Statistical Models: Theory and Practice*. Cambridge University Press.

Friedman, J., Hastie, T., & Tibshirani, R. (2001). *The Elements of Statistical Learning*. Springer, New York.

Fundamental theorem of calculus (n.d.). *Wikipedia*. https://en.wikipedia.org/wiki/Fundamental_theorem_of_calculus

Gelman, A., Carlin, J. B., Stern, H. S., Dunson, D. B., Vehtari, A., & Rubin, D. B. (2014). *Bayesian Data Analysis*. CRC Press, Boca Raton, FL.

Gelman, A., Lee, D., & Guo, J. (2015). Stan: A probabilistic programming language for Bayesian inference and optimization. *Journal of Educational and Behavioral Statistics*, 40, 530–543.

Gelman, A., & Loken, E. (2013). The garden of forking paths: Why multiple comparisons can be a problem, even when there is no "fishing expedition" or "p-hacking" and the research hypothesis was posited ahead of time. http://www.stat.columbia.edu/~gelman/research/unpublished/p_hacking.pdf

Gelman, A., & Shalizi, C. R. (2013). Philosophy and the practice of Bayesian statistics. *British Journal of Mathematical and Statistical Psychology*, 66, 8–38.

Gigerenzer, G., & Marewski, J. N. (2015). Surrogate science: The idol of a universal method for scientific inference. *Journal of Management*, 41, 421–440.

Good, P. I., & Hardin, J. W. (2012). *Common Errors in Statistics (and How to Avoid Them)*. Wiley, Hoboken, NJ.

Griffiths, T. L. & Tenenbaum, J. (2006). Statistics and the Bayesian mind. *Significance*, 3, 130–133.

Hájek, A. (2011). Interpretations of Probability. *Stanford Encyclopedia of Philosophy*. https://plato.stanford.edu/entries/probability-interpret/

Hoekstra, R., Morey, R. D., Rouder, J. N., & Wagenmakers, E. J. (2014). Robust misinterpretation of confidence intervals. *Psychonomic Bulletin & Review*, 21, 1157–1164.

Hoel, P. G., Port, S. C., & Stone, C. J. (1971). *Introduction to Probability Theory*. Houghton Mifflin, Boston, MA.

Hoffrage, U., & Gigerenzer, G. (1998). Using natural frequencies to improve diagnostic inferences. *Academic Medicine*, 73, 538–540.

Hogg, R. V., McKean, J., & Craig, A. T. (2005). *Introduction to Mathematical Statistics*. Pearson Education, Harlow, Essex.

Homer. Odyssey (Pope)/Book XII. (2016, July 3). Wikisource. https://en.wikisource.org/w/index.php?title=Odyssey_(Pope)/Book_XII&oldid=6311168

Hothorn, T., Hornik, K., Van De Wiel, M. A., & Zeileis, A. (2008). Implementing a class of permutation tests: the `coin` package. *Journal of Statistical Software*, 28, 1–23.

Hume, D. (1748, this edition 1999). *An Enquiry concerning Human Understanding*, ed. Beauchamp, T. L. Oxford University Press.

Imbens, G. W., & Rubin, D. B. (2015). *Causal Inference in Statistics, Social, and Biomedical Sciences*. Cambridge University Press.

Ioannidis, J. P. (2005). Why most published research findings are false. *PLoS Medicine*, 2, e124.

James, G., Witten, D., Hastie, T., & Tibshirani, R. (2013). *An Introduction to Statistical Learning*. Springer, New York, NY.

John, L. K., Loewenstein, G., & Prelec, D. (2012). Measuring the prevalence of questionable research practices with incentives for truth telling. *Psychological Science*, 23, 524–532.

Kabacoff, R. I. (2010). *R in Action*. Manning, Shelter Island, NY.

Kampstra, P. (2008). Beanplot: A boxplot alternative for visual comparison of distributions. *Journal of Statistical Software*, 28, 1–9.

Kass, R. E., & Raftery, A. E. (1995). Bayes factors. *Journal of the American Statistical Association*, 90, 773–795.

Klassen, T. P., & Hartling, L. (2005). Acyclovir for treating varicella in otherwise healthy children and adolescents. *Cochrane Database of Systematic Reviews*, 19, CD002980.

Koenker, R., Portnoy, S. Ng, P. T., Zeileis, A., Grosjean, P., & Ripley, B. D. (2017). Package quantreg: Quantile Regression. R Package.

Kolmogorov, A. N. (1933). *Grundbegriffe der Wahrscheinlichkeitrechnung*, Ergebnisse Der Mathematik. Translated as *Foundations of Probability*. Chelsea Publishing Company. New York.

Kraemer, H. C., & Blasey, C. (2015). *How Many Subjects? Statistical Power Analysis in Research*. Sage Publications, Thousand Oaks, CA.

Lehmann E. L. & Romano, J. P. (2005). *Testing Statistical Hypotheses*, 3rd Edition. Springer, New York.

Lindsey, J. (2013). RMUTIL: Utilities for nonlinear regression and repeated measurements. R package.

Louçã, F. (2009). Emancipation through interaction—How eugenics and statistics converged and diverged. *Journal of the History of Biology*, 42, 649–684.

Lyon, A. (2013). Why are normal distributions normal? *The British Journal for the Philosophy of Science*, 65, 621–649.

Martin, A. D., Quinn, K. M., & Park, J. H. (2011). MCMCpack: Markov chain Monte Carlo in R. *Journal of Statistical Software*, 42, 1–21.

Matloff, N. (2011). *The Art of R Programming: A Tour of Statistical Software Design*. No Starch Press, San Francisco, CA.

Mayo, D. G. (2018). *Statistical Inference as Severe Testing: How to Get Beyond the Statistics Wars*. Cambridge University Press.

McDowell, M. A., Fryar, C. D., Ogden, C. L., & Flegal, K. M. (2008). Anthropometric reference data for children and adults: United States, 2003–2006. *National Health Statistics Reports*, 10, 1–48.

McElreath, R. (2015). *Statistical Rethinking*. Chapman & Hall/CRC Press, Boca Raton, FL.

McElreath, R. (2017). Markov chains: Why walk when you can flow? Blog posted 28 November 2017. http://elevanth.org/blog/2017/11/28/build-a-better-markov-chain/

Meehl, P. E. (1978). Theoretical risks and tabular asterisks: Sir Karl, Sir Ronald, and the slow progress of soft psychology. *Journal of Consulting and Clinical Psychology*, 46, 806–834.

Meehl, P. E. (1990). Why summaries of research on psychological theories are often uninterpretable. *Psychological Reports*, 66, 195–244.

Muggeo, V. M., & Lovison, G. (2014). The "three plus one" likelihood-based test statistics: unified geometrical and graphical interpretations. *The American Statistician*, 68, 302–306.

Munafò, M. R., Nosek, B. A., Bishop, D. V., Button, K. S., Chambers, C. D., du Sert, N. P., Simonsohn, U., Wagenmakers, E.-J., Ware, J.-J., & Ioannidis, J. P. A. (2017). A manifesto for reproducible science. *Nature Human Behaviour*, 1, 0021.

Nolan, D., & Lang, D. T. (2015). *Data Science in R: A Case Studies Approach to Computational Reasoning and Problem Solving*. CRC Press, Boca Raton, FL.

Open Science Collaboration (2015). Estimating the reproducibility of psychological science. *Science*, 349, aac4716.

Pearl, J., Glymour, M., & Jewell, N. P. (2016). *Causal Inference in Statistics: A Primer*. Wiley, Hoboken, NJ.

Peña, E. A., & Slate, E. H. (2006). Global validation of linear model assumptions. *Journal of the American Statistical Association*, 101, 341–354.

Plummer, M. (2003). JAGS: A program for analysis of Bayesian graphical models using Gibbs sampling. In *Proceedings of the 3rd International Workshop on Distributed Statistical Computing (DSC 2003), March 20–22, 2003, Vienna, Austria*.

Plummer, M., Best, N., Cowles, K., & Vines, K. (2006). CODA: convergence diagnosis and output analysis for MCMC. *R News*, 6, 7–11.

R Core Team (2015). R: A Language and Environment for Statistical Computing. R Foundation for Statistical Computing. https://www.R-project.org

R Core Team (2017). foreign: Read Data Stored by Minitab, S, SAS, SPSS, Stata, Systat, Weka, dBase,.... R package version 0.8-69. https://CRAN.R-project.org/package=foreign

Rabe-Hesketh, S., & Skrondal, A. (2008). *Multilevel and Longitudinal Modeling Using Stata*. STATA Press, College Station, TX.

Rosenthal, R. (1979). The file drawer problem and tolerance for null results. *Psychological Bulletin*, 86, 638–641.

Ross, S. M. (2002). *A First Course in Probability*. Pearson, Upper Saddle River, NJ.

Ruppert, D., Wand, M. P., & Carroll, R. J. (2009). Semiparametric regression during 2003–2007. *Electronic Journal of Statistics*, 3, 1193–1256.

Shaffer, J. P. (1995). Multiple hypothesis testing. *Annual Review of Psychology*, 46, 561–584.

Shalizi, C. (to be published). *Advanced Data Analysis from an Elementary Point of View*.

Short, T. (2004). R reference card. https://cran.r-project.org/doc/contrib/Short-refcard.pdf

Silvey, S. D. (1975). *Statistical Inference*. Chapman & Hall, London.

Simmons, J. P., Nelson, L. D., & Simonsohn, U. (2011). False-positive psychology: Undisclosed flexibility in data collection and analysis allows presenting anything as significant. *Psychological Science*, 22, 1359–1366.

Singer, J. D., & Willett, J. B. (2003). *Applied Longitudinal Data Analysis: Modeling Change and Event Occurrence*. Oxford University Press.

Spiegelhalter, D., Thomas, A., Best, N., & Gilks, W. (1996). BUGS 0.5: Bayesian Inference Using Gibbs Sampling Manual (version ii). MRC Biostatistics Unit, Institute of Public Health, Cambridge, UK.

Stigler, S. M. (1983). Who discovered Bayes's theorem? *The American Statistician*, 37, 290–296.

Stigler, S. M. (2007). The epic story of maximum likelihood. *Statistical Science*, 22, 598–620.

Stigler, S. M. (2010). The changing history of robustness. *The American Statistician*, 64, 277–281.

Stuart, A., & Ord, J. K. (1987). *Kendall's Advanced Theory of Statistics*, Volume 1. Oxford University Press.

Székely, G. J., & Rizzo, M. L. (2017). The energy of data. *Annual Review of Statistics and Its Application*, 4, 447–479.

Taleb, N. N. (2008). The fourth quadrant: a map of the limits of statistics. *Edge*. https://www.edge.org/conversation/the-fourth-quadrant-a-map-of-the-limits-of-statistics

Taylor, J., & Tibshirani, R. J. (2015). Statistical learning and selective inference. *Proceedings of the National Academy of Sciences of the USA*, 112, 7629–7634.

Thompson, S. P., & Gardner, M. (1998). *Calculus Made Easy*. Macmillan, London.

Tufte, E. R. (1983). *The Visual Display of Quantitative Information*. Graphics Press, Cheshire, CT.

Venables, W. N. & Ripley, B. D. (2002). *Modern Applied Statistics with S*, 4th Edition. Springer, New York.

Wasserman, L. (2013). *All of Statistics: A Concise Course in Statistical Inference*, 2nd Edition. Springer, New York.

Wasserstein, R. L., & Lazar, N. A. (2016). The ASA's statement on *p*-values: context, process, and purpose. *The American Statistician*, 70, 129–133.

Westfall, J., & Yarkoni, T. (2016). Statistically controlling for confounding constructs is harder than you think. *PloS One*, 11, e0152719.

Whitehead, J. (1997). *The Design and Analysis of Sequential Clinical Trials*. Wiley, Hoboken, NJ.

Wickham, H. (2014). *Advanced R*. CRC Press, Boca Raton, FL.

Wickham, H. & Chang, W. (2016). devtools: Tools to make developing R packages easier. R package version 1.12.0. http://CRAN.R-project.org/package=devtools

Wilcox, R. R. (2011). *Introduction to Robust Estimation and Hypothesis Testing*. Academic Press, Cambridge, MA.

Xie, Y. (2013). animation: An R package for creating animations and demonstrating statistical methods. *Journal of Statistical Software*, 53, 1–27.

Young, G. A., & Smith, R. L. (2005). *Essentials of Statistical Inference*. Cambridge University Press.

Person index

Subject index

Printed and bound by CPI Group (UK) Ltd, Croydon, CR0 4YY